Endogenous Time Preferences in Social Networks

Endogenous Time Preferences in Social Networks

Marianna A. Klochko
Assistant Professor of Sociology
Ohio State University, Marion, USA

Peter C. Ordeshook
Professor of Political Science
California Institute of Technology, USA

Edward Elgar
Cheltenham, UK • Northampton, MA, USA

Published by
Edward Elgar Publishing Limited
Glensanda House
Montpellier Parade
Cheltenham
Glos GL50 1UA
UK

Edward Elgar Publishing, Inc.
136 West Street
Suite 202
Northampton
Massachusetts 01060
USA

A catalogue record for this book
is available from the British Library

ISBN 1 84542 598 7

Printed and bound in Great Britain by MPG Books Ltd, Bodmin, Cornwall

Contents

1 Cooperation and Time Discounting

Punishments and rewards that impact immediately have a stronger influence than punishments and rewards moved to the future, and the more they are moved the less their influence. (Pitirim Sorokin 1914, translated from the Russian 1992)

The view that preferences, norms, and, more generally, cultural attitudes should be considered as endogenous with respect to socio-economic systems has been now extensively motivated. In particular, various pieces of empirical evidence have been interpreted to suggest the relevance of the endogeneity of various elements of preferences, as, for example, the discount factor... (Bisin and Verdier 2001: 299)

1.1 The Critical Parameter

Regardless of whether the object of study is social norms (Lewis 1969, Bowles and Gintis 1998, Bendor and Swistak 2001, Mahoney and Sanchirico 2001, 2002), trust in interpersonal relations (Yamagishi and Cook 1993, Guth and Kliemt 1994, Yamagishi and Yamigishi 1994, Yamagishi et al. 1998, Strang and Macy 2000, Macy and Sato 2002), the mechanisms whereby the provisions of a political constitution are enforced (Hardin 1989, Ordeshook 1993), adherence to the law (Posner 2000), manifestations of altruism (Nagel 1970. Bergstrom and Stark 1993), public policies that might encourage sustainable economic development (Ostrom 1990), or the economic and political issues associated with the provision of public goods generally (Olson 1965, Hardin 1982, 1995), cooperation is problematical when the benefits of cooperation are such that individuals have an incentive to renege on agreements and 'free ride' on the actions of others. The most cited representation of this problem is the one shot Prisoners' Dilemma whereby each person's dominant strategy – the strategy they should choose regardless of anyone else's choice – is to defect from cooperative action and free ride on the efforts of others so as to yield a Pareto-inferior outcome. But social norms exist and are adhered to, political constitutions survive, the law constrains on individual action even when the severity or prospect of punishment is slight, people cooperate with others on the basis of beliefs that encompass notions such as trust and altruism, and public goods issues are sometimes resolved without force or coercion.

Within the rational choice paradigm, the usual way of understanding the emergence and sustainability of cooperation – of how people overcome the imperatives of the motives illustrated by the Prisoners' Dilemma – is conceptualizing individual choice as part of some ongoing social process. In addition to admitting a consideration of such things as learning and imitation, the formal acknowledgment that social processes are dynamic allows us to establish the existence of a class of equilibria that are not normally present in more limited (myopic or static) representations. Specifically, defectors from cooperation 'today' can be punished 'tomorrow' by those they injure or with whom they interact in the future and who act on the basis of reputations earned from earlier choices, while non-defection can be rewarded subsequently by allowing a person to share in the social gains from cooperation. Put simply, by expanding the set of strategies available to people, conceptualizing social processes as ongoing and dynamic expands the set of equilibria in our analyses to include cooperative as well as non-cooperative outcomes (Taylor 1976, Friedman 1977, Axelrod 1984, 1997).

This approach opens the door to a variety of issues. Since the strategy commonly used to illustrate the consequences of repetition, tit–for–tat, is not subgame perfect (i.e., it is not necessarily rational to implement a threat, thereby rendering the threat non-credible), what strategies can ensure full cooperation? What are the mechanisms whereby people monitor compliance with cooperative agreements? How does uncertainty, either about the nature of future interactions or the likelihood of alternative outcomes, impact the sustainability of cooperation? What is the role of networks in transmitting reputations? What is the role of leadership in coordinating us to one equilibrium as opposed to another? How can the notion of trust be represented formally using the concepts of belief and Bayesian updating? How do we conceptualize the notion of culture and the forms of social cooperation we associate with that term using the base ideas of preference, expectation and strategy? How can notions of learning and the dynamic adjustment of preferences be incorporated into our perspective? And how do we explain cooperation when social interactions between pairs of persons are not repeated (see, for example, Guth and Kliemt 1994, Bendor and Swistak 2001, and for a statement of the problem, Bowles and Gintis 1998)?

What warrants emphasis is that regardless of the questions asked, the credibility of punishment, the efficacy of monitoring, the promise of reward, the roles of reputation and trust, and the mechanisms whereby people form expectations about the future based on past experience, not to mention the nature of equilibria that repetition and associated processes allow, depend on the reality of a *shadow of the future* – on the extent to which people possess a 'sufficiently low' discount on payoffs that are realized tomorrow rather than today. If people are impatient and seek immediate gratification, cooperation is difficult to sustain and Pareto-inferior outcomes difficult to avoid, as opposed to when people's planning horizons are extended and they act not on the basis of immediate consequences but also on the basis of outcomes that may lie far in the future. If the threat of sanction and the promise of reward are to impact decisions in the present, then people must give the future sufficient weight in their strategic calculations. The emergence of cooperation as an equilibrium in social processes depends not only on the absolute payoffs people associate with alternative outcomes, but also on how they evaluate the prospect of future payoffs.

1.2 The Usual Representation

As we discuss later, there are objections to the supposition that a shadow of the future is essential to explaining cooperation, based largely on the observation that many forms of cooperation do not occur in the context of repeated interactions among the same persons. Why, for instance, do we tip a waiter or waitress at a restaurant we are unlikely to visit again? We address such concerns later with two arguments. First, time preferences can be critical to understanding cooperation even absent repeated interactions between the same set of persons (see, for instance, Bendor and Swistak 2001 and Gintis 2003). Second, although the details of the analysis that follows would change, the general perspectives we adopt can be applied to any component of individual preferences, including those things we might choose to label *taste*. However, once our attention focuses on individual time discounts a number of new issues and research challenges emerge, and here it is useful to review briefly the formal treatment commonly given to the shadow of the future in the literature. We begin by letting x_t correspond to an outcome that occurs at time t and $U(x_t)$ denote the value or utility a person associates with x_t upon its realization – the utility of x_t if it occurred today. Then the Present Value (PV) – the value a person assigns today (at $t =$ 0) to the (possibly infinite) stream of outcomes ($x_0, x_1, x_2, x_3,...$) is expressed as $PV = U(x_0) + d_1U(x_1) + d_2U(x_2) + d_3U(x_3) + ...$, where d_i is the discount he or she applies to any payoff realized in period $t = i$. The formal definition of time discounts, then, is that they represent the marginal rate of substitution between current and future consumption (Fisher 1930), and, in particular, the marginal rate along the 45° line in the current consumption/future consumption plane (Stigler 1984). This particular way of representing discounting was first introduced by Samuelson (1937) and subsequently formalized in axiomatic form by Koopmans (1960), and it is important to understand that discounting is distinct from an assessment of uncertainty in individual choice (for a survey of the origins of this formulation see Frederick, et al. 2002). It is true that people discount uncertain prospects, but here, by the 'shadow of the future' we refer to the weight people give to future outcomes, probabilistic or otherwise, under the assumption that, *ceteris paribus*, the present value of the outcome 'O today' is different from the present value of the outcome 'O tomorrow' (and thus, as defined by Fisher 1930, corresponds to the 'marginal rate of substitution between current and future consumption'). Thus, as Elster (2000: 10) argues when distinguishing between uncertainty and discounting in the context of drug addiction, 'an addict ... may have accurate beliefs about the disastrous effects of the drug on his or her body or purse, and yet ignore them because of an addiction-induced increase in the rate of time discounting.'

A number of assumptions concerning d_i are nearly universally applied in formal analyses of cooperation, repeated games, and the various models economists offer about investment and growth. The first is that future outcomes are weighted less heavily than more immediate ones (i.e., $0 < d_i < 1$, $t > 0$). A person should value \$100 today more than they value the certainty of that money a month or a year from now. The second is that discounting is exponential: 'to generate consistent plans, impervious to later contradiction due to the passage of time, requires that the discount function be of a special form, namely, that all future dates should be discounted at a constant rate of interest' (Winston and Woodbury 1991: 325, interior quote from Strotz 1955: 177) so that 'in the standard models ... the discounting is always assumed to be exponential' (Elster 2000: 43). The third common assumption is that the d_is are independent of the *x*s and $U(x)$s. For example, then, these latter two assumptions imply that if the sequence of outcomes $(x_0, x_1, x_2, x_3, ...)$ corresponds to an infinite constant stream $(x, x, x, x, ...)$, we can write

$$PV = U(x) + dU(x) + d^2U(x) + d^3U(x) + ... = U(x)/(1-d), \qquad (1.1)$$

which is the usual actuarial expression for present value. So if a person is concerned only with the present ($d = 0$), then PV equals the value of the first outcome, $U(x)$. But if a person values the future the same as the present ($d = 1$), then PV is infinite (or minus infinity if $U(x)$ is negative).

In the next section we review a series of behavioral anomalies which suggest that these assumptions about time preference are not universally valid. However, those anomalies and the generality of the preceding assumptions are not the primary concern of this volume. Instead, we are concerned with the fact that the applications of this representation commonly employ several additional assumptions in order to simplify analysis further. First, paralleling the economist's usual approach of assuming that individual preferences are fixed (Bowles 1998), d is generally assumed to be exogenously determined and fixed as well.[1] Although the researcher may explore the consequences of high or low values of d, the determinants of this parameter are unexplored, and wholly ignored is the

---

[1] Axelrod (1984), for example, explicitly adopts the standard construction of time discounting in his seminal analysis of evolutionary cooperation (see p. 12 and his discussion of the shadow of the future in pp. 126-32). For additional examples see Taylor 1976, Andvig and Moene 1990, Bicchieri and Rovelli 1995, Mauro 1997, Greif, Milgrom and Weingast 1998).

possibility that *d* is itself a function of the very process under investigation. Second, it is also often convenient to assume that *d* is common to all persons, or, when talking about strictly economic variables, that it equals the actual market rate of return. However, as even Baumol (1970, p. 273), as an economist, notes: '[d]espite the critical nature of [the discount rate], in some calculations it is assigned a value almost cavalierly, with little attempt to show that the selected figure is not chosen arbitrarily and capriciously,' and as Krahn and Gafni (1993, p. 415) note: 'Most analyses ... pull their discount rates either out of the air or off the shelf, and the lucky number is most often 5%.' Thus, although the preceding assumptions can be stated with mathematical precision, they are often justified only with the argument that they facilitate analysis, and their subsequent application occurs without regard for alternative formulations.

1.3 Anomalies

We emphasize that the preceding assumptions about discounting, especially that of exogenous and fixed discounts, are common to essentially all studies of cooperation, norms, trust and so on that depend on a repeated game formulation of cooperation and collective action. This is true despite the fact that a subsidiary literature gives ample evidence of anomalies which show that individual time discounts are a more complex parameter of individual choice than expression (1.1) suggests (for a general survey see Frederick et al. 2003, Camerer and Loewenstein 2004)

Myopic Discounting: Winston and Woodbury (1991), on the basis of hypothetical choices offered to 160 college students, suggest that some people change their time discount as a function of the uniqueness of events (duration) and how far in the future that event will occur. Briefly, 'a myopic discount function 'overvalues' the more proximate satisfactions relative to the more distant ones .... [and] as time passes and the rewards come closer, there is a choice reversal [so that] what is optimal when viewed from one time perspective ... is not optimal when viewed from another ... with myopia, the closer an event, X, the higher the interest rate used to discount its value' (p. 328). Admittedly, these results are not unchallenged. Roberts (1991), taking note of the fact that 'only 87 of the 360 [choices] exhibited straightforward myopia,' argues that 'in any experiment of hypothetical options there is always the question of whether individuals have an incentive to respond as they would when real money is at stake' (p. 343). Nevertheless, the suggestion of myopic discounting cautions us about the blind acceptance of classical time discounting assumptions.

Negative Discounting: In his seminal study, Loewenstein (1987) shows that if people value the anticipation of a favorable event for its own sake (or dread the anticipation of an unfavorable event), they will act as if they have a negative discount rate by postponing the favorable event and speeding up the realization of the unfavorable one: 'people will tend to delay desirable and fleeting consumption for purposes of savoring' (p. 678), in which case the usual assumption that individual time discounts vary between 0 and 1 is violated. Similarly, Nagin and Pogarsky (2001), in a study of college student attitudes towards and evaluations of the risks and rewards of driving while intoxicated find that fully 21% of their sample reveal preferences consistent with negative discount rates: 'For these individuals, a six-month suspension [of driving privileges] beginning immediately corresponds to a suspension of less than six months in the future, which suggests that these subjects would prefer to endure the punishment as soon as possible' (p. 880). One might attempt, of course, to circumvent this issue by an appropriate re-conceptualization of outcomes (e.g., incorporating the time of its realization into an outcome's description), but the parsimony that formulations such as expression (1.1) provides soon disappear insofar as no outcome can ever be repeated and all outcomes are unique.

Hyperbolic Discounting: A phenomenon closely related to negative discounting is discount rates that are non-exponential. Specifically, 'there is a substantial body of evidence that preferences are dynamically inconsistent. Research on animal and human behavior has led psychologists to conclude that discount functions are approximately hyperbolic (Ainslie 1992) [where] hyperbolic discount functions are characterized by a relatively high discount rate over short horizons and a relatively low discount rate over long horizons' (Laibson 1997: 445; see also Ainslie and Haendel 1982, Ainslie 1991, Sally 2000). For example, inconsistency is revealed when people, as is often the case in experimental research, express a preference for '\$100 now versus \$110 tomorrow' but at the same time prefer '\$110 a year and a day from now over \$100 a year from now' (Ok and Masatlioglu 2003). Noting that '[i]n terms of criminal decision making, hyperbolic discounting implies a pronounced diminution in the impact of sanctions over initial, small delays, with such diminution becoming rapidly less discernible over time' Nagin and Pogarsky (2001: 885) suggest that the notion of hyperbolic discounting can explain the criminologist's difficulty in validating the hypothesis of a 'celerity effect' (whereby the immediacy of punishment contributes to its deterrence value). The general consequence of hyperbolic discounting, however, is what behavioralists refer to as time inconsistency: 'Individuals who discount the future hyperbolically will tend to deviate from their plans unless they take precautionary measures' (Elster 2000: 29). That is, not only might contracts across individuals be difficult to sustain, but even those 'contracts' one makes with oneself might be difficult to uphold as well.

Intransitive Time Preference: Consider a person whose discount over some period exceeds his or her discount over subintervals of that period. That is, suppose the utility of the outcome '\$100 today' exceeds that of '\$250 two weeks from now' and that '\$200 in a week' is preferred to '\$100 today'. To this point there is nothing inconsistent with this preference and the classical exponential model (e.g., if we set $d = .6$). However, as Read (2001) shows, such choices do not necessarily preclude a preference for '\$250 two weeks from now' over '\$200 a week from now' in which case we have the intransitive preference $U(\$100 \text{ at } t = 0) > U(\$250 \text{ at } t = 2) > U(\$200 \text{ at } t = 1) > U(\$100 \text{ at } t = 0)$.

Non-independence: Essentially all models of social processes that use discounting to understand cooperation, norms, and the evolution of trust impose an independence condition, defined as follows: Let A, B and C be three outcomes occurring at times $t = 0$, 1 and 2, and the present value of this sequence be $PV(A,B,C) = U(A) + dU(B) + d^2U(C)$. Now if the order in which A and B are realized is reversed so we can write $PV(B,A,C) = U(B) + dU(A) + d^2U(C)$, let $PV(A,B,C) > PV(B,A,C)$. The independence assumption is that if we substitute the outcome D for C, then $PV(A,B,D) > PV(B,A,D)$. Because C is common to the first two sequences, and D common to the third and fourth, this substitution should play no role in a person's evaluation of the order in which A and B occur. However, Loewenstein (1987, 1992) offers empirical evidence that this assumption is not universally satisfied, and Becker and Murphy (1988) explicitly acknowledge the possibility of non-interdependence in the case of (drug) addiction, whereby one's consumption in one period impacts the absolute utility of consumption in subsequent periods.

1.4 Empirical Correlates

Such anomalies are merely suggestive of complexity in the character of individual time discounts, and some of them, as we note, can be accommodated and the traditional assumptions about discounting sustained by an 'appropriate' conceptualization of outcomes. In the case of negative discounting, for example, we can conceptualize outcomes so that intermediate ones contain the description 'consumption of X awaits'. Similarly, violations of non-independence might be treated by conceptualizing outcomes to include what has already occurred. Such an approach, however, threatens the parsimony of our models (if every outcome must include the prospect of all conceivable futures and all realized histories), and perhaps even requires that our 'abstract' descriptions of things become as complex as the processes under investigation so as to undermine the ultimate intent of modeling, the development of general theory. It may be that the solution is some fundamental re-conceptualization of time preferences and preferences generally. But our concern here is not the discomfort such anomalies might cause in the character of economic theory. Rather, that concern is the exogenous treatment given to whatever time discount parameter might be required to account for individual preference. There is, nevertheless, a lesson to be learned here. These anomalies suggest that people's revealed time preferences depend on the specific context of choice and that a discount, and even the model of discounting, that applies in one context may not apply universally.

However, before we address the matter of endogenous versus exogenous time preference, it is useful to review briefly a literature which suggests that it is inappropriate to assume that the same set of parameter values apply universally. Even within the classical framework, time preferences vary across populations in seemingly consistent and understandable ways.

Age and Gender: Davids and Falkof (1975), in their study of delinquent youth, show, among other things, that females tend to value the future more highly than males (have a higher propensity to save than males) and that the younger cohort of their study tend to value the future less than the older cohort. In addition, there appears to have been a general decline between 1959 and 1974 in the weight given the future by all groups in their analysis. Essentially equivalent results are derived theoretically from a biological evolutionary approach to the determination of d by Rogers (1994), who concludes that 'young adults should discount at a higher rate than their elders, especially over short delays' (p. 477). Equivalently, 'once an individual has acquired sufficient wealth to support a mate and children, maximizing the expected value of biological fitness would be expected to imply risk aversion and income seeking behavior' (Rubin and Paul 1979: 594).

Fertility: With an evolutionary perspective, Becker and Barro (1988) offer a model that rationalizes the prediction that a person's valuation of the future will increase with fertility – a result that sustains the relationship between age and time discounts. Moreover, the relationship is bi-directional: 'If the costs of rearing descendants is constant over time, fertility depends only on interest rates ... the time preference factor ... and the degree of altruism' (p. 23). Admittedly, however, this argument presumes considerable 'rationality' on the part of parents in the decision about whether or not to have children, and wholly ignored is any of the social or cultural determinants of fertility.

Religion: Becker and Mulligan (1997) offer the hypothesis that 'anything that raises future utility without raising marginal utility of current consumption will tend to lower the equilibrium discount of the future' (p. 739). This hypothesis is not the mere reflection of an observational equivalence between low discounting and high absolute valuations of future outcomes. It is a statement about equilibrium consumption patterns when 'consumers' are free to invest energy and resources in increasing their appreciation of the future – when people are allowed to invest in increasing d. The corollary they offer, however, is especially interesting: 'if consumers expect positive utility after death – perhaps because they believe they will go to heaven – this raises their investments in future-oriented capital ... [and] in equilibrium, therefore, consumers who expect to go to heaven will discount the future less' (741). As Becker (1996: 11) summarizes the argument, 'religion often increases the weight attached to future utilities, especially when it promises an attractive afterlife.' Thus social processes and participation in social groups that raise people's 'religious consciousness' can contribute to raising *d*.

Income: In his classic study of the economics of investment, Fisher (1930: 73) notes that 'A small income, other things being equal, tends to produce a high rate of impatience, partly from the thought that provision for the present is necessary both for the present itself and for the future as well, and partly from the lack of foresight and self-control.' Similarly, Becker and Mulligan's (1997) model holds implications for the relationship between wealth and time discounting. If people are allowed to invest in 'appreciating the future', 'the rich will have the greatest incentive to [invest] because they have both low marginal utility of wealth and high future utilities' (p. 745). It follows that those with higher income will be more patient than those with lower income. It needs to be emphasized, moreover, that the derived causal relationship is not simply that patience implies wealth (via savings and postponed consumption), but rather that 'it is possible to detect the importance of the effect of wealth on patience rather than the reverse causality' (p. 752). This analysis is wholly theoretical, but while a study of foraging creatures such as pigeons might seem far removed from our concerns, Kagel, Green and Caraco (1986) reinforce this view: 'hungrier foragers discount the future more steeply than better-fed individuals' (p. 273) and 'birds were more likely to choose small, more immediate rewards when hungry than when partially satiated ... a forager might reasonably be expected to discount the future more steeply as its energy reserve falls toward critical lower levels' (p. 275). These results, although they illustrate discount rates that fail to satisfy the independence condition, seem unsurprising. But we should also note their relevance to the relationship between discounting and age: 'experience may influence the way a forager interactively evaluates reward size and delay. The experimental subjects first learned to wait for a larger reward and, as a result of this experience, were more apt to choose the more delayed, larger reward than birds lacking the same experience' (p. 274). Thus, we not only see here an illustration of an endogenously determined discount rate, but also a reason for supposing that individual discount rates will vary with age: 'experience may result in age-dependent discounting behavior. For example, children have a more pronounced preference for immediate rewards than adults do' (Kagel et al, 1986: 274). Insofar as the interaction of parameters here is concerned, 'studies on cognitive abilities of persons from different social status have pointed out that … long-tem planning is better developed in middle than in low status people' (Oevermann 1969). Similarly, Teahan (1963) demonstrates a strong relation between present social standing and future time perspective and that children who had greater success in school thought more about the future and had wider temporal perspectives than those at the bottom of the class. Schneider and Lysgaard (1953) argue that middle class children are more willing than working class children to postpone a reward, while Trommsdorf and Lamm (1975: 350) note that the 'greater capacity of middle class children to delay rewards may be attributed to their ability to look further into the future than lower class children, and may be related to high achievement motivation and the ability to tolerate frustration.'

Criminals: In their classic *A General Theory of Crime*, Gottfredson and Hirschi (1990) conclude that, as Nagin and Pogarsky (2001: 866) summarize it, 'persons who engage in crime can be distinguished by their 'here and now' orientation.' Similarly, Wilson and Herrnstein (1985) draw our attention to the impulsiveness of criminals in a way that we can interpret their discussion as referring to those who discount the future virtually to the limit. Nagin and Pogarsky (2001) build on this argument by measuring the time discounts of 252 college students (for the methodology used to measure time preference non-monetary contexts see Cropper et al. 1994) in order to assess their attitudes towards various penalties for driving while intoxicated. Their findings are twofold. First, they see 'immense variation in [discount rates], which range from –175% to +364%' (p. 879). And second, 'negative discounters report lower probabilities of driving while drunk. This negative discounting effect is also large – *ceteris paribus*, negative discounters are 9.8% points less likely to drive while drunk, a 22% reduction from the sample average of 34%' (p. 882). And after controlling for such variables as gender, past experiences with alcohol and traffic related accidents, experimentally adjusted severity of punishments, they find that 'although the discounting variable ... falls short of statistical significance at conventional levels' the overall positive impact of inferred time discounts on the likelihood of driving while intoxicated 'comports precisely with the Gottfredson and Hirschi and Wilson and Herrnstein views of present orientation' (p. 882).

Addiction: The seminal study of time discounts and their variability across a population appears in the pair of essays by Becker and Murphy (1988) and Becker and Mulligan (1997) and the application of their analysis to addiction. Here the idea of the adjacent complementarity of goods is employed (where the value of a good's consumption today depends on how much was consumed in the past) and a model offered in which not only is it that 'present-oriented individuals are potentially more addicted to harmful goods than future oriented individuals' (Becker and Murphy, p. 682), but also, in terms of reversing the causal relationship, that 'harmful addictions induce even rational persons to discount the future more heavily' (Becker and Mulligan, p. 744). Or, as Elster (2000: 10) suggests, when confronting the argument that addiction changes absolute utility, 'an addict ... may have accurate beliefs about the disastrous effects of the drug on his or her body or purse, and yet ignore them because of an addiction-induced increase in the rate of time discounting.' This conclusion, like Becker and Mulligan's, is not reached simply by assuming that past consumption of an addictive substance (drug) increases the utility of consumption today. Rather, it is the addict's time preference itself that undergoes change (see also Ainslie and Haendel 1982, Orphanides and Zervos 1995, 1998, O'Donoghue and Rabin 1999, and especially Bickel and Johnson 2003).

Other Correlates: The literature offers a number of additional socio-economic variables that correlate with time preference. Lawrance (1991: 69; see also Cropper et al. 1994) offers evidence in support of the argument that 'time preference rates for white household are slightly lower than those for nonwhite households in all but the highest income levels,' while Elster (2000: 27) mentions 'investment in schooling' and a number of studies mention circumstance at the home (illnesses, presence of minor children, etc.). As summaries of several studies on time preferences, we have: 'subjects who are more concerned with future and attach greater importance to it are also more intelligent and successful ... and achievement oriented' (Agarval and Tripathi 1980: 144); 'subjects who have suffered high prolonged deprivation are indeed predominantly and significantly more past oriented and lack a vision of future' (p. 149); and 'the capacity to choose a larger reward, delayed for a relatively short and specified period of time, over a smaller reward available immediately is related to the degree to which personal future events in general appear to be endowed with a sense of reality, as well as to the degree of everyday preoccupation with future rather than present events' (Klineberg, 1968: 256; see also Benzion et al. 1989, Chapman and Winquist 1998, Frederick 2002 and for a comprehensive survey and classification of correlations, Godoy et al. 2004).

1.5 Vignettes On Endogenous Discounting

The preceding two sections establish that the concept of individual time discounting is not a simple fixed parameter and that discounting can vary across and within individuals depending on circumstances and choice context. For the most part, however, the research we review, especially that which concerns anomalies, concentrates on modeling individual choice outside of any social context (with the exception of those aggregate evolutionary models that justify time discounts as the product of natural selection and process of intergenerational transfer of resources). Reported anomalies, moreover, seem more a problem for economic theory or models of social process that borrow directly from economics than anything else. There are, however, circumstances in which within-individual discounts change owing to the specifics of social processes in such a way as to suggest that a variety of processes and policies ranging from anti-drug programs, to environmental protection, to the proliferation of political – economic corruption, to the spread of street gangs that cannot be addressed adequately until we better understand the endogenous nature of time preferences – of how individual time discounts are learned and change as a consequence of social interaction and the structure of social networks, and how social welfare is impacted by that learning and change.

Sustainable Development and the Ogallala Aquifer: It is by now well understood that one of the crucial problems of the 21st century is that of resolving the 'contradiction between the growth imperative of the modern world system and the constraints of a finite planet ... [whereby] sustainable development meets the needs of the present without compromising the ability of future generations to meet their own needs' (Gallopin et al. 1997: 2). Correspondingly, students of ecology construct scenarios of the future development of the planet, given the different decisions about overall investment in new technologies and normal consumer choices, which achieve different end points, depending on the technologies chosen, the policies pursued by governments, and so on. Naturally, the preferred scenario is that of sustainable development in which people exhibit greater care about future generations in everyday behavior. However, before this social transformation can occur 'two legitimate moral and social imperatives must be reconciled: the needs of the present and the needs of the future. The living standards of the billions today who do not enjoy the benefits of human progress must be improved. At the same time, development patterns in both rich and poor countries must be altered so as not to leave a bitter social and environmental legacy to future generations' (Gallopin et al. 1997:2). The particular difficulty here is twofold. First, sustainability confronts the standard problem of collective action whereby individuals have an incentive to free ride on the contributions to conservation made by others. And second, the weight given to the future, even if adequate for sustainability among developed economies, may be inappropriate for meeting the needs of the developing world. In other words, different persons in different economic circumstances will weight current needs against the future differently.

An environmentally conscious social planner, then, might ask whether, to what extent, and how individual time discounts can be changed so as to encourage sustainable development, and here it is useful to survey recent experiences with respect to the Ogallala aquifer. Briefly, the Ogallala aquifer provides much of the groundwater for irrigation and ordinary consumption in Kansas, Nebraska, Wyoming, Colorado, eastern New Mexico and western Texas and Oklahoma. It is in fact one of the largest sources of fresh water in North America. But like other depletable resources, it is a public good subject to exploitation and over-use. As the *National Geographic* (Zwimgle, March 1993: 99 – 100) noted, although it was not until the 1950s that the aquifer was tapped as a significant source of irrigation by new technologies, 'by the end of the 1970s, farmers and geologists could not ignore a pattern of declining local water levels. In 1982 the Department of Commerce undertook a study of the entire region: It predicted unvarnished disaster ... groundwater levels in northwestern Kansas were dropping an average of one foot per year ... In some places the Ogallala's 'overdraft' ... had reached 95 percent – [or] for every gallon of water pumped out a teacupful was being replaced by natural processes.' There were any number of solutions available to individual farmers: adoption of dryland farming techniques, closer monitoring of soil wetness relative to optimal levels, water recycling systems, and precision application of water. Adoption of any of these solutions, though, required the adoption of new techniques and technologies at private expense, thus occasioning a classic example of the Prisoners' Dilemma. Various approaches to stem overuse were attempted, including government intervention via regulation and fines. But especially here farmers resisted government regulation and interference. One approach, though, proved fruitful: education. Not merely the education of farmers with respect to the long -term dangers of current rates of usage, the benefits of alternative technologies, and heightened awareness of their 'civic responsibilities' in preserving the aquifer, but education directed at children that alerts them to the problem. Although it might have been assumed that such programs were designed to educate the 'farmers of the future,' they in fact had a more immediate impact, and as one of the originators of this effort is quoted by the *National Geographic* (p. 103), 'a worried grandchild is worth five visits from a regulatory official.'

The question here is what precisely in the grandfather's calculation of self-interest is being changed by the grandchild? What is the grandchild's impact on the grandfather's beliefs and preferences that leads to choices that depart from the status quo? There are several possibilities, including his sense of civic responsibility (altruism), heightened sensitivity to the private costs of current practice, and an appreciation of the long-term private benefits of changing one's practice and technology. But there is yet a fourth possibility. The 1982 Department of Commerce report forecast that at the current rate of usage, water levels would have fallen by 23 percent by 2020, with Texas alone having depleted two- thirds of its total. Thus, confronted with a warning that applied some 40 years into the future, a farmer might reasonably ignore the problem altogether if he applied a 'reasonable' time discount and took into account only the technical information provided by the government (for example, with the commonly applied discount rate of 5%, $1000 forty years from now is worth a mere $120 today). However, by drawing attention to the concerns of a future generation – those most likely to inherit the farm currently being tilled – we can speculate that the grandfather's private discount is lowered and his planning horizon extended over decades rather than years. Rather than be concerned only with the farm's productivity in the current and following year, a grandchild asking if the farm will be his when he grows up extends that horizon nearly indefinitely, and this extension as scenarios such as the repeated Prisoners' Dilemma tell us, make cooperative (water conserving) equilibria more sustainable.

In support of this argument we can offer a quote from a recent issue of *Time* devoted to the problems of sustainable development, where the author emphasizes how having to plan his children's future had changed his own perspective on the sustainability of the environment:

> Children have a wonderful ability to focus their parents' attention on the essentials. Before our twin sons were born in 1987, I had often heard about all the environmental problems projected to come to a head toward the middle of this century. But I was born in 1937, so I would surely be dead before 2050 as a real date and I couldn't grasp that the environmental risks were real. After the birth of our kids, my wife and I proceeded to obsess about the things most parents obsess about – schools, our wills, life insurance. Then I realized with a jolt: my kids will reach my present age of 65 in 2052. That's a real date, not an unimaginable one! My kids' lives will depend on the state of the world in 2052, not just on our decisions about life insurance and schools (Diamond 2002: A54)

Drug and Alcohol Abuse, Alcoholics Anonymous: If our hypothesis is correct as to the impact of the 'worried grandchild' on the 'grandfather', we have an example of how interactions can impact time discount rates. We can speculate, however, that the process whereby time preference changed involved group processes as well. As 'grandfathers' met and socialized via their normal networks, they impacted each other's beliefs, perceptions and preferences – reinforcing the effects of their 'worried grandchildren'. For a more explicit example, though, of how time discounts can be impacted in a group context, and with the research of Becker and Murphy (1988) in mind along with their argument that addiction impacts time preference (thereby rendering those rates endogenous to whatever other factors encourage or discourage addiction), consider the following sample of quotes from the *Alcoholics Anonymous Handbook* (2001, 4th ed) that concern the typical experiences of members (all italics added):

1. *I know my goals, dreams, values*, and boundaries, and I know how to protect, nurture and validate them (p. 327).
2. Yes, there is a substitute [to drinking liquor in companionship] and it is vastly more than that. It is a fellowship in AA ... Your imagination will be fired. Life will mean something at last. *The most satisfactory years of your existence lie ahead.* (p. 152).
3. When I wake up today, there are lots of possibilities. *I can hardly wait to see what's going to happen next* (p. 374).
4. We have a sign at the AA meetings .. that says *Expect a miracle* (p. 380).

5 My marriage is no longer on the brink ... *And the best is yet to come* (p. 406).
6 Every action, word, prayer, and 12th Step call is *an investment* in my spiritual freedom and fulfillment (p. 469).
7 At my first meeting the people fulfilled their responsibility – they made me welcome. I met others like me and it felt good. Maybe this thing was for real. So I went to another meeting, and I got the same feeling. Then another meeting ... (p. 509).
8 I learned how to be a good AA member by watching good AA members and doing what they do. I learned how to have a good marriage by watching people with good marriages and doing what they do. I learned how to be a parent by watching good parents and doing what they do (p. 521).
9 You are going to meet new friends in your own [AA] community ... Among them you will make lifelong friends. You will be bound to them with new and wonderful ties, for you will escape the disaster together and you will commence shoulder to shoulder your common journey (p. 153).
10 Of far more importance was the fact that he was the first living human with whom I had ever talked, who knew what he was talking about in regard to alcoholism from actual experience (p. 180).

The value of organizations such as Alcoholics Anonymous (including Cocaine Anonymous, Marijuana Anonymous, Narcotics Anonymous, and so on) is well documented. But the question is what, in generalized theoretical terms, is it that this organization does to the motives of its members? What values, perceptions and beliefs are changed by participation in its 12-step process of recovery and by sustained participation in the program? Answers are suggested by the preceding quotes. The italicized parts of the first six quotes reveal the heightened awareness of and value given to the future. This, in turn has two complementary interpretations: (1) the beneficial value of future outcomes is made more vivid and (2) members are encouraged to take greater cognizance of those benefits by giving the future greater weight in their day-to-day existence. That is, in a manner wholly consistent with Becker and Murphy's argument that addicts are likely to be those with limited patience, rather than discount the future so that only short-term immediate needs are valued (presumably the need to quickly resolve the need for alcohol and the personal circumstances that might have led one to drink) and long-term costs and consequences ignored (lost wages, lost relationships, medical problems), AA members are encouraged to extend their horizons and look to the future in anticipation of the benefits that result from sobriety (restored relationships, stable employment, health).

The last five quotations give us some insight as to how AA implements this change: through the establishment of a peer group (network) that reinforces this change in perspective – a peer group with a similar long-term view (see Burkett and Jensen 1975, Johnson et al. 1987). We can only speculate as to how this peer group exerts influence, although we can take Nee and Lian's (1994) suggestion with respect to how opportunism spreads in a communist system ('the inevitable opportunism among high-level party officials and economic bureaucrats sets a 'demonstration effect' that only induces others to become opportunists' (p. 278)) and hypothesize that active membership in AA demonstrates the benefits of sobriety and reinforces patience among recovering alcoholics. The consequence is that private discounts are endogenous to the group, with each member reinforcing the heightened value given to the future of every other member (and presumably encouraging new members to do the same).

Crime, Gang Membership: Much of the research into crime, especially urban crime among youth and the phenomena of urban gangs, focuses on social correlates – education, poverty, race, ethnicity, unemployment rates, domestic problems, and so on. There are hints in the literature, however, of a dynamic that takes us beyond simple correlations. First, if we combine the fact that gang activity largely involves persons between the ages of 13 and 21 (i.e., 'youth'), the general conclusion that youth are more impatient that adults, and the argument that propensity to commit crimes correlates with a 'here and now' impulsiveness (Wilson and Herrnstein 1985, Gottfredson and Hirschi 1990), we can conjecture that impatience (a high discount on the future) contributes to a person's willingness to join a gang so as to realize the immediate rewards of criminal activity and peer group support at the expense of long term consequences (jail, rendering oneself less employable, foregoing educational opportunities, etc). Indeed, the potential relevance of time discounting is explicitly recognized in Gordon's (2000) study of street gangs in Vancouver: 'It may well be that experiencing the frustrations of attempting but failing to achieve material success in a highly materialistic society, results in a rejection of conventional ways of succeeding and an embracing of the unconventional ways. This may be especially prevalent amongst adolescents and young adults who are bombarded with achievement messages and who are raised on the idea of instant gratification for minimal effort' (p. 55).

Gordon's (2000) recommendations for combating gang formation builds on this perspective: 'Anti-gang programming appears to be most effective when it is aimed at the supply of new gang and group members ... Programs in high school can reduce fear and intimidation, dry up the source of gang personnel, and help generate negative perspective of gang membership' (p. 57). Similarly, Maxson and Whitlock's (2002) study of street gangs reports that participation in religious activities as opposed to other community or school-based groups best discriminated gang and non-gang memberships. Once again, we can ask what parameters in the motives of potential members would such programs impact. Taking our cue from the inferred strategy of AA and its explicitly spiritual or religious (but non-denominational) overtones, we might reasonably guess that it is encouraging the application of a lower discount on the future by supplying youth with alternative peer networks and finding ways to sustain relatively higher valuations of the future – or finding ways to disrupt the high discounting that characterizes gang membership.

With respect to sustaining gang membership, again we find parallels with Alcoholics Anonymous. The methods differ (e.g., 'The gang member who fails to live up to the gang's requirements ... incurs immediate loss of status, removal from the gang, physical harm, or even the risk of being killed' (Ruble and Turner 2000: 247)), but just as AA supplies a social network in which a recovering addict can overcome temptation, 'the gang provides rules to live by and a code of conduct' (Allender 2001: 4). And just as Alcoholics Anonymous discourages its members from having daily contacts with non-recovering addicts, and attempts to reinforce its impact on individual motives by establishing new peer groups, 'street gangs tend to have rigid boundaries, which denotes that boundaries are relatively impermeable. Gangs typically do not approve of active members having friends, especially close friends, outside of the gang' (Ruble and Turner 2000: 247).

Also in support of the idea that peer associations within gangs account for motives and actions, Thornberry, et al. (1993) assess why crime rates among gang members are higher than among non-members with otherwise similar socio-economic characteristics. As Howell (1994) summarizes their findings:

> they examined three explanatory hypotheses: (a) a selection ...model, (b) a social facilitation ... model, and (c) an enhancement model that combines aspects of the selection and social facilitation ... [a] posits that gangs recruit adolescents who are already delinquent, whereas ... [b] holds that group processes and normative structure of gangs bring about high rates of delinquency among initially non-delinquent youths who join. Thornberry et al. find that social facilitation or kind of group model to receive the strongest support ... [and that] the normative structure and dynamics of the gang must be the central focus (p. 503)

Left unexplored are specific parameters of individual choice (preference, belief and time discount) that provide for the fundamental description of and explanation for a gang's influence. But one reasonable hypothesis is that gangs enforce impatience – a willingness among members to forego a concern with long-term costs and benefits in favor of the immediate rewards of criminal activity.

1.6 Corruption

To this point our argument is simply that individual time discount rates are not wholly exogenous, fixed, and uniform across a population, with the suggestion that they are often determined by social processes we can label *socialization* or *learning* (as well as *evolutionary*). Certainly we concur with Bowles's (1998) argument that 'most writers have invoked a kind of functionalist correspondence between economic structures on the one hand and values, customs and tastes on the other, without explaining the mechanisms by which the former might affect the latter ... Nonetheless, the argument that economic institutions influence motivations and values is plausible, and the amount of evidence consistent with the hypothesis is impressive' (p. 76). We would only add that this argument should not be restricted to economic contexts, but applied to individual time preferences and social processes generally. Thus, to make use of these ideas we turn, by way of an extended and elaborated example, to an assessment of corruption as a specific mode of interpersonal cooperation and exchange. We do this not because of any immediate apparent connection between the different forms of corruption found in the world and the dynamics of individual discount rates. But if we interpret corruption as a particular form of cooperation, we can gain some insight into variations in its pervasiveness across societies as well as changes in the degree of corruption within a society.

Monitoring and Cooperation: Corruption exists in almost all countries but flourishes perhaps with special force in the newly independent states of the former Soviet Union, where it is commonly associated with market failure, economic stagnation, and an ever-present threat to democratization. Corruption is not new to the territory of the former USSR, but today it seems especially pervasive: 'In Russia in 1991, for example, getting a business started often required bribing the local legislature, the central ministry, the local executive branch, the fire authorities, the water authorities, etc.' (Shleifer and Vishny 1993: 605) and 'Ports and customs check point are notorious. Stories like that of the American importer who had to pay $6000 in cash to release a container of goods sitting on the dock in Northern Russia circulate within the business community. Unfortunately, corruption seems to be largely tolerated by both government and business' (Wilson and Donaldson 1996: 159). It is also worth noting that the general level of corruption does not appear to have stabilized (as reported in ITAR-TASS, 19 November 1999), and that we can observe considerable variation in apparent levels of corruption across the different parts of the former Soviet empire, where corruption levels and bribery are much higher in component states of the USSR than in those countries that belonged to its Eastern bloc. It is not the existence of corruption *per se*, then, that focuses our attention, but its apparent dynamic and variability across states.

To draw a connection between corruption and the potential dynamics of individual time discounts that might account for this variability, note first that corruption requires cooperation. In its most common form, an instance of corruption requires an exchange: money for services, or services for services, that, ostensibly, the law doesn't or shouldn't allow. However, regardless of whether we deem a particular exchange corrupt, the most theoretically interesting aspects are the mechanisms employed to enforce agreements and the particulars of an exchange. The problem with which cooperation of any type must contend is that few if any examples of exchange allow for an instantaneous event. If, for example, money is paid for some service, then that service is either performed before the full cash value is exchanged or it is performed subsequently. In either case, whoever moves second has an incentive to renege on their payment or provision of the agreed upon service. Efficient exchange, then, requires an enforcement mechanism that ensures against or at least discourages defection: 'although cooperation produces mutual gains for those who participate, in the absence of an enforcement mechanism cooperation will not be possible' (Posner 2000: 13).

As for enforcement mechanisms, there are the courts and the remedies the judicial process establishes. There are also those social norms that try to enforce notions of justice and fair exchange through informal sanctions and the value people place on their reputations for fairness and honesty. Here enforcement takes the form of a refusal to deal subsequently with a person who has been judged untrustworthy in the past. Such enforcement mechanisms are illustrated by the literature on temporal embeddedness (see, for instance, Rooks et al. 2000) whereby it is noted that the effort that needs to be spent in negotiations is decreased if transactions are embedded in a temporal network that allow for a repeated sequence of interactions. Thus, as Rooks et al. (2001: 124-6) note, temporal embeddedness consists both of a shadow of the past as well as of the future, where 'shadow of the past' refers to the duration, frequency, volume and success of previous transactions. In turn, the notion of a shadow of the future is a function of this past insofar as it establishes expectations whereby 'firms learn to trust [or distrust] each other over time' (Blumberg 2001: 842).[2]

Note, however, that regardless of form, these mechanisms require two things for their operation: monitoring and people who do not discount the future 'too greatly'. A contract will be adhered to – indeed, a contract will be forged – only if the threat of punishment for a defection or promise of reward for compliance is credible. And credible threats and promises require two things. First, it must be possible to monitor actions so that parties to an exchange can discriminate between a defection and compliance. Monitoring may be imperfect as when a person is a member of some large group and only the product of the group's combined effort is observable. In this instance, sanctioning threats may have to be applied against every member of the group or to some member chosen randomly, but the principle is the same: monitoring allows a defection today to be punished tomorrow. The second thing required for credibility is that implementing a threat or promise must correspond to a subgame perfect equilibrium in the game that describes the circumstance in question. That is, those empowered to

---

[2] For additional literature on temporal embeddedness see Bergadaa (1990), Tybout and Artz (1994) and Buskens and Weesie (2000).

administer, say, a punishment in the event of a defection must find it in their self-interest to apply the appropriate sanction even if doing so is personally costly. Absent subgame perfection, a threat becomes non-credible – both contracting parties know that it is individually irrational to implement the threat. But in this case there cannot be an agreement, since the contracting parties know that whoever moves second will have little incentive to hold up their end of the bargain and deliver on their promise.

The credibility of future actions is critical to sustaining cooperation in personal relations, even when the word 'threat' entails nothing more than a damaged reputation and a refusal to cooperate with the defector later. But having placed responsibility for cooperation on credible threats and promises, we must then explore the requirements for subgame perfection and the role of monitoring, and here it is useful to consider the classic illustration of the problem of cooperation, the repeated Prisoners' Dilemma as portrayed in Table 1.1, which corresponds to a standard Dilemma provided only that K exceed 1, in which case 'defect' dominates 'cooperate' and the unique equilibrium outcome is (0, 0). Now, however, consider an especially simple repeated form of this game that allows only two strategies: (D) defect always; and (GT) a grim-trigger strategy that chooses C as long as the 'opponent' cooperates, but which

Table 1.1. Standard Prisoners' Dilemma

	cooperate	defect
cooperate	1, 1	–2, K
defect	K, –2	0, 0

switches to D for all time upon the detection of a single defection. Rather than analyze this situation by simply letting d be the discount factor a person applies to future payoffs, suppose detection is uncertain: specifically, let q be the probability of detecting a defection. Hence, if a person defects from GT to D when their opponent continues to play GT, they get a payoff of K for that period, a payoff of zero thereafter with probability q (the probability that their defection is detected and they are subsequently punished in all future rounds of play), and a chance to take advantage of their colleague in the next round with probability $1 - q$. Thus, assuming an exponential pattern of time discounting that satisfies the independence assumption, the time discounted value of defecting to the strategy D when played against GT is:

$$K + Kd(1 - q) + K d^2(1 - q)^2 + Kd^3(1 - q)^3 + \ldots = K/[1 - d(1 - q)]$$

so that the repeated game in strategic form becomes as shown in Table 1.2.

Table 1.2: Repeated grim trigger reduced game

	G: Grim Trigger	D: Defect Always
GT	$1/(1-d)$, $1/(1-d)$	$-2/[1-d(1-q)]$, $K/[1-d(1-q)]$
D	$K/[1-d(1-q)]$, $-2/[1-d(1-q)]$	0, 0

Note now that the strategy pair (GT, GT), which uniquely yields cooperation in this limited version of the repeated Dilemma, is an equilibrium if and only if

$$1/(1-d) > K/[1-d(1-q)],$$

or, equivalently, if

$$d > (K-1)/(K-1+q) \qquad (1.2)$$

Thus, if detection is certain (if $q = 1$), we get the usual condition for a cooperative equilibrium in the repeated dilemma, $d > (K-1)/K$. That is, the 'shadow of the future' must be sufficiently great so that the cost of future punishments outweighs whatever short term gain a person might realize from defection. Conversely, if $q = 0$, the mathematical identity $1/(1-a) = 1 + a + a^2 + a^3 + ...$ no longer holds, and cooperation is impossible.

The implications of this discussion are intuitive. First, as K increases so that the relative value of defection increases, the requirement on q and d for cooperation becomes more severe. And second, as monitoring becomes more difficult (as q decreases), the value of d required to sustain cooperation increases. Alternatively, we can say that as the weight given to the future, d, decreases, *cooperation will be sustained in only those exchanges in which monitoring is near-perfect.* Moreover, there is nothing exceptional about this analysis. We can readily substitute a cost of monitoring for q, or allow a more complex array of strategies that permit punishments in various forms (e.g., tit–for–tat or grim trigger strategies that punish for a prespecified number of rounds), but the substance of our conclusions remains unchanged.

Corruption as Cooperation: Our task now is to establish the relevance of this analysis to corruption, and here it is useful to consider the fact that corruption, especially in its political form, typically entails bilateral or at least small-scale relations – between the bribe giver and the bribe taker or between a bribe giver, an agent for the bribe taker (e.g., a bureaucrat who acts as an agent from some higher-level official with the authority to deliver on a promise), and the bribe taker. Scale is dictated here by the fact that owing to the ostensible illegality of the exchange, a certain amount of secrecy is required: 'the illegality of bribery induces participants to spend resources keeping the transaction secret' (Rose-Ackerman 1999: 12). Indeed, without a well-defined and firmly established judicial system to enforce contracts and with little more than social acquiesce available to ensure adherence to social norms, 'gift giving and bribery will be more common ... [and] if legal guarantees are not possible, trust is correspondingly more important ... both sides to the deal have an interest in blurring the meaning of the payment in the eyes of the outside world while keeping it quite explicit between themselves.' (Rose-Ackerman 1999: 98). But trust is based on reputation – 'unless there are repeated transactions, how is an official to guarantee performance or a payee to guarantee payment if the service is provided prior to payment?' (Jain 2001: 90) – which, if a degree of secrecy is to be maintained, requires a more atomistic and less anonymous form of exchange than the forms of market cooperation we assume correspond to, say, international investments or corporate finance. Corrupt officials simply do not have the luxury of establishing a brand label (a reputation for fulfilling their corrupt obligations) in the same way a manufacturer of baby food might seek to develop a reputation for a safe and nutritious product: 'How can an official announce that he or she is willing to enter a corrupt deal? While open dissemination of this information would attract the best buyer, it would also attract attention from controlling authorities and rivals or superiors who might wish to share in the proceeds' (Jain 2001: 90). Put differently, 'Because legal advertising and official certification are out, corrupt officials need to communicate by other means. The difficulty of establishing a reputation may lead crooks to deal only with known partners... Establishing a corrupt relationship between strangers is risky because the first one who is explicit about wanting to make or accept a payoff is at the mercy of the others who can threaten to report the illegal act ... Stable corrupt systems will be easier to maintain in small, local markets than in large national or international ones' (Rose-Ackerman 1999: 101).

This argument fits well with Raiser's (1999) discussion of the absence of extended trust in post-Soviet societies and Grief and Kandel's (1994) assessment of the difficulties of contract enforcement in Russia. Briefly, having emerged from a political-economic system that, in effect, discouraged the development of extended trust (where by extended trust we mean the basis of establishing and maintaining contracts between individuals who have only limited information about each other's attributes), economic exchange must rely instead on trust based on kinship or explicitly repeated interaction – hence the pervasiveness of corruption in post-Soviet states at the expense of other forms of economic cooperation. This is not to say that all reputational cooperation is necessarily individualistic or even that all legal exchange occurs wholly in the shadow of a legal system: witness the role of intergenerational reputations in Asia as a substitute for contract law enforcement, as well as the role of reputation in private exchanges between repeat customers and, say, a small corner grocery store (Posner 2000). However, reputation based enforcement, combined with the absence of both a legal backup, necessitates closer monitoring by all parties to an exchange and, correspondingly, a more atomistic structure to cooperation generally.

One implication of this argument is that *the more atomistic the exchange necessitated by the requirements of monitoring, the more likely are those exchanges,* ceteris paribus, *to fall into the category we call corruption*. If this is correct, then we have here an alternative approach to understanding differing levels of corruption across societies as well as variations in rates of corruption within a society. Specifically, we have an argument for hypothesizing that as people's time discounts increase – as $(1 - d)$ increases – the forms of cooperation we observe will more likely correspond to corruption. Briefly, as $(1 - d)$ increases, cooperation will occur only under conditions of closer and more perfect mechanisms of monitoring than when $(1 - d)$ decreases. This, in turn, will tend to drive out of the political-economic market those forms of cooperation that rely on impersonal and judicial modes of monitoring (e.g., extended trust) and substitute instead those that require close ties, kinship and the like. And to the extent (but only to the extent) that this form of monitoring is dominated by forms of cooperation we label 'corrupt', our argument thereby rationalizes the view that if people give the future little weight, their actions will correspond to the rule 'get what you can while the getting is good,' which, in turn, seems more suitable to an economy driven by corrupt business and political practices and short-term objectives rather than the long-term investment criteria often employed when evaluating a more standard business venture.[3]

The Endogeneity of Corruption and Discounting: Because not all forms of cooperation enforced by informal yet efficacious monitoring can be labeled corrupt and because there may be substitutes for impersonal and judicial modes of monitoring upon which to build a viable political-economy, our discussion here should be viewed as limited to tendencies that apply 'more or less' to different societies and circumstance. We also appreciate that it may be reasonable to conceptualize people as if they operate with a variety of time discounts: long-term for easily monitored relationships and short-term for less easily monitored exchanges, notably those that occur outside the sphere of their immediate social connections. As the external environment grows more risky so that operating with a limited planning horizon becomes a necessity in this sphere, people

---

[3] Not all forms of cooperation that require close monitoring should be labeled 'corrupt' (e.g., that based on family or kinship trust). The implicit assumption of our argument, then, is that as discounts increase, the proportion of 'corrupt' forms of cooperation do not decrease relative to these other forms. This is a fragile assumption and ought to be viewed as part of the description of the substantive boundary of any model based on our argument.

come to rely more on those relationships that allow for long-term commitments and exchanges, which, if our earlier argument is correct, implies substituting exchanges that might be deemed instances of corruption at the expense of those less-personal and more anonymous exchanges that characterize markets elsewhere. Nevertheless, to complete the circle of our argument, consider this observation with respect to political elites: '[a] ruler's very venality may make him insecure and subject to overthrow. This insecurity induces him to steal more, making him even more insecure, and so forth. As a consequence he or she will have a relatively high discount rate for government projects and will support projects with quick short-term payoffs and costs spread far into the future' (Rose-Ackerman 1999: 34). The suggestion here is that corruption feeds impatience (high discounting), which in turn feeds corruption. That is, both the level of corruption in society and people's discount rates are endogenous to the political-economic system, which is itself impacted by the level of corruption in society (see Klochko and Ordeshook 2003 for elaboration of this argument).

Aside from the specific case of the political leader referred to in Rose-Ackerman's discussion, the precise nature of this endogeneity is unspecified – specifically, the individual and social processes whereby levels of corruption impact discounting. But we can offer some speculations. First, if as we noted earlier, high discounting is encouraged by low incomes and even poverty, and if corruption is detrimental to overall social welfare, then corruption can induce higher rates of discounting directly through its impact on overall economic performance. Second, we should consider those 'demonstration' or learning effects whereby people with 'inappropriate' low discounts mimic the behavior of those with high discounts who, because they give the appearance of economic success (e.g., high rates of conspicuous consumption and low, less visible rates of saving), become 'role models' for behavior and normative development. And finally, we should not ignore the possibility that people can exert a direct influence (as in Alcoholics Anonymous and street gangs) on the time preferences of others via various systems of reward and punishment as applied within the social networks that describe society.

The relevance of understanding how private discounts might change extend beyond such issues as drug addiction, street crime and corruption. They go to the very heart of understanding how democratic polities function and are sustained. Suppose we ask why some persons might choose to serve, for example, as members of a national supreme court and forgo far more lucrative legal practices, why persons of wealth might forgo a further husbanding of their fortunes and seek instead public office with remunerations that are a mere fraction of what they can earn elsewhere, why private fortunes are sometimes dissipated in the quest for public office, or, returning to the issue of corruption, why public office holders might resist the temptation to use their official positions solely for private gain? What, in other words, is the source of these forms of 'democratic patriotism'? The answer, in part, must be a desire to 'live for the history books' and to be remembered as statesmen, jurists, and legislators rather than 'mere' entrepreneurs, inheritors of wealth, or 'politicos'. But such motives can operate only if two conditions are satisfied. First, there must be some reasonable guarantee that history will be there to cast its judgment. Such motives matter little to a jurist if there is little prospect of the court surviving into the next decade, or to the legislator if political instability characterizes the state. What happens to the incentives to become the *n*th president of some state or why function in a democratic way once that office is secured when there is little likelihood that the office will survive into future generations? And second, individuals themselves must value the future, perhaps even a future that extends beyond their physical existence. The time preferences of a society, then, are often critical in determining the performance and fate of states. But note the complexities here: Individual time preferences are likely to discount the future heavily if the state is unstable and performs poorly, whereas poor performance and instability are the likely consequences of a society in which individuals give the future little weight. Uncertainty, then, feeds impatience and impatience feeds uncertainty. But now the critical question: Is the stability of the state, insofar as it depends on individual time preferences, predetermined by some set of cultural parameters and determinants of uncertainty, or are uncertainty, a state's stability and individual time preferences all endogenous and changeable? And more to the point, is the discount people apply to the future determined strictly by the uncertainties they perceive in combination with some set of fixed cultural determinants? Or do those discounts have other sources, perhaps even ones that can be manipulated by human design, quite apart from the prevailing insecurities of everyday existence? In understanding the time preferences people apply to their actions, is it sufficient to understand how they perceive and accommodate uncertainty, or are there other processes that impact individual time discounts so that those uncertainties themselves are endogenous and inexplicable without an understanding of these processes?

This volume cannot provide answers to all such questions. But in the next chapter we offer some evidence in support of the idea that time discounts follow a logic independent of perceived uncertainty. And in the chapters that follow we pursue a model consistent with this idea, if only so that we can begin to disentangle what seems a complex nexus of relationships in the social decision calculus of individuals.

2 Evidence of the Endogenous Determination of Discount Rates

An addict ... may have accurate beliefs about the disastrous effects of the drug on his or her body or purse, and yet ignore them because of an addiction-induced increase in the rate of time discounting (Elster 2000: 10)

2.1 Introduction

To this point the evidence we offer in support of the hypothesis of endogenously determined and potentially changeable private time discount rates is largely circumstantial. The anomalies we review intimate that time preferences adhere to no simple pattern and may indeed be idiosyncratic, empirical correlates highlight changes that simply correlate with individual life cycles, and the vignettes we consider are merely suggestive but hardly conclusive. In this chapter, then, we offer further evidence against the case of exogenously set and permanently fixed time preferences. That evidence comes from two sources: A survey of time preferences among recovering drug addicts and a survey that contrasts the time preferences of Ukrainian students who have studied or are studying in the West against a similar cohort of students who remain in Ukraine. Once again, of course, that evidence does not 'prove' the case any more than we can ever establish any empirical hypothesis as fact. But hopefully it will be strong enough to convince the reader that the theoretical work that follows in subsequent chapters treats a real phenomenon and not one of our imagination.

2.2 Recovering Addicts

We begin our discussion of drug recovery and time discounts by recalling first that immediate surroundings play a crucial role in a person's descent into and recovery from self-destructive behavior such as drug or alcohol addiction. Friends, acquaintances and significant others serve as role models and provide support and encouragement for patterns of behavior, good or bad. Thus, 'drinking ... is a social phenomena in two related ways: (1) what one does with and thinks about alcohol is a function of his or her group membership and identification, and (2) almost all drinking takes place in social group settings which the drinker 'believes makes his drinking socially approved by the people who matter to him' (Akers 1973: 122). Social imperatives apply with special force in the case of drugs and drug addiction: 'The peer group is now widely regarded as an important social environment capable of influencing both the initiation of substance use and the maintenance of patterns involved' (Cotterell, 1996: 129).

It is well understood that the problem is sustaining the recovery process even following an initial 'commitment' to sobriety. Too often people relapse back to their self-destructive choices, and here we acknowledge the success of recovery and support programs such as Alcoholics Anonymous (AA), Cocaine Anonymous (CA), Narcotics Anonymous and so on. The structure and methods of these programs are well documented and certainly consistent with the academic literature on addiction, peer groups and social learning. But a question remains. If Elster is correct in the argument that introduces this chapter and if we extend that argument to say that the recovering addict has accurate perceptions of the 'pleasures' of addiction, then what is it, in generalized theoretical terms, that these organizations, by their readjustment of peer group associations, change in the motives of members? If it is not an understanding of the objective costs and benefits of addiction and recovery, what other values, perceptions and beliefs are changed by a sustained participation in a 12-step recovery program?

Clearly, the changed social environment these programs attempt among participants is crucial for achieving and sustaining sobriety: 'differential peer associations played the largest role in explaining cessation of drug use, particularly that of alcohol, marijuana, and stronger drugs' (Winfree et al. 1993: 103). In this way programs such as AA and CA do more than merely communicate costs and benefits. In addition, they immerse the recovering addict in a sober peer group using a variety of social activities – picnics, camping trips, and so on. But once again the question that remains is: What precisely is it in the parameters of an individual's decision calculus that changes when peer groups change? The hypothesis we consider here with a simple interview of recovering addicts in a CA program is that one parameter in particular that is changed is the time discount people apply to the future.[1] That is, regular participation in a voluntary recovery program extends people's horizons such that they give future outcomes greater weight and more immediate outcomes less weight in day–to–day decisions. We also suggest that sustained participation in a program is crucial in the sense that the critical change occurs sometime after five years

[1] The relevance of time discounting in drug recovery is suggested as well by the literature on crime and gang membership that points to the importance of peer group associations. See, for example, Miller 1975, Johnstone 1983, Wilson and Herrnstein 1985, Gottfredson and Hirschi 1990, Winfree et al. 1994, Gordon 2000, Nagin and Pogarsky 2001, Maxson and Whitlock 2002, and Klein 2002. The results reported in this section, however, are taken from Klochko (2005).

of regular participation and sobriety. Our analysis here, then, is wholly consistent with Bickel and Johnson's (2003) conclusions that 'delay discounting appears to be a fundamental behavioral process underlying drug dependence' and that 'individuals with sufficient experience in pursuing and accomplishing relatively long-term goals may be less susceptible to allowing the brief, intense reinforcing effects of drugs to commander a large part of their behavior' (p. 435).

One might infer an implicit understanding of these effects in the very philosophy of a '12-step program'. Requiring years to complete, such programs set achievement milestones that motivate individuals by future goals. If one completes step 3, there is a degree of instant gratification, but *at the same time* one is encouraged to anticipate the rewards of accomplishing steps 4 through 12. Clearly, one of the objectives of a program such as AA is not simply to change the addict's peers, but to change their degree of impatience as well. To explore this relationship, a sample of 141 attendees at a series of Cocaine Anonymous meetings in the Los Angeles California area were administered a simple anonymous questionnaire, where, in addition to each respondent's age and gender, respondent's were asked to answer three questions.[2]

Q1: Would you say that your attendance at CA meetings is
____ : regular (not less than once a week)
____ : intermittent (on average at least once a month)
____ : sporadic (a few times a year)
____ : this is relatively new to me

Q2: How many months/years of sobriety can you claim?
____ : less than 6
____ : between 6 and 12 months
____ : between 1 and 2 years
____ : between 2 and 5 years
____ : more than 5 years

Q3: Would you prefer to receive $200 today or $500 in six months.[3]
____ : $200 today
____ : $500 six months from today

---

[2] No significant differences exist between male and female respondents or between those 40 years or younger vs. over 40. Of 87 males, 36 (41.4%) chose $200 whereas of the54 females, 26 (48.1%) chose $200. Of the 70 respondents under age 40, 30 (42.9%) chose $200 whereas of the 70 respondents over 40, 31 (44.3%) chose $200.

[3] For other attempts at measuring time preference see Benzion et al. 1989, Horowitz 1991, Holcomb and Nelson 1992, Green et al. 1994, Nagin and Pogarsky 2001, Frederick 2002, Frederick et al. 2003.

Question 3 provides only a crude ordinal estimate of time preference, but the results are illuminating. First, consider the data in Table 2.1, which divides our sample approximately in half according to the reported length of sobriety. The data here are striking and the imputed relationship statistically significant at the .01 level (Chi^2.= 19.17). Specifically, we see a nearly inverse relationship between length of sobriety and one's choice of an immediate reward of $200 over the more distant $500.

Table 2.1: Length of sobriety vs. discounting

	$200 today	$500 in 6 months
sober < 5 years	45 (53%)	39 (47%)
sober ≥ 5 years	18 (31%)	40 (69%)

A somewhat finer breakdown of the data is even more revealing. Table 2.2 divides those who report being sober less than five years into two subgroups: sober less than a year and sober for a year or more but less than five. With a relationship significant at the .01 level (Chi^2 = 19.46), the suggestion here (but only a 'suggestion', given the size of our sample) is that the critical change in individual discount rates occurs at some point after five years of sobriety. That is, during the first few years of abstinence, recovery may be slow, at least as measured by impatience and a willingness to postpone gratification. It is only after a full (even lifetime) commitment to sobriety (and the recovery program) that the fruits of participation are fully realized.

Table 2.2: Length of sobriety vs. discounting

	$200 today	$500 in 6 months
sober < 1 year	31 (53.4%)	27 (46.6%)
1 ≤ sober < 5 years	13 (52.0%)	12 (48.8%)
sober ≥ 5 years	18 (31.0%)	40 (69.0%)

We use the word 'commitment' meaningfully, since we should differentiate between those who make a drug recovery program a central feature of their lives versus those for whom participation is a tenuous thing. If we consider Question 1about the regularity of participation, then, of those indicating regular attendance, 21 (39%) chose $200 while 33 (61%) chose $500, whereas irregular attendees divide nearly equally between $200 and $500 (although only 17 (23%) respondents reported less than regular attendance and a bare majority chose $200, the relationship is in the anticipated direction and significant at the 0.01 level: Chi^2 = 12.48). What is perhaps more interesting is the suggestion of an interactive effect for regularity and years of sobriety. Looking at the data in

Table 2.3, notice the strong relationship between years of sobriety and inferred time discount among those reporting less than regular attendance and the weaker but still statistically significant relationship among those reporting regular attendance (at the .05 level, $Chi^2 = 4.450$). We appreciate that with only twenty respondents reporting irregular attendance, conclusions must be tentative, but it appears that, although years of sobriety has the stronger impact, regular attendance can 'substitute' somewhat for duration of sobriety.

Table 2.3: Regularity, length of sobriety and time discounting

		\$200	\$500
irregular attendance	< 5 yrs sobriety	15 (68.2%)	7 (31.8%)
	≥ 5 yrs sobriety	4 (30.8%)	9 (69.2%)
		19 (54.3%)	15 (45.7%)
regular	< 5 yrs sobriety	29 (47.5%)	32 (52.5%)
	≥ 5 yrs sobriety	14 (29.5%)	31 (70.5%)
		43 (40.0%)	63 (60.0%)

Certainly it would be valuable to apply similar questions to addicts who have not sought the aid of a recovery program or whose participation failed to yield the desired sobriety. The data here, moreover, warrant an important note of caution. Specifically, the bimodal distribution in the duration of reported sobriety portrayed in Table 2.2 suggests the possibility of selection bias whereby those who persevere in the program are those with low discounts initially. That is, rather than impacting this parameter of choice directly, we cannot reject the hypothesis that a 12-step program merely filters on it so that those with high discounts relapse before five years. Speaking against this interpretation is the fact that those with intermediate terms of sobriety do not appear, on the bases of their responses to Question 3, to be in the process of 'filtering' and indeed are not statistically different in that respect from those beginning their 'journey of sobriety'. Nevertheless, as always correlation cannot imply cause, especially in a cross-sectional study. But our data do lend weight, tentative as it might be, to the idea that not only is the discount people apply to the future subject to endogenous change, but that that change can be effected by an effective policy of drug recovery.

2.3 Culture, Selection and Socialization

Although the data in the preceding section are consistent with the assumptions we employ in the chapters that follow, we cannot at the same time reject the hypothesis that time preferences are culturally determined and fixed and that it is simply the composition of various subpopulations in our data that change, where by culture we mean a 'set of stable, habitual preferences and priorities in men's attention, and behavior, as well as in their thoughts and feelings' (Deutsch 1962: 62). In this section, then, we consider some additional evidence in support of the hypothesis of dynamic time preference using data from another source – students from the former Soviet Union (notably, Russia and Ukraine) – that allow us to more explicitly address the issue of whether those preferences are 'hard wired' or whether observed differences in subpopulations are a function of simple selection.[4]

Data: It is generally agreed that human personality traits are most readily subject to change in formative years (puberty through ages 18–25). The older one gets, the more difficult it is to change core aspects of personality without invoking major cognitive dissonance or imbalance. Hence, the targeted population of our analysis is youth, ranging in age from approximately 17 to 35. One component of that population consists of people whose countries of origin were part of the former Soviet Union (primarily Ukraine and Russia) but who came to the United States and other Western countries as students. Some persons in this sample came with immigrating parents, several have completed their education and found jobs in the West after graduating, and a few returned to their respective countries of origin. Of special importance to us here is the fact that we have a sample of 172 respondents whose length of stay in the West varies. The second component of our data is our sample of 324 high school graduates, university students, young professionals, and graduate students currently living in Ukraine.

For the purposes of describing the methods of collecting data, we call the population originating from former USSR and currently living in the West, or who studied in the West and returned, 'immigrants'. A questionnaire consisting of thirteen questions was e-mailed to approximately 400 such individuals along with an introductory note stating the goal of research, a privacy statement and a promise to publish the results for those who express an interest in our findings.[5] Our rate of return here was approximately 43% and yielded a total of

4 The analysis reported here is taken from Klochko (2006).

5 We used several sources to contact 'immigrants': (1) participants in the 'Muskie' exchange program supported by the US government, which provides funding for graduate studies for young citizens of the former USSR; graduates of the US government-funded FSA (Freedom Support Act) exchange program, which provides funding for undergraduate non-degree one-year studies; undergraduate and graduate students from Cornell University; graduates and current participants of a non-profit US organization that helps Ukrainian citizens apply to US educational institutions;

172 completed questionnaires. The survey for Ukraine was conducted among students enrolled in several Kharkiv high schools, the National Pharmaceutical Academy, the Kharkiv Police Academy, and the Kharkiv Pedagogical University. The questionnaire itself was translated into Russian (the most common language in eastern Ukraine [Kharkiv] where the questionnaires were collected)[6] and conducted in person due to the limited access of the general population to internet technologies. Questions 1 through 4, however, were modified, with monetary terms scaled and expressed in national Ukrainian currency (hryvnya), and a total of 324 interviews conducted. The appendix to Klochko (2006) gives the English text of both questionnaires along with the introductory note that accompanied its e-mailed version.

The Questionnaire: There are some slight differences in the questions asked of both samples that reflect the different context and experiences of respondents. However, most of the questions are identical or at least compatible. Briefly, **Question 1**, which asks whether the respondent prefers $200 today or a guaranteed $500 six months hence replicates the question asked of our sample of recovering drug addicts.[7] **Question 2** is a modification of Question 1, set in a slightly different context, and asks respondents what would they do with a fixed amount of money, where presumably the choice of saving for a vacation indicates future orientation. **Question 3** may pose a confusion of terms, where the choice between an average versus an expensive restaurant may address issues of social status more than any view of the future. Nevertheless, this question represents yet a third approach to measuring qualitatively time discounts. **Question 4** attempts to measure in quantitative terms the discount people place on time. We understand that responses here may be skewed since in the former Soviet Union it was considered 'immoral' to charge interest on borrowed money. It is therefore interesting that within our Ukrainian sample we find people who indicate a willingness to charge up to rates of 300 and 400 percent, whereas the highest rate we observe among our 'immigrant' sample is 25 percent. It would appear, then, that this question taps directly the connection between an uncertain political/economic environment and people's valuation of the future. **Question 5** operationalizes the 'religiousness' of respondents as their reported frequency of church attendance, and we include it here because the literature on time discounting hypothesizes that religion correlates with time

---

and members of the Ukrainian Student Organization in the US.

6 The survey was also used in Western Ukraine, although with fewer people (20).

7 For other attempts at measuring time discounts and discussion of these rates see Benzion, Rapoport and Yagil (1989), Horowitz (1991), Holcomb and Nelson (1992), Loewenstein and Prelec (1993), Green et al. (1994), Chapman and Winquist (1998), and Frederick (1999, 2002).

preferences to the extent that the expectation of a 'life after life' extends the 'shadow of the future.' At the same time, this question can serve as one of our controls to ensure comparability across our two samples. **Question 7** asks respondents about their professional plans. Presumably, the existence of a plan, however vague, indicates an ability to project into the future. **Question 8** concerns marital status, and we consider it here since the literature offers the hypothesis that married people tend to discount the future less. As a control variable, we would not want to infer that time preferences change merely because our sample of 'immigrants' contains a higher proportion of married respondents than does our sample of respondents in Ukraine. **Question 9** concerns gender, and although we have no specific hypotheses here, there is some evidence in the literature that women discount the future less than men given their clearer focus on family and child rearing. **Question 10** asks the age of respondents, and we include it here not merely as a general control across our two samples, but also because of the considerable evidence (see Chapter 1) that adolescents are more impatient than older generations. Knowing age, moreover, also allows us to compute the percentage of one's life 'immigrant' respondents have spent in the West.

The remaining questions differ across our two samples although they are structured to maximize compatibility. With respect to our 'Western' sample, **Question 6** asks respondents the number of years they have studied in the West (undergraduate and graduate). **Question 11** simply asks country of origin ranging from 0 = USSR to 4 = Eastern Europe, and where 1 = Russia, 2 = Ukraine, 3 = other former USSR countries such as Armenia, Georgia, Latvia. Note that even though this question is not asked in our Ukrainian sample, we assign all respondents there a score of 2 [Ukraine].[8] **Question 12** asks whether the respondent is currently living in a different country from her country of origin, and if so, for how long. This question, then, provides an estimate not only of residence change but also a period of time spent in a 'new' country. This information is useful for those cases when somebody immigrated without attending a higher educational institution. In terms of asking current country of residence, coding is similar to that of question 11 after adding the categories 5 = United Kingdom and 6 = USA. **Question 13** attempts to assess a respondent's current peer group or set of social acquaintances by asking whether he or she communicates primarily with people who are citizens of the current country of residence versus other foreign nationals.

As for our Ukrainian sample, **Question 6** asks educational level, ranging from incomplete high school to complete college degree. The questionnaire

---

[8] Mobility is low and you must be a citizen of Ukraine to go to Ukrainian schools and colleges. Hence, we can safely assume that our sample uniformly contains only Ukrainian citizens.

administered in our Western sample includes this question, but since our target audience consists exclusively of university students we know a priori that all are or were undergraduate or graduate students who can be differentiated by age. **Question 11** parallels Question 13 in our Western sample, except that here we ask the respondent to estimate the number of people they associate with from the West as measured by number of people they know for whom the native language is English, German or French. **Question 12** asks whether respondents have plans to leave the country to study or work. Our hypothesis here is that people who answer 'yes' would discount the future less than those who answer 'no'.

Descriptive Statistics: We should not be so naive as to assume that respondents who emigrate to the West as students or for other reasons are a random sample of the college-age population in the former USSR. Certainly, just as selection occurred in all previous waves of emigration to North America – the risk takers as they are sometimes called – selection doubtlessly continues today. And we would be surprised to find that those who do emigrate value the future the same or less than those who remain at home – that they are less prone to take risks than those who preceded them. Thus, the same selection issues that arise in our study of drug recovery – namely, the possibility that instead of observing any change in time preferences, our study shows only that those who value the future highly, or at least are not subject to the same level of impatience that describes a propensity to drug addiction, remain in a recovery program – arise here as well. Our sample of 'immigrants' however, gives us some hope of sorting through this effect, since it is one thing to drop out of a drug rehabilitation program and another to return to Ukraine after years of study in the West. Thus, years of study gives us some handle on the processes of socialization that might occur after any initial selection. But first let us review some of the summary statistics of our respondents to assess the comparability of our two samples.

We begin by looking at the overall differences between the two samples, and in Table 2.4 we see that, generally, the two are approximately comparable with respect to the demographic characteristics of age and gender although the 'immigrant' sample is slightly older on average owing to the larger share of those between ages 26 and 30 (but note that if we combined the categories 21– 25 and 26 – 30, our two samples would look quite similar). The mean age of 'immigrants' is 25.4 years, whereas that for their Ukrainian counterparts is 21.7. Our 'immigrants' also contains a slightly higher proportion of females: 49 percent as compared to 39 percent of their peers remaining in Ukraine. Other variables differ more significantly. First, most respondents who live or have lived in the West have professional plans (58%) or choose economics or the law as their profession (30%). This total of 88 percent compares with only 39 percent having similar plans in the Ukraine. Again, we most likely see here how political-economic instability and the perceived absence of opportunity takes its toll in

Ukraine, but there are other demographic and attitudinal differences as well. A greater proportion of our 'immigrant' sample is married (31% versus 13%), 'immigrants' naturally tend to communicate more with a mixed population than do those who live in Ukraine, and the proportion of 'immigrant' respondents who claim to have no religious interests whatsoever is more than twice that of our Ukrainian sample (41.4% versus 20.2%).

Table 2.4: Descriptive Statistics

	'Immigrants'	Ukrainians
Age		
15-20	19.2%	34.5
21-25	34.3	58.5
26-30	35.5	3.9
31-35	8.1	2.7
36 or more	2.9	0.3
Gender		
male	51.0%	61.0
female	49.0	39.0
% married	31.0	14.0
Occupation and career goal		
none	4.1	45.5
student	8.1	2.1
professional	56.4	39.5
econ or law	31.4	13.0
Church attendance		
not at all (0)	41.4	20.2
religious but don't attend (1)	19.5	45.5
holidays only (2)	25.4	26.8
once a month (3)	5.3	3.6
weekly (4)	8.3	3.9
Communication		
citizens of current residence	12.2	7.8
mixed	72.7	26.8
home country (or no foreign)	15.1	65.4

Greater still are some of the differences between the two samples with respect to the first four questions that seek to tap future orientation and time preference. Table 2.5 gives the breakdown of these two samples and although we see little appreciable difference on Questions 2 and 3 (there is a tendency among 'immigrants' to save any bonus as opposed to partaking of immediate consumption), we find sharp differentiation on Questions 1 and 4. First, with respect to Question 4, approximately the same percentage of respondents would charge no interest (48.2% versus 40.2%), but if they were to charge, Ukrainians would clearly charge a higher rate (with several respondents suggesting rates of up to 200, and in one case 300 percent). It seems that once Soviet-era ideas about the 'immorality' of monetary interest are rejected or otherwise subverted, students who live in Ukraine expect a far higher rate of return than do their counterparts in the West. This, of course, may reflect not only a different preference for time, but also a general distrust of people borne of an uncertain political-economic climate and the countless stories one reads there of criminal activity, corruption, dishonesty, etc.

Table 2.5: Basic Comparisons on time preference

	'Immigrants'	Ukrainians
Question 1		
$200 (1000 h) today	18.0	58.7
$500 (2500 h) in 5 months	82.0	41.3
Question 2		
luxury or party	47.7	51.4
save	52.3	38.6
Question 3		
average restaurant	29.1	27.4
expensive restaurant	70.9	72.6
Interest charged on loan		
0	48.2	40.2
0 – 5%	18.0	0.0
5 – 10%	9.6	1.2
10 – 25%	21.7	14.2
25 – 30%	1.8	24.5
50 – 100%	0.6	13.0
> 100%	0.0	1.9

The most interesting difference, however, in terms of the stark contrast it offers, concerns Question 1 (the question that was the focus of our study in the preceding section) for here we see over 80% of our 'immigrant' sample, like drug addicts who have spent five years or more in a drug recovery program, choosing the greater sum of money some six months hence. In contrast, less than half our sample in Ukraine make that same choice (41.3 percent). This contrast becomes even more pronounced if we look at whether a respondent studied in the west, since there are six respondents who emigrated to the US but not for purposes of college attendance (and two who failed to answer the question). Here, as Table 2.6 shows, the percentage choosing $500 six months from now among those who attended a Western university rises slightly from 82 to 84 percent, while the percentage choosing $200 today among those who have never attended a Western university is virtually unchanged owing to differences in sample size.

Table 2.6: Question 1

	never studied in West	studied in West
$200 now	198 (58.6%)	27 (16.5%)
$500 later	140 (41.4%)	137 (84%)

The Argument for Selection: The data Tables 2.5 and 2.6 summarize suggest a considerable difference between Ukrainian and 'immigrant' samples in both their willingness to postpone immediate gratification (Q1) and the desire to be compensated for postponed repayment of debts (Q4), and both questions suggest that respondents in Ukraine discount the future much more heavily than do their counterparts who have traveled to the West. Moreover, any variation we find in these two questions when we combine our two samples of respondents and conduct a series of logistic regressions tells a simple story, regardless of which variables we include as independent variables; namely, having studied or spent time in the West is significant and critical to explaining any variation in our two dependent variables. Table 2.7 reviews a series of logistic regressions that make little attempt at screening independent variables according to any criteria since our general conclusions are unchanged by which variables we include or delete (bold faced coefficients are significant at 0.01 whereas '*' denotes a coefficient significant at 0.05, and standard errors in parenthesis). As this table reveals, although religion and education are significant at the 0.01 and .05 levels of significance respectively with respect to Question 1, whether or not a person studied in the West is significant at the 0.01 level for both dependent variables regardless of the configuration of independent variables.

Asserting that our data confirm the impact on time preferences of experience in the West ignores the extent to which we can attribute the differences we observe in our two samples on Questions 1and 4 to mere selection. Indeed, the argument for attributing most if not all of these differences to selection has a reasonable justification and receives some support in our data. The mere act of applying to a graduate program in the United States (researching universities, filling out forms, taking SAT and TOEFL exams, submitting transcripts and so on) and giving up the security of one's family entails a certain acceptance of risk and a willingness to incur immediate costs and forego that security for potentially greater but postponed rewards. Indeed, it has long been argued that America benefitted from its immigrants not merely because of the labor, ideas and energy they bring with them, but also because immigrants themselves are not a random draw from their native populations. Instead, they are the risk takers, people willing to trade security and certainty for potentially greater rewards not only for themselves but for their children, grandchildren and so on.

Table 2.7: Logistic regressions for combined sample

	Q1: (n = 502), linear			Q4: (n = 472) linear		
intercept	–0.40 (.34)	–0.01 (.19)	–.35 (.11)	**33.3** (13.3)	**31.5** (2.92)	**27.5** (2.51)
studied in west	**1.43** (.58)	**1.97** (.26)	**1.97** (.24)	**–18.6** (6.1)	**–16.3** (4.85)	**–21.3** (4.5)
career	0.16 (.108)	0.07 (.10)	–	–1.77 (2.26)	–	–
religion	**–0.24** (.10)	**–0.24** (.10)	–	–0.40 (1.20)	–	–
gender	–0.41 (.23)	–0.35 (.22)	–	–0.45 (4.69)	–	–
marriage	–0.02 (.29)	–	–	3.18 (5.86)	–	–
communication	–0.30 (.18)	–	–	– 7.9* (3.69)	**–9.04** (3.44)	–
education	.037* (.17)	–	–	4.20 (3.70)	–	–
age	–0.31 (.17)	–	–	–0.34 (.68)	–	–
adj R^2	0.25	0.23	0.21	0.05	0.06	0.04

Our data are not inconsistent with this view. First, consider Table 2.8, which offers a simple cross-tabulation of Question 1 against 7. The suggestion here, of course, is that respondents who state that they have no interest in leaving Ukraine have a (statistically significant) greater likelihood of choosing 1000 hryvnya today as opposed to 2500 hryvnya six months from now. Thus, if those who move to the West to work or study are drawn only from those who indicate a willingness to leave Ukraine or who at least are not committed to staying, there is an immediate upward bias in the propensity to reveal a preference for future reward over the present among those who are in the West.

Table 2.8: Question 1 versus 7, Ukraine

	1000 h	2500 h
stay in Ukraine	134 (66%)	70 (34%)
difficult to say or leave	58 (48%)	62 (52%)

Consider also Table 2.9, which offers a simple cross-tabulation between Questions 1 and 5 (religion), and note the decided relationship between the frequency of church attendance and one's propensity to choose a payment immediately: Slightly more than half of those who stated that they do not attend church and have no religious interests whatsoever (coded 0 here) choose the postponed payment of 2500 h. But as we move to higher categories of religious conviction or commitment, the proportion choosing an immediate payment of 1000 h increases appreciably.

Table 2.9: Question 1 versus 5 (religion), Ukraine

religious attendance/interest	1000 h	2500 h
0	32 (48%)	35 (52%)
1	88 (58.3%)	63 (41.7%)
2	55 (61.8%)	34 (38.2%)
3 + 4	20 (80%)	5 (20%)

It might seem that the data in this table are at odds with the usual relationship observed between impatience and religiosity and tangential to the issue of selection bias. But Table 2.9 imposes no additional controls for such potentially intervening variables as age and indeed is almost certainly a stand-in for other factors. And with respect to selection bias, recall from Table 2.4 that the proportion of respondents coded 0 on this question in our 'immigrant' sample is twice that of its Ukrainian counterpart (41.4 versus 20.2 percent). Thus, if we assume that the relationship between religion and one's propensity to choose 1000 h today over 2500 h six months from now is constant, and if we assume as well that respondents who have moved to the West were initially chosen disproportionally from those with no religious interests (as Table 2.4 suggests), then again the preceding table is consistent with the selection hypothesis.

Finally, consider the variable age and Table 2.10, which reports the relationship of age to Question 1. Although the differences here are not spectacular, there is a tendency on the part of younger Ukrainians to prefer the later reward compared to older respondents. As with religion, this relationship is the opposite of the one usually reported in the literature, but is understandable in terms of people's experiences with the uncertainties associated with the transition from the Soviet to post-Soviet era. Certainly youth are far less likely to have been impacted in their lifestyles by the collapse of the USSR and all the uncertainties that accompanied that event. At the same time, the potential role models people might have confronted, as generated by Western media, would also have had a differential impact, depending on age. Insofar as the selection hypothesis is concerned, however, we can calculate that the average age at which our 'immigrant' respondents arrived in the West was 19.5 (age minus number of years living in the West). If we assume that time preferences are thereafter 'frozen' – that whatever the cultural or political/economic influences might be of living in the former USSR might no longer operate after emigrating – then there would be a decided bias in favor of choosing to postpone rewards among our 'immigrant' sample as compared to respondents who, for one reason or another, chose to remain in Ukraine.

Table 2.10: Age and Question 1, Ukraine

	1000 h	2500 h	*n*
age 17 – 21	55.3%	44.7%	141
22 – 25	59.3%	40.7%	140
26 – 45	70%	30%	20

Changing Time Preferences: Despite the preceding evidence for selection, there are several anomalies in the data that do not wholly fit this hypothesis. Consider Table 2.11, which reports the answers respondents from our Ukrainian sample give to Question 1, broken down by gender and the extent to which they indicate that they communicate regularly with people other than those who speak their native language (Question 11). The first thing to note is the virtual absence of any relationship between gender and responses to Question 1and, more surprisingly, the negative (though admittedly weak) relationship of this question to our question about communication. If we were to look at our 'immigrant' sample, we would first see that males outnumber females approximately 60 to 40. More interestingly, men in that sample are more likely to choose a postponed reward in answering Question 1 than are females (90 versus 74 percent).

Thus, we cannot argue that the 'immigrant' sample appears to discount time less than our Ukrainian respondents merely because men are more likely to move to the West than are women. Nor can we argue that those who emigrate bring with them a higher valuation of time because they communicate more regularly with non-Ukrainian or non-Russian speakers. Indeed, we have here the somewhat unexpected finding that our Ukrainian respondents are more likely to choose an immediate reward if they report a circle of foreign speaking acquaintances than if their acquaintances are uniquely other Ukrainians.

We cannot assert that the data in this table alone wholly subverts the selection hypothesis. In fact, as we show later, selection almost certainly accounts for a good share of the differences we see across samples. But such relationships (or their absence) raise questions about the precise nature and extent of selection and whether it can assume that it accounts for all of the differences we observe.

Table 2.11: Gender, communication and Question 1, Ukraine

	1000 h	2500 h
Gender		
male (n = 197)	58.4%	41.6%
female (n = 127)	60.6%	39.4%
Communication		
0 (n = 215)	57.2%	42.8%
1 (n = 86)	60.0%	40%
2 (n = 23)	74%	26%

To begin our untangling of selection from dynamically changing time preferences, Table 2.12 reports the results of a series of logistic regressions with Questions 1 and 4 again as dependent variables, but using only data from our 'immigrant' sample in conjunction with the several measures of length of time spent in the West in the set of independent variables. The first thing to note here is that the sole demographic variable that remains significant in every regression – indeed, it is the unique demographic variable that achieves any reasonable level of statistical significance – is gender. But more importantly, time spent in the West – measured either as number of years in college, number of years in absolute terms, or number of years as a percentage of age – is significant (generally at the 0.01 level) throughout. Gender and time alone explain much of the variance in our dependent variable, and looking at the last column of this table, it is also evident that it is not the case that time is a substitute for any demographic variable or set of variables. Time in the West has its own impact on an immigrant's propensity to choose \$500 six months from now rather than \$200 today, and by inference a significant impact on time preferences. That is, not only does time spent in the West discriminate strongly the responses to Question 1 for our combined sample (see Table 2.7), it does the same even within the 'immigrant' only sample.

***Table 2.12**: Logit regressions, Question 1, 'immigrants' only*

	1	2	3	4	5	6	7
intercept	2.22 (1.49)	**1.42** (.53)	2.20 (1.63)	**1.78** (.50)	1.54 (1.67)	**1.82** (.49)	2.4 (1.5)
years West	–	–	**0.17** (.07)	**0.15** (.07)	–	–	–
years West %	–	–	–	–	**0.04** (.02)	**0.04** (.02)	–
years college	0.34* (.14)	**0.34** (.13)	–	–	–	–	–
career	0.12 (.30)	–	0.07 (.32)	–	0.06 (.32)	–	0.11 (.30)
religion	– 0.00 (.18)	–	0.00 (.18)	–	0.01 (.18)	–	– 0.12 (.16)
gender	**–1.45** (.47)	**–1.47** (.46)	**–1.52** (.51)	**–1.57** (.50)	**–1.51** (.51)	**–1.60** (.50)	**–1.30** (.45)
marriage	0.22 (.51)	–	0.11 (.52)	–	0.13 (.51)	–	– 0.01 (.49)
comm.	– 0.44 (.44)	–	– 0.57 (.49)	–	-0.59 (.49)	–	– 0.47 (.43)
age	– 0.03 (.05)	–	– 0.01 (.05)	–	0.02 (.05)	–	0.01 (.05)
adj R^2	0.18	0.17	0.20	0.18	0.20	0.18	0.11

It is interesting now to predict what our immigrant sample might have looked like upon arrival in the West in order to estimate how much of the difference in responses to Question 1 between our two samples can be attributed to socialization versus selection. Since we are interested here in orders of magnitude we use simple linear regression and consider three regression models:

Model A: time and gender (since gender is the uniquely significant demographic variable);

Model B: time, gender and age (since the average age of our 'immigrant' sample is a bit more than 25 years whereas the age of emigration is approximately 19.4 years; and

Model C: time, gender, age and career plans (since we cannot exclude the possibility that those plans change as well after emigration).

To illustrate our calculations after each of these models is estimated, consider the following simple linear regression:

$$\underset{}{\textit{Question 1}} = \underset{(0.172)}{.786} + \underset{(0.01)}{.0325}\textit{YrsCollege} - \underset{(0.06)}{.202}\textit{Gender} + \underset{(0.006)}{.0006}\textit{Age} + \underset{(0.039)}{.01}\textit{Plans}$$

Suppose we first set *YrsCollege* = 0, since that would be the appropriate value of this independent variable upon emigration. Since the actual average of this variable in our 'immigrant' sample is 3.266, this would decrease our estimate of average responses to Question 1 by 3.266 times 0.0325 = 0.1061. The average age of our sample is 25.24 years whereas the average age at which they arrived in the west is 19.4 years. If we multiply this difference (5.84) by 0.0006 we get an additional reduction in our estimate of responses to Question 1 equal to 0.0038. Finally, if we also assume that career plans change after emigration, we can multiply .01 by the difference in the average value of this variable for our two samples (in this case 2.10 minus 1.37) for an additional change of 0.0073. Summing these differences, we get 0.1061 + 0.0038 + 0.007 = 0.1172. Since the actual mean value of our dependent variable is 0.8200, we would estimate that upon arrival in the West, 82 – 11.72 = 70.28 percent of our sample of 'immigrant' respondents would have chosen $500 six months later over $200 today. But because only 41.3 percent of our Ukrainian sample made this choice, we would thereby estimate that of the difference between samples (82 – 41.3 = 40.7), 11.72 of this difference can be attributed to socialization and 40.7 – 11.72 = 28.98 to selection. Table 2.13 presents the result of similar calculations for the three regression models listed above using each of the three measures of time in the West available from our survey, and as it shows, time spent in the

West accounts for anywhere between 18 and 33 percent of the differences in responses to Question 1 across our two samples, with selection accounting for the rest.[9]

Table 2.13: Three models compared

model		yrs college	yrs in West	% life in West
A	socialization	10.7	7.4	8.1
	selection	30.0	33.3	32.6
B	socialization	10.9	7.5	13.4
	selection	29.8	33.2	27.3
C	socialization	11.7	9.3	13.5
	selection	29.0	31.4	27.2

2.4 Conclusions

Even if we deem the issue of selection versus socialization resolved, there remains the matter of determining whether and to what extent we can attribute different responses to Question 1 to variations in perceived uncertainty as opposed to some independently determined change in decision parameters. Unfortunately we do not have a direct measure of perceived uncertainty. However, we can examine the relationship between responses to that question and Question 4 – the rate of interest a respondent would charge for repayment of a loan. Certainly it is not unreasonable to suppose that presupposing that one is entitled to a high rate of interest derives from the perception of risk. And given the clear propensity on the part of respondents in our Ukrainian sample to charge significantly higher rates of interest than those in our 'immigrant' sample, along with the clear differentiation between samples with respect to Question 1, there is an immediate suggestion that perceptions of risk and a willingness to forgo the immediate reward of $200 or 1000 h in favor of a larger but postponed payment are related. To explore this one explanation for the differences between samples in their responses to Question 1, we can begin by noting once again that answers to Question 4 are most likely confounded by a Soviet-era attitude toward monetary interest, especially when applied to friends and acquaintances. This attitude

[9] We appreciate that our assumption that career plans change after emigration is suspect, but this variable never accounts for more than 0.73 percent of the difference, and it is worthwhile bearing in mind that we have made no adjustment for marriage which would raise slightly the share of the difference attributable to socialization.

doubtlessly accounts for the fact that an equal proportion of both samples (more than 40 percent) indicate that they would charge no interest. Suppose, however, that we look only at those who offer a non-zero answer to Question 4 and consider respondents in Ukraine, who exhibit the widest variation in their responses to both Questions 1and 4. Table 2.14 gives the results of three linear regressions with Question 4 as the dependent variable among this restricted subsample, as well as two logit regressions using answers to Question 1among this same sample. In each case we include a regression that includes all demographic variables, followed by subsequent regressions that incrementally restrict the list of independent variables to those that promise significance.

The two variables with the most evident impact on a willingness to charge interest on a personal loan are gender (with women more willing to charge interest than men) and whether or not a respondent has foreign acquaintances. Note first that when Question 4 is our dependent variable, the variance explained in the regression approaches zero; there is virtually no explanatory power in either Question 1 or its parallel, Question 2. It is true that gender and the likelihood that the respondent has foreign acquaintances yield significant coefficients, but we can only speculate as to their meaning. Similarly, when we set responses to Question 1 as our dependent variable, although variance explained increases considerably, Question 4 has no explanatory power. Thus, if we interpret responses to Question 4 as a measure of perceived risk, we cannot argue that the variation in answers to Question 1 is the consequence of variations in the perceptions of risk or vice versa, at least within our sample of Ukrainian respondents.

This is not to say that changing perceptions of risk and political uncertainty do not play a role in changing one's attitudes about the discount one ought to apply to the future across our two samples. Simply because we cannot attribute the different answers respondents give to Question 1on the basis of some inferred measure of perceived risk does not mean that this feature of a respondent's environment does not impact attitudes upon emigration. Nevertheless, if we take our two empirical studies together – both the one offered in this section and our study of recovering drug addicts from Section 2.1 where it is more difficult to argue that perceptions of risk plays any role in one's propensity to accept postponed rewards – we have an emerging picture of people with dynamically changing time preferences. Certainly, a good proportion of the differences we see in subpopulations can be attributed to the various processes of selection that occur within each study. But at the same time, although neither study 'proves' the hypothesis (and as we know, empirical research can only 'disprove', never 'prove') of dynamically changing time discounts, both give strong evidence in favor of the hypothesis of endogenously changing preference as opposed to one that sees time preferences as fixed and culturally (exogenously) determined.

It may be true that changing preferences are in part a product of changing

Table 2.14: Linear and logit regression., reduced Ukranian sample (n = 192)

	Linear for Q4			Logit for Q1	
intercept	25.92 (42.33)	**53.40** (8.03)	**55.00** (7.66)	2.49 (1.67)	**3.03** (1.49)
Q1	6.36 (10.28)	6.46 (9.72)	–	–	–
Q2	– 9.85 (5.81)	– 10.01 (5.64)	– 8.84 (5.34)	**0.92** (.21)	**0.90** (.21)
Q4	–	–	–	0.00 (.00)	0.00 (.00)
religion	– 1.05 (5.49)	–	–	–0.343* (.20)	–0.35* (.20)
education	– 1.19 (10.52)	–	–	0.752* (.38)	0.72* (.37)
married	6.53 (13.74)	–	–	–0.31 (.52)	–
gender	20.89 (13.18)	**22.84** (11.68)	**23.46** (11.63)	–0.19 (.48)	–
age	1.20 (2.21)	-	-	**–0.22** (.09)	**–0.24** (.08)
communication	**– 18.41** (8.39)	**– 16.7** (8.69)	**– 17.45** (8.61)	**–0.85** (.37)	**–0.83** (.33)
where to work	7.04 (7.84)	–	–	**0.64** (.28)	**0.64** (.28)
career	1.45 (4.93)	–	–	.16 (.17)	–
adj R^2	0.004	0.03	0.03	0.256	0.250

perceptions of risk. But these perceptions are likely to be a product of the social interactions that subsequent chapters attempt to model – one's experiences with and observations of other people and the information gathered through the various sub-networks of which one is a member. But this much seems certain: economic, political, and sociological analyses in which time preferences are a critical component but which impose a static and exogenously given 'shadow of the future' are built on a tenuous foundation. We should not assume, a priori, that patterns of cooperation, coordination, learning, and social network development can be satisfactorily understood absent a failure to model and simultaneously account for the value people give to the future as opposed to the present whereby that value is itself a function of the phenomena we otherwise seek to address. If cooperation depends on the time preferences of participants, it may also be true that those time preferences depend on the forms of cooperation that emerge in some dynamically changing simultaneous system of relationships.

One final comment is warranted. Attempts at assessing the value of international student exchange programs typically rely on questions that ask whether participants think they derived any benefit from the program – whether the experience was, in some vague way, deemed a positive one. There is also, among proponents of such programs, an element of faith that they facilitate democratic development in the home countries if and when students return to them. The preceding analysis of Ukrainian students suggests that in fact preferences are changed in a fundamental way – in this case, in the parameter that represents the value people place on the future. And to the extent that this change is maintained upon returning to one's home country, then if giving the future 'sufficient' weight facilitates the various forms of cooperation required to develop and sustain a democracy, our analysis gives some measure of the value of these exchanges.

3 The Evolutionary Perspective

In human society almost any cultural entity can be seen as a meme: religions, language, fashions, songs, techniques, scientific theories and concepts, conventions, traditions, etc. The defining characteristic of memes as informational patterns is that they can be replicated in unlimited amounts by communication between individuals, independently of any replication at the level of the genes (Heylighen 1992: 71)

3.1 Introduction

Chapters 1 and 2 discuss the definition and complex character of individual time discount rates with a review of the evidence suggesting that those rates vary systematically across populations and within individuals as a function of age, experience and so on. In this chapter we move away from specific empirical arguments to a more formal analysis of endogenous time discounting. This analysis is, in fact, largely a speculative combination of two somewhat separate modules. First, consider this argument: '[w]e cannot expect people to take steps to reduce their rate of time discounting, because *to want to be motivated by that long-term concern ipso facto is to be motivated by that long-term concern*' (Elster 2000: 28, italics in original; see also Raub and Weesie 1990). The fact is that time preferences do change, and in lieu of requiring that they do by some well defined conscious and explicitly utility-maximizing choice, we implement the suggestion by Ainslie (1992) that individual time preferences are subject to (perhaps unconscious) evolutionary selection. Framing our presentation in the scenarios set forth in Chapter 1, we employ Maynard-Smith's (1974, 1982) notions of *Evolutionary Stability* (ES) and of an *Evolutionary Stable Strategy* (ESS) and Dawkins's (1976) notion of a *meme* to describe alternative equilibria for time preference. Our assumption is that an individual discount is like a behavioral heuristic that prospers or perishes on the basis of its relative performance, given the social context (a society vulnerable to corruption, a neighborhood populated by street gangs, a drug rehabilitation program that impacts peer associations, or a decision to move to another country).

The analysis that follows, then, is in the tradition of, or at least consistent with, those models that attempt to explain cooperative action as a 'trait' that is subject to evolutionary selection (see, for example, Boyd and Richerson 1992,

Lomborg 1996, Bester and Guth 1998, Bisin and Verdier 2001, Gintis 2003, Guttman 2003). However, rather than assume that the thing that evolves is simply a propensity to cooperate or not cooperate or corresponds to a specific preference, set in the context of some 2x2 game, we in effect dig further into the structure of individual preferences to the extent that individual time discounts are a basic general parameter of preferences that may or may not apply to decisions about cooperation.

The analysis that follows in this volume is in fact consistent in principle with studies that assume that what evolves is either a propensity to cooperate or not cooperate or preferences that render cooperation an equilibrium, as well as with those analyses that find repeated interactions among the same persons inessential for the evolution of cooperation but which, nonetheless, see time preferences as critical (see, for instance, Bendor and Swistak 2001and Gintis 2003). However, our analysis differs from previous formulations in several key respects. Briefly, the discussion in this chapter is intended merely to illustrate the types of parameters that can determine the composition of time preferences in a group – in this case, the proportion p of those who discount the future 'greatly' as opposed to those who do not. In evolutionary game terms, p is the ES proportion of high discounting types. However, rather than fully explore the structure of a specific evolutionary game, in the next chapter, we embed our analysis in a model of investment and growth that seeks to describe patterns of development in society's overall welfare as a function of the dynamic adjustment of individual time discounts in the direction of 'appropriate' ES values. And in subsequent chapters we allow p itself to vary endogenously as a function of the relative performance of alternative types.

To see now how a specific value of p can be rationalized as the ES outcome of some evolutionary game, suppose as for example Macy and Skvoretz (1998) do when taking the payoffs to a two-person Prisoners' Dilemma as the measure of genetic fitness ('In our model, fitness is a function of cumulative payoffs in repeated PD games. Relative fitness determines the probability that each strategy will propagate' p. 644; see also Macy 1996), we can take the situational specific payoffs achieved in pairwise interactions between people (of similar or different types) to be a measure of the fitness of those types. Those payoffs, of course, are a function of the situation under investigation , but in Section 3.2 we first and more generally apply the methodology of Maynard-Smith (1974, 1982) to derive p, the evolutionary stable proportion of different types (ES) in a population. The analysis that follows, then, is consistent with the view that 'homogeneity [in social groups] is achieved, not only by the assortative interactions of individuals with fixed phenotypes ... but by social norms that cause phenotypically plastic individuals to converge on a single behavior' (Wilson 1997: 352–53).The particular innovation we offer in this chapter is to depart from the usual assumption that 'strategies' in evolutionary contexts are backwards looking (i.e., that their fitness depends only on prior experience). As suggested by

Shapiro's (2002) formulation of adaptive computations in neural networks, whereby the genetic algorithm 'pick[s] the state which is predicted to yield the highest value of (discounted) future rewards' (p. 23; see also Sutton and Barto 1998) we allow for the possibility that time discounting memes take as their measure of fitness the prospect of future rewards, discounted accordingly. In Section 3.3 we consider a number of scenarios of endogenous discounting derived from our discussion in Chapter 1 so that we can relate the calculation of p to specific payoffs. Here we argue that a polymorphic population in which memes of different types continue to interact in equilibrium seems most likely in some situations, whereas in other situations there are two distinct ESS, thereby suggesting the development of clique or distinct 'subcultures' whereby the members of each subculture interact only with themselves.

3.2 A Simple Evolutionary Model

To introduce the evolutionary component of our analysis, we proceed by first reviewing the notion of an evolutionary stable strategy (ESS) as introduced by Maynard-Smith (1974, 1982) set in an example commonly referred to as a Hawk-Dove interaction (Dixit and Skeath 1999: 341–45; for a more detailed presentation of evolutionary games, see Samuelson 1997 and Weibull 1995). However, here we assume that the two competing types in the population are those who discount the future greatly and those who discount it less. Specifically, in accordance with the exponential model of time discounting, let d be the weight a person employs when evaluating future outcomes so that we say a *person discounts the future greatly as* d *approaches 0.* Referring now to Table 3.1, suppose, when two 'people' interact in whatever social context we ascribe to them, they are in effect, playing an evolutionary game in which: (1) if two 'high ds' meet, they engage in a mutually beneficial exchange that yields each a payoff of 1; (2) if two 'low d' people meet, they try to cheat each other (owing to the low value they place on the future) and thereby realize a payoff of 0; and, (3) if a high d and low d meet, their respective payoffs are the parameters $-v$ and u (with the idea here being that a low d can take advantage of a high d).

Table 3.1: Generic evolutionary game

	high d	low d
high d	1, 1	$-v$, u
low d	u, $-v$	0, 0

The payoffs u and v can model a variety of situations. If $u > 1$ and $v > 0$, the game is a Prisoners' Dilemma; if $u > 1$ $v < 0$, the situation corresponds to the game of Chicken; if $u > 1$ and $v < -1$, the situation is a Battle of the Sexes; and if $v > 1$ and $0 < u < 1$, the situation is an Assurance game. However, rather than treat this structure as a two-person game, we want to use it as a model of an evolutionary process in which cell payoffs correspond to 'fitness' (Macy 1996) so that we can assess whether a population consisting exclusively of one type or the other, or some mixture of both, confers an advantage on one type or the other. That is, for a particular distribution across types, will one type or the other replicate itself at a greater rate than the alternative? Evolutionary stability, then, corresponds to a 'strategy' or distribution $p = (p, 1-p)$ across the two types such that genetic 'defects' in the form of one type or the other can 'invade' a population that corresponds to that distribution. Defined generally,

> *Definition*: If X is a set of pure strategies and p is a probability distribution across X, then p is ES if for every alternative q across X,
>
> $$e(p, p) \geq e(q, p) \qquad \text{(A)}$$
>
> and if $e(p, p) = e(q, p)$ then
>
> $$e(p, q) > e(q, q) \qquad \text{(B)}$$
>
> where e denotes fitness (payoff).

(Note that if p is an ES strategy, then (p, p) is a Nash equilibrium; but because of condition (B) the opposite need not be true). Of course, p may equal zero or one, in which case the equilibrium corresponds to a situation in which only one type or the other can exist in equilibrium and cannot be 'invaded' by its opposite number. But if the strategy that satisfies conditions (A) and (B) had $0 < p < 1$, the equilibrium is said to be polymorphic in which the two types, in the proportion p and $1 - p$, coexist. Which situation prevails is a function of the parameters u and v, and since this game is well-analyzed elsewhere, we simply offer a description of its ES outcome for various values of u and v (see Thomas 1984, Samuelson 1997, Weibull 1997, Dixit and Skeath 1999):

- if $u > 1$ and $v > 0$, the unique Nash equilibrium is (0, 0), so the ESS is 'all low *d*s'. A population that contains high *d*s will evolve to one that contains only low *d*s.
- if $u < 1$ and $v < 0$, (1, 1) is the Nash equilibrium so 'all high *d*s' is ES,
- if $u > 1$ and $v < 0$, there are two pure strategy Nash Equilibria, $(u, -v)$ and $(-v, u)$, and one mixed strategy equilibrium ($p = -v/(u-1-v)$, $(1-p) = (u-1)/(u-1-v)$).[1] But only the mixed strategy is ESS.[2] The population is

---

1. To see this, suppose $(p, 1-p)$ is a mixed strategy equilibrium for column chooser. Then the expected payoff to row choose from the strategy 'high *d*', $p - v(1-p)$, must equal the expected payoff, up, from the 'strategy' low *d*. Setting $up = p - v(1-p)$ and

polymorphic with low and high *d*s.
- if $u < 1$ and $v > 0$, both (1,1) and (0,0) are Nash equilibria and ES. That the mixed strategy cannot be ES follows from the result offered by Thomas (1984: 181– 2): if *x* is ES and if we define $W(x)$ as the pure strategies that make up *x*, then if *y* is also ESS, $W(x)$ cannot be contained in $W(y)$. So if the game is 2 × 2 – if only two types are possible – then if there is a pure ESS, there cannot be a mixed ESS since *W* of the pure strategy would be contained in *W* of the mixed strategy.

Although the analysis of situations such as Hawk-Dove – the source of our example – is most commonly applied to modeling evolutionary stability among 'hardwired' types in which the processes of mutation, extinction and reproduction occur by the physical replacement of one type by others, 'sociologists are often interested in extending evolutionary models to the spread of cultural rules, such as heuristic routines, conventions, rituals, moral sentiments, habits, and normative codes ... [and] unlike instinctual rules, cultural rules can be 'softwired,' that is, they can be modified and passed from one organism to another without the need to replace the host' (Macy 1996: 104–5). Put differently, at issue here is what it is that evolves with respect to discounting since we are not so bold as to assume that people with high or low discounts literally die out in any of the processes we describe (as in, for example, Kacelnik 2003). If that were true, it would be difficult to account for the rapid changes in levels of corruption we observe in, say, Russia or Ukraine that appear to have required less than a decade to reach fruition or changes in time discounts within people as we suggest is the case with recovering addicts, 'grandpas with worried grandchildren on their laps', new members of gangs, or students who emigrate to the West. For example, if, as the literature suggests, delinquency is encouraged by high discounts on the future and if gang membership reinforces such discounts, then we need to assume that those discounts can change *quickly* within a person in order to explain facts such as that 'when [youths] are active members of the gang, their rate of delinquency is substantially higher than it was

---

solving for *p* yields $-v/(u-1-v)$.

2. Let the initial population contains mostly low *d*s in the proportion $(1\text{-}h)$ where *h* is nearly 0. A low *d* meets another with probability $(1–h)$ and a high *d* with probability *h*, for an expected payoff of $(1-h)0 + hu$, whereas a 'mutant' high *d* will get $h–v(1–h)$. But since *h* is small, the relevant comparison is with the terms involving $(1–h)$, in which case $(1–h)0 > –v(1–h)$. Thus, high *d*s can invade a population of predominantly low *d*s. Conversely, suppose the population is predominantly high *d*s in the proportion $(1–h)$, *h*. Then, a high *d* gets $(1–h) – vh$ whereas a mutant low d gets $(1-h)u + 0h$. Again we can ignore the terms with *h* alone to see that $(1–h) < (1–h)u$, so the mutant low *d* can invade a population of predominantly high *d*s. Thus, only the mixture is ES. It is considerations of this sort that occasion the necessity for condition (B) in the definition of evolutionary stability.

either before or after gang membership' (Thornberry et al. 1993: 80).

Here, then, we can appeal to Dawkins's (1976) notion of a *meme*, which he describes thus: 'Just as genes propagate themselves in the gene pool by leaping from body to body via sperms or eggs, so memes propagate themselves in the meme pool by leaping from brain to brain via a process which, in the broad sense, can be called imitation' (p. 192) wherein 'selection favors memes that exploit their cultural environment to their own advantage' (p. 199). Our hypothesis, then, is that private discount rates are like memes. If one rate or the other appears to be relatively advantageous either because of that person's personal experiences or because of what he or she observes, then either incrementally (or probabilistically) or instantaneously, that person will switch to making decisions in accordance with the advantaged rate.

There are, in fact, two alternative interpretations of the analysis that follows. In one interpretation we can view the decision-making entities of our models as single individuals, each possessing a meme that renders them high or low discounters of the future. In another interpretation, we can view those entities as the same person, but one with competing memes. That is, we can suppose that in day to day actions, the individual samples at random from a 'population' of high and low discounts (as with a mixed strategy) and incrementally adjusts his or her sampling probabilities in accordance with experience or on the basis of what they observe among other decision making entities. In subsequent chapters, then, when we consider networks, the nodes of a network can be interpreted either as populations of individuals, each characterized by a single discounting meme, or as a single individual with a 'population' of memes.

Setting aside this distinction for the moment, at this point it is important to appreciate that this evolutionary approach does not require abandonment of the rational choice paradigm. Specifically,

> an evolutionary model of the spread of a behavioral pattern over an entire population is not in conflict with an explanation of its emergence in terms of individual learning ... Voting, contributing to public charities, and refraining from littering or polluting are choices that are not easily amenable to a rational explanation. They need not, however, be thought of as counterexamples to rational choice theory. They may result from compliance with norms of cooperation that emerged in other contexts out of rational behavior, and were subsequently extended to the entire population through pressures that are analogous to selection in the wild. The advantages of supplementing a rational choice explanation with an evolutionary approach are twofold. On the one hand, an evolutionary model does not require sophisticated reasoning and learning in circumstances, such as large-group interactions, in which it would be unrealistic to assume them. We may rather suppose that some behavioral patterns borne out of strategic interactions spread and evolve in a large population out of simple adaptive mecha-

nisms. It is not too far-fetched to assume that strategies that make a person do better than others will be retained, while strategies that lead to failure will be abandoned (Bicchieri, 1993: 248 – 49).

Like strategies, the memes of which we speak can be simple or complex in terms of the determinants of relative fitness. But because we are not speaking of fitness in a biological sense whereby only current and past experience pertain, but also of things that allow for cognition, beliefs and speculation, we follow Shapiro's (2002) suggestion for adaptive neural networks by supposing that Table 3.1, rather than representing fitness, merely describes the possible payoffs experienced in a single event or interaction. Instead, we let a 'player' or meme's fitness be determined by whatever payoff accrues to it in that interaction under the assumption that this payoff is but the first realization of an indefinite sequence of equivalent payoffs whenever it encounters the corresponding type. Thus, as a generalization of Table 3.1 (since we can always let $d_{low} = d_{high} = 0$) 'fitness' becomes

Table 3.2: Meme's fitness discounted

	high d	low d
high d	$1/(1-d_{high})$, $1/(1-d_{high})$	$-v/(1-d_{high})$, $u/(1-d_{low})$
low d	$u/(1-d_{low})$, $-v/(1-d_{high})$	0, 0

Next, let high and low d's be in the population in the proportion p and $(1-p)$ respectively, in which case the expected present value of being 'high d' equals

$$PV(high\ d) = p/(1-d_{high}) - (1-p)v/(1-d_{high}),$$

whereas that of being 'low d' is

$$PV(low\ d) = pu/(1-d_{low}).$$

That is, if the population proportion is p and $1-p$, then a high d meme expects a payoff of 1 with probability p and a payoff of $-v$ with probability $1-p$ in the current period and in all subsequent periods. Thus, *ceteris paribus*, its expected present value is

$$[p - v(1-p)] + d_{high}[p - v(1-p)] + d_{high}^2[p - v(1-p)] + \ldots$$
$$= [p - v(1-p)]/(1-d_{high}),$$

which is what our calculation states. So, if a polymorphic population is evolutionarily stable (ES) – if (high d, low d) and (low d, high d) are equilibria to Table 3.2 interpreted as a game – then to solve for the ES equilibrium value of p we set PV(high d) equal to PV(low d) since otherwise a high d meme or low d meme will give way to its opposite number. That is, we set

$$p/(1-d_{high}) - (1-p)v/(1-d_{high}) = pu/(1-d_{low})$$

and solve for p thus:

$$p = -v/[u(1-d_{high}) - (1-d_{low}) - v]$$

Note that if both discount rates equal 0, then this expression is equivalent to the ES value of p when we equate fitness with immediate payoffs from a single interaction. Note also the equivalence here between this formulation and one in which fitness is measured by the discounted value of past experiences as opposed to a projection of future experience. Specifically, if experience in period $t = (0, -1, -2, -3, ...)$ is weighted by the discount d^{-t} (so that recent events are given more weight than less recent ones), we achieve an analytically identical formulation.

It is useful to offer a generalized version of Table 3.2 that does not contain specific payoffs but which nevertheless retains the symmetry of that representation. Thus, if for the entries in Table 3.3, the ES population is polymorphic in the proportions p and $(1-p)$, we can solve for the ES value of p thus

$$p = [D\,(1-d_{high}) - B\,(1-d_{low})]\,/\,[(A-B)(1-d_{low}) - (C-D)/(1-d_{high})]$$

Of course, this calculation of p applies only when the following inequalities hold:

$$B/(1-d_{high}) > D/(1-d_{low}) \text{ and } C/(1-d_{low}) > A/(1-d_{high})$$

which ensures that (low d, high d) and (high d, low d) are Nash equilibria to Table 3.3 interpreted as a 2-person non-cooperative game. Otherwise, either all 'high d' and/or all 'low d' are ES. For example, if $B < 0$ (if low ds can impose a cost on high ds) and $A > 0$, then for some value of d_{high} and d_{low} there will be two ES outcomes: all high d and all low d.

This case is noteworthy in that it suggests the possibility of an equilibrium in which two distinct subpopulations exist side by side by, but with no member of one group interacting with a member of another group. In fact, we are reminded here of those various instances of distinct cliques forming within, say, a school population where one clique emphasizes scholastic attainment, another athletic prowess and a third drugs or some such debilitating addiction. All three cliques exist within geographic proximity of each other, but members of one group are isolated from the other two – or if they do interact it is only infrequently.

Table 3.3: Generalized version of a meme's fitness

	high d	high d
high d	$A/(1-d_{high}), A/(1-d_{high})$	$B/(1-d_{high}), C/(1-d_{low})$
low d	$C/(1-d_{low}), B/(1-d_{high})$	$D/(1-d_{low}), /(1-d_{low})$

We can generalize this observation somewhat by noting, with respect to Table 3.3, that

Case 1: If $B < 0$ and $D > 0$, then (high d, high d) and (low d, low d) are simultaneously ES, although if $A << C$ (i.e., if $A/(1-d_{high}) < C/(1-d_{low})$), the unique ES is (low d, low d)

Case 2: If B and A are both positive, there exists a d_{high} such that (high d, high d) is the unique ES outcome

Case 3: If $B > 0$, the ES outcome will be polymorphic provided that A is not 'too large' relative to C.

There are, of course, other imaginable and seemingly reasonable time discount-dependent formulations of an ES game. For example, suppose high d memes, in addition to their day-to-day payoffs, 'anticipate' a reward R some time in the future, say t periods from the present, as long as they remain high ds. High ds, for instance, may be a type that 'plans for college' and a reward R that accrues some time in the future to those who pursue higher education. That is, a meme's fitness need not be a function solely of current and past experiences, or even the mere projection of those experiences into the future, but also a function of the potential qualitatively different futures it anticipates. Then we would add d^tR to high d payoffs. Note that because this added payoff is not realized until some future date, it is inappropriate to incorporate it in a purely biological interpretation of genetic fitness. But here we are speaking of a cultural, sociological, or psychological trait – a norm, a convention, a decision or evaluative criterion, and so on – that presumably can 'anticipate' events and establish its fitness on the basis of its presumed relative *future* performance.

However, before we turn to the various scenarios we might use to model endogenous time discounting, there is a generalization of the notion of ES outcomes that can be especially useful in modeling social processes such as the spread of corruption. Specifically, corruption might not involve a simple competition between individuals of the same type with competing memes, but rather individuals of two types – (type 1) 'honest versus dishonest' entrepreneurs and (type 2) 'honest versus corrupt' public officials – each with their own competing memes. Evolutionary stability, then, would require an appropriate matching of types – dishonest and corrupt, honest and honest – in the 'proper' proportions. Fortunately, the notion of ES can be extended to this case. Again limiting our discussion to situations in which only two memes compete 'within an individual', let $p = (p, 1-p)$ be the probability that a type 1 'plays' its first versus second strategy, and let $q = (q, 1-q)$ be the probability that a type 2 'plays' its first versus second strategy. Define the fitness (payoff) of a 'deviant' type 1 playing p' within the population (p, q) as $e_1[p', (p, q)]$ and the fitness of a type 2 playing q' as $e_2[q', (p, q)]$. Then,[3]

> *Definition:* The pair (p^*, q^*) is ES if, for any p not equal to p^* or q not equal to q^*,
>
> $$e_1[p^*,(p^*,q^*)] + e_2[q^*,(p^*,q^*)] \geq e_1[p,(p^*,q^*)] + e_2[q,(p^*,q^*)]$$
>
> and if equality holds,
>
> $$e_1[p^*,(p, q)] + e_2[q^*,(p, q)] > e_1[p,(p, q)] + e_2[p,(p, q)].$$

This definition is illustrated later, but note that if we divide the first inequality into two halves and require that

$$e_1[p^*,(p^*, q^*)] \geq e_1[p,(p^*, q^*)]$$
$$e_2[q^*,(p^*, q^*)] \geq e_2[q,(p^*, q^*)]$$

which a non-cooperative equilibrium to a game played between the two types must satisfy, we see that in a sense, (p^*, q^*) is an equilibrium, although again, not all equilibria correspond to ES strategies owing to the second condition in

3. If there is no interaction between players of the same type, we can define e_1 and e_2 thus: $e_1(p',(p,q)) = p'Mq$ and $e_2(q',(p,q) = q'Np$ where M is a 2×2 fitness matrix for type 1's when interacting with type 2's and N is the 2×2 fitness matrix for type 2s when interacting with type 1s. The e's can be defined to allow fitness to depend on their interactions with each other, but we ignore that complication. For example, then, $p'Mq$ becomes the familiar $pqM_{11} + p(1\text{-}q)M_{12} + (1\text{-}p)qM_{21} + (1\text{-}p)(1\text{-}q)M_{22}$.

the preceding definition. One other fact is worth mentioning here. For the case of a 'symmetric' 2 × 2 game it is well known that there necessarily exists an ES outcome. If the game is not symmetric in the sense that the interacting players are of distinct types, then there is no guarantee that the game will have an ES outcome (see Taylor 1979, Thomas 1984, Weibull 1998).

3.3 Alternative Scenarios

To be of empirical value the evolutionary games offered in the previous section require a substantive interpretation of parameters. But first it is useful to look at a version of the game in Table 3.3 without a consideration of future payoffs. Looking then at Table 3.4, suppose for purposes of discussion that $D > B$ and $C > A$. Then unambiguously, the unique ES outcome is (low *d*, low *d*) and in fact there is nothing to preclude the possibility that the payoffs parallel that of a one-shot Prisoners' Dilemma. But now, looking back at Table 3.3, notice that if $B > 0$ and d_{high} is sufficiently high and/or d_{low} sufficiently low, then we can readily move to a situation in which the evolutionary stable population is either polymorphic or (high *d*, high *d*). This shows us how a one period (or undiscounted) perspective on payoffs formed in accordance with a biological interpretation as opposed to psycho-sociological interpretation of meme fitness can distort conclusions about equilibrium social states. With this idea in mind we can now turn to our scenarios describing fitness in alternative situations.

Table 3.4: Undiscounted payoffs

	high *d*	low *d*
high *d*	*A, A*	*B, C*
low *d*	*C, B*	*D, D*

Corruption Scenario #1: Consider the argument 'if everyone or nearly everyone places a low value on the future, a person would be a fool to act with a long-term horizon.' This attitude mirrors the following monologue of the head of a Russian farming unit: 'I would live as the law dictates if those in power would do so. But they put us in conditions where we are forced to cheat ... I have to sell my pigs without official papers so that I can have at least some profit. At least 50% of produce goes towards avoiding the tax' (Klyamkin and Timofeev, 2000: 335). Alternatively, 'if I believe that everyone or nearly everyone will take advantage of me by failing to fulfill the terms of a contract unless the sanctions I can impose are large and detection certain, then it would be irrational to act as if I held a long term horizon and assumed that the terms of exchange would necessarily be honored as a matter of course.' Such arguments suggest a dynamic in which corruption feeds on itself via its impact on people's valuations of the future (for a discussion of the dynamics of corruption in the former Soviet Union following its dissolution, see Kneen 2000). Indeed, a more explicit appreciation of the endogeneity of discount rates in the context of corruption can be found in Rose-Ackerman's (1999) discussion of the effects of corruption on firms that somehow become immersed in corrupt practices: 'The claim of no impact on firm behavior is, however, too simple ... the corrupt nature of [a] deal may give the firm a short run orientation' (p. 34).

Now consider two individuals who, if left to themselves without state interference, can each earn X per time period, indefinitely into the future. However, in accordance with the idea that the society in question is vulnerable to corruption, suppose these ventures are subject to government 'oversight' in the form of excessive taxation, fines, extortionary regulation, and even, as has been boldly illustrated by the case of the dismantling and what amounts to the state-sponsored theft of the assets of the oil company Yukos in Putin's Russia, confiscation. That is, suppose that each time the state is called upon to 'inspect' the enterprise in question, it can impose a cost of C on our unfortunate entrepreneur. However, only when one encounters an entrepreneur who chooses 'defect' and abides by a strategy other than live-and-let-live will the state be provided with the excuse to thus interfere and reduce our unfortunate entrepreneur's payoff to $X - C$. On the other hand, we can suppose that the 'defector' who acts as a government informant earns the reward R from the state for each such discovery, but only if he forgos a measure of profit from the more cooperative live-and-let-live strategy in the form of a bribe to state officials (presumably, so that they don't inspect him). So suppose a defecting player earns $A < X$. However, if both actors are defectors, there is no reward and both entrepreneurs earn the reduced payoff A. Table 3.5, then portrays this situation as a single play 2-person 'game' which, if interpreted as the payoffs to an evolutionary scenario and if C and R are sufficiently great, yields (defect, defect) as the (presumably Pareto-inferior) unique ES outcome.

Table 3.5: Corruption Scenario # 1

	live and let live	defect
live and let live	X, X	$X - C, R + A$
defect	$R + A, X - C$	A, A

However, rather than simply take the payoffs in Table 3.5 as the input to an evolutionary model in the same way as some researchers interpret the payoffs to a Prisoners' Dilemma as a measure of a strategy's genetic fitness, let us suppose, in accordance with what we know about defectors from cooperation in repeated games as well as in accordance with Rose-Ackerman's suggestion that corruption (here in the form of accepting a payoff from the state for 'turning in' a business associate) breeds low valuations of the future, that defectors are characterized by low values of d, live-and-let-live types by high values, and that the fitness of either type is determined by a discounted expected value under the assumption that interactions will be repeated indefinitely into the future. Then Table 3.5 becomes as shown in Table 3.6.

Table 3.6: Modified corruption scenario

	high d	low d
high d	$X/(1-d_{high})$, $X/(1-d_{high})$	$(X-C)/(1-d_{high})$. $(R+A)/(1-d_{low})$
low d	$(R+A)/(1-d_{low})$, $(X-C)/(1-d_{high})$	$A/(1-d_{low})$, $A/(1-d_{low})$

Now the ES strategies depend not only on the magnitudes of X, C, R and A, but also on d_{high} and d_{low}. As long as $X > C$ (as long as the 'fine' extracted from the hapless entrepreneur does not yield negative profits), then regardless of whether a defector can earn any reward, there exists a value for d_{high} that renders (high d, high d) the unique ES strategy pair. That is, as long as an 'honest' entrepreneur is not literally threatened with economic extinction, corruption need not overtake a society. Second, if $C > X$, then both (high d, high d) and (low d, low d) are ES, and we might expect a population in which cooperators attempt to isolate themselves from their opposite type. Finally, note that it is also possible to have a polymorphic population. This occurs whenever

$$(R+A)/(1-d_{low}) > X/(1-d_{high}) \text{ and } (X-C)/(1-d_{high}) > A/(1-d_{low})$$

These two inequalities can be rewritten as

$$R/(1-d_{low}) + A/(1-d_{low}) > X/(1-d_{high}) > C/(1-d_{high}) + A/(1-d_{low})$$

or simply

$$R/C > (1-d_{low})/(1-d_{high})$$

Thus, a necessary condition for a polymorphic population is that $1\text{-}d_{low}$ not be too high and $1\text{-}d_{high}$ not be too low relative to the ratio of reward to imposed costs, in which case the proportion of high *d* types in equilibrium will equal

$$p = \frac{A(1 - dhigh) - (X - C)(1 - dlow)}{C(1 - dlow) - R(1 - dhigh)}$$

Corruption scenario #2: Perhaps a more straightforward example of corruption is formed by assuming that high and low *d* types characterize persons who are predisposed to keep and not keep bargains. If we take the Prisoners' Dilemma as our model of this circumstance, but assume payoffs are repeated and discounted, we can use the 'game' in Table 3.3 as our model, provided we assume that $C > A > D > B$. If, in addition, $B > 0$, then as before we can have (low *d*, low *d*), (high *d*, high *d*) or a polymorphic population be ES, depending on the values of d_{high} and d_{low} that are assumed to apply. For a bit more content, suppose people can be either one of two types: a person who when engaged in a cooperative venture, will put full effort into that venture, versus a person who will exert minimal effort. If both give full effort, their interaction earns *E*, which they share equally, at a cost of e to each person. If only one puts in full effort, their 'venture' earns $E' < E$ (which again is shared equally). This scenario is described by the matrix of payoffs inTable 3.7.

Table 3.7: Corruption Scenario # 2

	full effort	no effort
full effort	$E/2- e, E/2- e$	$E'/2- e, E'/2$
no effort	$E'/2, E'/2- e$	0, 0

Thus far, as long as $E'/2 > E/2- e > 0 > E' - e$ the situation corresponds to a Prisoners' Dilemma. However, assume that if the profits of their venture are too great, then as is often the case in the former USSR, they will both be taxed at near-confiscatory rates (or the mafia will move in to take control of their venture). Finally, as before, assume that a behavioral heuristic, in choosing between whether to be one that gives full effort or not, applies a discount to the future. Then the structure of payoffs is the one shown in Table 3.8a:

Table 3.8a: Discounted payoffs for corruption scenario #2

	full effort	no effort
full effort	$- e/(1- d_1), - e/(1- d_2)$	$(E'/2-e)/(1-d_1), E'/2(1-d_2)$
no effort	$E'/2(1- d_1), (E'/2- e)/(1-d_2)$	0, 0

Note that if Table 3.8a is treated as a game, we have a Battle of the Sexes in which there are two pure strategy equilibria and a mixed one (with full effort chosen with probability $\frac{1}{2} - e/E'$). If the players cooperate by randomizing between the two pure equilibria, each gets the expected payoff $(E' - e)/2(1 - d_i)$. But also suppose we allow the grim trigger strategy of liquidating their venture so that each person receives a one-time payoff of $B/2$: the asset value of their enterprise. The situation confronting these two persons can then be represented by the extensive form game in Figure 3.1.

Figure 3.1: Extensive form of the game in Table 3.8a

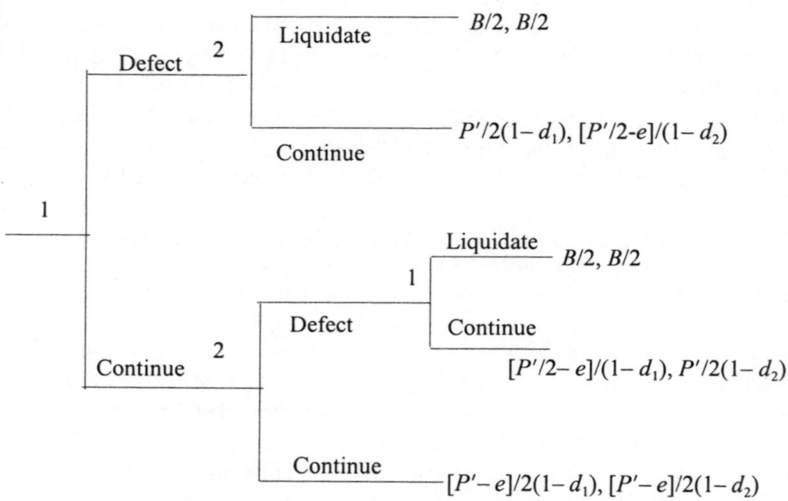

How this situation will unfold depends now on private discount rates. For example, if d is low for both persons, by which we mean that

$$(E - e)/2(1 - d_{low}) < B/2$$

then a defection by either one of them will induce the other to liquidate. Moreover, liquidating is preferred to continuing their cooperation, so liquidation occurs if both players are characterized by low d's. On the other hand, if both have high ds so that the previous inequality is reversed, they will continue their cooperative arrangement and realize the payoff $[E - e]/2(1 - d_{high})$. Finally, suppose person 1 has a low d and 2 a high d, by which we mean that if 1 defects,

2 will not liquidate, that is, $[E/2 - e]/(1 - d_{high}) > B/2$. The payoffs in Table 3.8b, then, follow from the play of the preceding game in extensive form if we assume that people have a choice between high and low valuations of the future.

Table 3.8b: Corruption Scenario # 2 (Modified)

	high *d*	low *d*
high *d*	$(E' - e)/2(1 - d_{high})$, $[E' - e]/2(1 - d_{high})$	$(E'/2 - e)/(1 - d_{high})$, $E'/2(1 - d_{low})$
low *d*	$E'/2(1 - d_{low})$, $(E'/2 - e)/(1 - d_{high})$	$B/2$, $B/2$

Interpreted as ES payoffs, note that if $(E' - e)/2(1 - d_{high}) < E'/2(1 - d_{low})$ and $(E'/2 - e)/(1 - d_{high}) > B/2$, then as before the ES outcome describes a polymorphic population. However, the second of these inequalities is in fact part of our definition of high versus low *d*. So all we require, in effect, for a mixed population to be evolutionary stable is the first inequality, which is satisfied simply by having a high enough cost of effort.

Street Crime Scenario: Consider this observation about street crime and gangs: 'joining the gang may be viewed as normal and respectable, particularly by the youth, even when the consequence is a series of delinquent and violent acts. Stealing, aggression, and vandalism may be secondary to the excitement of interacting with other peers of similar class, interest, need, and persuasion. The *consequences* of joining a gang and participating in delinquent acts *may not be recognized* by adolescents and even young adults' (Spergel 1990: 223, emphasis added). This last sentence targets a core idea; namely, that of a 'delinquent's' inability to foresee the future or account for the consequences of immediate decisions, which in turn is suggestive of a low valuation of the future. So suppose we construct a matrix of payoffs as shown in Table 3.9 that depend on whether a person is characterized as having a high or low valuation of the future:

Table 3.9: Street crime scenario

	high *d*	low *d*
high *d*	$A + Bd_{high}^{t}$, $A + Bd_{high}^{t}$	$-C + Bd_{high}^{t}$, $P - Id_{low}^{t}$
low *d*	$P - Id_{low}^{t}$, $-C + Bd_{high}^{t}$	$F - Id_{low}^{t}$, $F - Id_{low}^{t}$

The payoffs entered here are as follows: first, high d types associate a payoff B (e.g., stable employment, the rewards of education, escape from current surroundings) some distance in the future, at time t, the present value of which equals Bd_{high}^{t}. In addition, the interaction of two such types yields an unspecified immediate payoff of $A \geq 0$ (i.e., is minimally costless). On the other hand, when two low d types interact they realize the 'identity and self-esteem they cannot find elsewhere' (Spergel 1990: 223). 'Although the prospect of material gain cannot be discounted ... the members of gangs were more likely to become involved as a result of peer group attraction. They wanted to belong to a friendly supportive group ... the peer group attraction that is evidently responsible for young males becoming involved with street gangs is probably no different than the attraction of other socially legitimate groups' (Gordon 2000: 51–2). There is, then, the reward of fellowship, F, among gang members, but the cost of this association is the discounted (underappreciated) potential for incarceration if caught for carrying out illegal activities, Id_{low}^{t}. Finally, consider the situation whereby opposite types interact. Here '[t]he gang member can "control" turf, school, park ... when he cannot perform adequately and achieve respect for himself through legitimate means' (Spergel 1990: 224). This benefit we denote by P. On the other hand, outsiders to a gang or defectors from a gang can incur significant costs: '[a] youth may wish to leave the gang but be unable to, particularly if he remains in close physical proximity to other gang youths in the neighborhood or prison. The threat of violence may also induce him to remain.' (Spergel 1990:225). Less dramatically, 'gang members reward certain behaviors in their peers and punish others' (Winfree et al. 1994: 149). This punishment, or cost of not being a member but nevertheless interacting with them we denote by $-C$.

With payoffs as in Table 3.9 it is somewhat easy to appreciate the problems associated with the control of and spread of gangs and gang violence in 'vulnerable' social environments and why some researchers conclude that the primary strategy for gang control is 'to concentrate on drying up the sources of recruitment' (Johnstone 1983: 298). Specifically, note that a polymorphic population is ES only if

$$-C + Bd_{high}^{t} > F - Id_{low}^{t}$$

and

$$P - Id_{low}^{t} > A + Bd_{high}^{t}$$

Combining these inequalities so that each is expressed as an inequality that Bd_{high}^{t} must satisfy, we get as a necessary condition for polymorphic ES outcome:

$$P - F > A - C.$$

Thus, as represented, adjusting d_{high} or d_{low} cannot determine whether or not this condition is satisfied. This condition, moreover, may be especially difficult to satisfy when the psychological rewards of gang fellowship, *F*, or the rewards of remaining an outsider, *A*, are large. The value of gang fellowship may in fact be considerable in neighborhoods we deem 'vulnerable'. This is not to say that d_{high} and d_{low} play no role in determining whether (high *d*, high *d*) or (low *d*, low *d*) are ESS, but only that, as presented, they play no role in determining whether a mixed population can survive. Thus, as we suggest earlier when populations exclusively of either type are ESS, this is perhaps one reason why we might see cliques or networks developing in otherwise mixed populations whereby persons of one type fail to mix with persons of another type. It may be that a youth with his sights set on college and beyond will attempt to minimize his interaction with gangs, but we also need to appreciate that '[g]angs typically do not encourage active members to have friends, especially close friends outside of the gang' (Winfree et. al. 1994: 150).

What is especially interesting about this scenario, however, is that matters are only slightly changed if we take full account of time discounting by also allowing the payoffs *P, E, A* and *C* to accumulate in discounted fashion. That repetition is an appropriate representation is suggested by at least one researcher who defines a gang as 'a group of *recurrently associating* individuals with identifiable leadership and internal organization, identifying with or claiming control over territory in the community, and engaging either individually or collectively in violent or other forms of illegal behavior' (Miller 1975: 9, emphasis added). But if we assume repeated interactions to some indefinite future and add the appropriate discount factors to the earlier inequality for a polymorphic population, we get

$$[P - E]/(1 - d_{low}) > [A - C]/(1 - d_{high}).$$

Polymorphism now becomes even more difficult to sustain to the extent that both $[P - E]$ and $[A - C]$ are positive since those discounts can be expected to decrease the left side of this inequality and increase the right side. Only if both sides of this inequality are negative can we guarantee that an 'appropriate' choice of discounts will ensure a polymorphic population. That is, a society will most likely become exclusively all high *d*s or all low *d*s unless the fellowship from gang activity and the costs of mixed interactions are great.

Addiction Scenario: Consider this classic observation about alcoholics: '[the] social-psychological scheme for explaining alcoholism has the following parts: personality features set the stage, making one a candidate, [and] qualities of drinking-centered groups that uniquely attract and reward the use of alcohol for such persons, linking 'readiness' with alcohol' (Trice 1966: 43). Akers (1973) expands on Trice's observation by noting that 'drinking phenomena are a social phenomena in two related ways: (1) what one does with and thinks about alcohol is a function of his or her group membership and identification, and (2) almost all drinking takes place in social group settings which the drinker 'believes makes his drinking socially approved ... a drinker's consumption ... closely matches that of his or her drinking companions ... several features of many drinking groups are attractive and rewarding for the drinker. They accept and legitimate the use of alcohol' (p. 152-62). The social imperatives and context of alcoholism seem to apply with special force to drugs and drug addiction: 'The peer group is now widely regarded as an important social environment capable of influencing both the initiation of substance use and the maintenance of patterns involved' (Cotterell 1996: 129) and 'when drugs are actually used, it is almost always in a peer context. Peers initiate the youth into drugs. Peers help provide drugs. Peers talk with each other about drugs and model drug-using behaviors for each other and in doing so shape attitudes about drugs and drug-using behaviors' (Oetting and Beauvais 1987: 137).

We can generalize then that 'birds of a feather enjoy flocking together' – alcoholics (and by extension, drug addicts) gain greater pleasure from drinking when associated with other alcoholics, and less pleasure when in the company of non-alcoholics; conversely, recovering alcoholics gain more pleasure from being in the company of other recovering addicts, and less when in the company of those who drink heavily. In fact, if, in accordance with Becker and Mulligan (1997) and Becker and Murphy's (1988) arguments about the relationship between addicts and time discounting, we associate a recovering addict with a high d and a non-recovering one with a low d, then the situation, portrayed in Table 3.10 is not much different from the one we associate with street gangs.

Table 3.10: Addiction scenario

	high d	low d
high d	$F/(1-d_{high}) + Bd_{high}^t$, $A/(1-d_{high}) + Bd_{high}^t$	$G/(1-d_{high}) + Bd_{high}^t$, $D/(1-d_{low}) - Id_{low}^t$
low d	$D/(1-d_{low}) - Id_{low}^t$, $G/1-d_{high}) + Bd_{high}^t$	$F^*/(1-d_{low}) - Id_{low}^t$, $F^*/(1-d_{low}) - Id_{low}^t$

The critical difference here between our representation of the payoffs for addiction and street crime is allowing for two forms of fellowship – that among recovering addicts of the sort they might get from weekly meetings in Alcoholics Anonymous, F ('At my first meeting the people fulfilled their responsibility – they made me welcome. I met others like me and it felt good. Maybe this thing was for real. So I went to another meeting, and I got the same feeling. Then another meeting', as quoted from the AA Manual in Chapter 1), and the fellowship addicts get by being among themselves, F^* ('Like smoking, drinking is seen by youth as a marker of social sophistication, with additional (if dubious) benefit of helping the youth adapt to social situations' (Cotterell 1996:140). We also allow for two somewhat more ambiguously defined payoffs, G and D. Here G is the utility a recovering alcoholic associates with an interaction with a non-recovering counterpart, whereas D is the utility a non-recovering alcoholic derives from an interaction with a recovering addict. We can imagine both D and G assuming different signs and magnitudes. On the one hand G might be negative as when a person in recovery feels uncomfortable in the presence of someone who openly drinks; but it might also be positive when other considerations pertain in interpersonal relations. The same can be said about D. Nevertheless, we can see here the logic of Alcoholics Anonymous success. Granting that $F^* > G$, it must establish an environment in which $F > D$, in which case there are, at worst, two alternative ES equilibria. But if in addition it can render G negative and simultaneously encourage a high d among those who join, then 'high d' becomes the unique ES outcome.

Depletable Resources: If one maintains a strict biological interpretation of an ES game, it is difficult to translate the analysis to systems in which only two entities interact. But if we assume that the entities in question are memes and that an individual is characterized by a probabilistic decision rule whereby he chooses a meme 'at random' according to the distribution $(p, 1-p)$ upon each encounter with another individual, we can take a situation that is limited to only two persons and apply the notion of an ES outcome to describe the equilibrium nature of that encounter. With this in mind, let us now consider a scenario that concerns some depletable resource, where high d implies acting to conserve the resource and low d implies otherwise. In particular, suppose that if both persons 'conserve', each can earn 1 unit indefinitely into the future and realize the present value $1/(1-d_{high})$. On the other hand, if one person raises his level of consumption by $y > 0$ per period, the resource in question depreciates at the rate r. In this case, the individual conserving realizes the diminishing stream

$$1, r, r^2, r^3, \ldots$$

whereas his less ecologically conscious counterpart realizes

$$(1+y), (1+y)r, (1+y)r^2, (1+y)r^3, \ldots$$

If both persons fail to conserve suppose the resource depreciates at twice the rate, in which case each realizes the stream

$$(1+y), (1+y)r/2, (1+y)(r/2)^2, (1+y)(r/2)^3, \ldots$$

Now consider the 'game' in Table 3.11.

Table 3.11: Depletable resources

	high d	low d
high d	$1/(1-d_{high})$, $1/(1-d_{high})$	$1/(1-rd_{high})$, $(1+y)/(1-rd_{low})$
low d	$(1+y)/(1-rd_{low})$, $1/(1-rd_{high})$	$(1+y)/(1-rd_{low}/2)$, $(1+y)/(1-rd_{low})$

As before, (high d, high d), (low d, low d) or a polymorphic population can be ES, depending on parameters, but the more plausible possibility is when one strategy or the other prevails. We are, after all, speaking of iterated interactions between the same two people and it is difficult to conceptualize what it means for them to randomize over time in choosing between conservation and non-conservation. Certainly their calculation of the rate of depletion should be adjusted accordingly, where it is not simply a function of the period in which it occurs but rather of how many times in the past a non-conserving strategy has been employed. So excluding polymorphism, consider when (high d, high d) versus (low d, low d) are ES. Suppose for our example that $d_{low} = 0$, in which case (high d, high d) is stable when $1/(1-d_{high}) > 1+y$. This tells us that there is an ES solution that implies conservation regardless of the rate of resource depreciation. All we require is that d_{high} be 'sufficiently' close to 1. In other words, a policy (e.g., the 'worried grandchild') that targets private discount rates is sufficient to establish conservation as ES. Next, consider the condition for low d to be ESS, which here requires $y/(1+y) > rd_{high}$. Now the rate of resource depreciation enters into the equation and yields a calculation that is intuitive; namely, (low d, low d) cannot be evolutionary stable unless r or d_{high} are 'sufficiently' low. That is, if the depletable resource diminishes rapidly (if r is small) upon non-conservation and if even high d types discount the future too greatly, then (low d, low d) can be ES as well. Notice finally that there is nothing in these inequalities that precludes both pure strategies from existing simultaneously. (high d, high d) is stable if $d_{high} > y/(1+y)$ and (low d, low d) is stable if $d_{high} < y/r(1+y)$, so both are stable if $y/r(1+y) > d_{high} > y/(1+y)$.

Corruption Scenario #3: We conclude this chapter with a scenario in which there are two types of actors: entrepreneurs and government officials, each of which can be either corrupt or honest, holding a high valuation for the future or a low one. First, not taking account of time valuations, consider the scenario depicted in Table 3.12, wherein our notation is as follows: P is the normal profit an honest entrepreneur receives in a period if not harassed by a corrupt government official and W is that official's formal wage; S is the side-payment an honest entrepreneur must give to a corrupt official, who simply adds that payment to his wage; $-K$ is the value of attempting to engage in corruption with an honest official (loss of income and presumably imprisonment), whereas the honest official earns a normal wage, but enjoys a reward R for having turned in the attempt at bribery; finally, D denotes the profit each can earn by being jointly corrupt (presumably from others not party to the exchange).

Table 3.12: Corruption scenario # 3

		government official: high d (honest)	government official: low d (corrupt)
entrepreneur:	high d (honest)	P, W	$P-S, W+S$
	low d (corrupt)	$-K, W+R$	$P+D, W+D$

Note that Table 3.12, viewed as a game possesses a unique pure strategy equilibrium with any reasonable assignment of payoffs whereby both persons are corrupt so 'low *d*' is ESS for both types. But now, as before, consider the time discounted version of this scenario (Table 3.13).

Table 3.13: Discounted version of corruption scenario # 3

		government official: high d (honest)	government official: low d (corrupt)
entrepreneur	high d (honest)	$P/(1-d_{high})$, $W/(1-d_{high})$	$(P-S)/(1-d_{high})$, $(W+S)/(1-d_{low})$
	low d (corrupt)	$-K/(1-d_{low})$, $(W+R)/(1-d_{high})$	$(P+D)/(1-d_{low})$, $(W+D)/(1-d_{low})$

Aside from an entrepreneur preferring not to be corrupt when the official is honest, there is no natural set of inequalities that is invariant with parameters and discount rates. And at this point there are simply far too many parameters and permutations of values to conduct any meaningful analysis. So to simplify matters and gain a sense of possibilities, suppose $K = 0$, $d_{low} = 0$, $d_{high} = 1/2$, $P = W = 1$, and $S = ½$. Table 3.14 gives the resulting matrix.

Table 3.14: Possible Outcomes of the Game with Defined Parameters

		government official: high d (honest)	government official: low d (corrupt)
entrepreneur:	high d (honest)	2, 2	1, 1.5
	low d (corrupt)	0, $2(1 + R)$	$1 + D$, $1 + D$

Clearly, there is now the potential for two pure strategy equilibria (provided that $(1+D) > 2(1+R)$), and the question is whether the situation can sustain a polymorphic population of both honest and dishonest people. Solving for a mixed strategy equilibrium, we get

$$p(\textit{entrepreneur is honest}) = (2D - 4R - 2)/(2D - 4R - 1)$$

$$q(\textit{official is honest}) = D/(2 + D)$$

To assert that these mixed strategies are ES we must check the second condition in the definition of an ES outcome. However, since we merely want to illustrate a specific result, we give our example still greater analytic tractability by assuming that $D = 4$ and $R = 1$, in which case Table 3.14 is transformed to become Table 3.15

Table 3.15: Corruption scenario # 3 with additional defined parameters

		government official	
		high *d* (honest)	low *d* (corrupt)
entrepreneur:	high *d* (honest)	2, 2	1, 1.5
	low *d* (corrupt)	0, 4	5, 5

and $p = q = 2/3$. However, if entrepreneur plays (2/3, 1/3) against a corrupt official, the entrepreneur's expected payoff is 7/3, while if an official plays (2/3, 1/3) against a corrupt entrepreneur, that official's expected payoff is 13/3. The sum of payoffs, then, is 20/3. But this is less than when both are corrupt, so $p = q = 2/3$ is not ES. In fact, for the payoffs shown, there are only two pure evolutionary stable outcomes – either all are high *d*s or all are low *d*s. The impact of discounting here, then, is not to induce polymorphism, but merely to open the door to the possibility of a second equilibrium of honest players such that if d_{high} is large enough (and $S < P$), honesty can correspond to the unique equilibrium.

The only general circumstance, in fact, in which we can have a polymorphic ES outcome is if (low *d*, high *d*) and (high *d*, low *d*) are simultaneously equilibria to the situation analyzed as a 2-person game. For this to be true, however, it must be that a corrupt entrepreneur can take advantage of an honest official and a corrupt official needs to prefer to take advantage of an honest entrepreneur rather than collude with a dishonest one. Otherwise, as in our example, the system will converge either to everyone being honest or to everyone attempting to be dishonest.

4 Investment, Consumption and Change

For social species such as humans ... an especially important form of learning is social learning. Observing the activities of others and learning about the world through or from them enables individuals to acquire information with less effort and risk that if they were forced to learn on their own (Tomasello 2004: 51)

4.1 A Simple Linear Model

The scenarios of the previous chapter give us some understanding of how individual time discounts might be modeled using an evolutionary perspective. However, many if not all of the parameters that are part of those scenarios and that determine a scenario's ESS are themselves endogenous to social circumstances. Things such as economic uncertainty, population mobility, residence patterns, the presence or absence of alternative peer groups, the role of the state and the viability of its legal code, while possibly depending on individual attitudes, also impact those parameters. One thing seems clear, however: if the scenarios we offer that pertain to, say, the relationship between corruption and time discounts are viable, and if corruption impacts social welfare generally, then time discounts impact social welfare as well. And in this way, perhaps, we can begin to disentangle the complex interdependence of corruption and social welfare and address, in a theoretically viable way, whether and when forms of cooperation we might deem 'corrupt' facilitate or hinder economic development. Moreover, as the scenarios in the preceding chapters suggest, our analyses need not be limited to any assumed connection between corruption and cooperation but can be applied to other substantive domains as well.

However, an essential step in developing the requisite analysis is confronting the fact that absent a more comprehensive model, letting the ES proportion of types be the product of some simple 2x2 evolutionary game as in our scenarios merely pushes the problem back a step to those parameters taken as measures of 'fitness'. Stopping there accomplishes little in the way of making discount rates endogenous. What we require is a model of the process whereby individuals interact and converge to some (potentially endogenously changing) ES proportion. To that end (and as but a step on the way to the development of a complete model), we let ourselves slip temporarily into the economist's world and review a simple model of investment and growth. Ultimately we want to allow for the substitution of simple monetary variables with more complex social notions *a la* Bourdieu (1985) and Coleman (1987) that pertain to cultural and social capital, but for now it is convenient to assume purely monetary rewards and losses. We begin, then, by letting there be $N = \{1, 2,..., n\}$ persons, each endowed initially with K_o units of 'resources' or capital. Each person also begins by being characterized as holding memes that correspond to either one of two time discounts, $d = d_{high}$ and $d = d_{low}$ such that $0 < d_{low} < d_{high} < 1$. People are not conscious of these memes or the mechanisms of their adaptation to the specifics of a social circumstance. Instead, they act as if their time discount is 'hard wired'. Thus, in the initial period $t = 0$, each person j in N must choose an allocation of their initial endowment K_o between immediate consumption (enjoyment), C_0, and investment, I_o, which produces future capital for consumption and investment. Immediately thereafter, we suppose that each person's memes 'participate' in an evolutionary 'game' G that determines their survival rates into the next period. Postponing discussion of that process, for now we assume simply that (1) every unit of resource 'spent' on consumption yields one unit of value or utility for the person in question; (2) all persons in N play the same game G though all periods; and (3) in any period t, each person assumes, for purposes of making their investment-consumption decision, that all subsequent periods have the same characteristics as the current period.

Turning to the model of investment and consumption in which G is embedded, and dropping the subscript j for convenience, suppose investment yields a one period return of r, so that X invested in period t yields $(1+r)X$ in period $t+1$. Let C_0 and I_0 be a person's consumption and investment in the initial period ($t = 0$), so that for $t = 1$ they must divide $(1+ r)I_0$ between period 1consumption and investment. Let that decision be C_1 and I_1 respectively, where $C_1 + I_1 = (1+r)I_0$. Similarly, in period $t = 2$, that person must then allocate $(1+ r)I_1$ in a similar fashion between C_2 and I_2, where $C_2 + I_2 = (1+ r)I_1$, and so on. Since we are concerned with steady states, we make the simplifying assumption that this state is reached at $t = 1$ (given our model's simplicity, this is a feasible consumption-investment path). Thus, $C^*_1 = C^*_2 = C^*_3 = ...$ etc. In period 0, then, a person foresees (perhaps incorrectly) the present value of the consumption stream as

$$C^*_0 + dC^*_1 + d^2C^*_1 + d^3C^*_2 + ... =$$

$$C^*_0 + C^*_1 - C^*_1 + dC^*_1 + d^2C^*_1 + d^3C^*_1 + ... =$$

$$C^*_0 - C^*_1 + C^*_1/(1-d) =$$

$$C^*_0 + dC^*_1/(1\text{-}d) \tag{4.1}$$

If utility is linear, expression (4.1) describes what j maximizes when making an initial investment-consumption decision. This is not necessarily what j experiences since her assumptions about the game G may be erroneous (primarily because private time discounts may change), but this is the function that dictates a person's initial consumption-investment decision. We can, then, solve for the maximum pattern of investment over time by observing that at steady state, if $K_0 - C_0$ is invested in period 0, the period 1 'budget' is:

$$K_1 = (1 + r)(K_0 - C_0).$$

Thus, in period 1, if j consumes C_1, then the amount $(1+r)(K_0 - C_0) - C_1$ is invested, which yields:

$$(1+r)[(1+r)(K_0 - C_0) - C_1]$$

to be allocated in period $t = 2$. But since we are describing a presumed steady state, a person will assume that $C_2 = C_1$ and invest

$$(1+r)[(1+r)(K_0 - C_0) - C_1] - C_1.$$

It also follows that this quantity, the amount invested in period 2, must equal the amount invested in period 1. Thus,

$$(1+r)(K_0 - C_0) - C_1 = (1+r)[(1+r)(K_0 - C_0) - C_1] - C_1$$

Simple algebraic manipulation reduces this equation to $C_1 = R(K_0 - C_0)$. Substituting this identity into (4.1) gives

$$C_0(1 - d - dr) + drK \qquad (4.2)$$

as the thing a person maximizes by an appropriate selection of C_0 (and, from expression (4.1), of C_1, etc.). But note that (4.2) is linear in C_0, so

- if $(1 - d - dr) > 0$, a person's utility stream, as perceived in period 0, is maximized by setting C_o at the maximum, K_0, leaving nothing for the future, in which case the person's anticipated stream of consumption takes the form $(K_0, 0, 0, \ldots)$, whereas[1]
- if $(1 - d - dr) \leq 0$, the stream of utility is maximized by setting C_0 at the

[1] Note that if we award people with an exogenously supplied supplemental income in each period, say k, so that low ds do not disappear, all investment consumption decisions with respect to endowments are unchanged: high ds invest $K + k$ rather than simply K, their return on investment from the prior period, whereas low ds consume all of k rather than 0, and the impact of other parameters is unaffected.

minimum, 0, so that all consumption lies in the future in accordance with expression (4.2) and the anticipated stream takes the form $(0, rK_0, rK_o, ...)$.

At this point we should comment again on the seemingly restrictive economic content of this analysis. Specifically, it is important to appreciate that the idea of 'capital' or 'resource' can be given a broad interpretation. We can think of capital, *a la* Bourdieu (1985) and Coleman (1987), as encompassing a variety of 'resources' such as education, physical health, and social class. The general idea is the same: a person's stock of capital needs to be allocated between current pleasures and future rewards, where postponed pleasure produces rewards later at the exogenously determined 'interest' rate $r \geq 0$ but which the individual discounts using his or her private discount, d. For example, the time spent at physical exercise can be considered an investment in a healthier future at the cost of alternative forgone pleasures in the present (assuming, as is the case with some of us, that exercising is not altogether pleasurable), but where one prefers some optimal mix between exercising and other things. That mix will be a function of the "rate of return" of exercising (the presumed added life expectancy and quality of life years hence) and the rate at which one discounts the future and prefers immediate gratifications.

Thus far, though, our model is disjoint from our discussion of endogenous time discounting in that it takes time preference to be the same in each period. What we want is a mechanism whereby that preference adjusts according to some endogenous criterion. So suppose that in playing the evolutionary game G, each player j in N is given the weight C^*_{tj} in period t, which we take to mean that the game will be analyzed 'as if' there are $C^*_{t1} + C^*_{t2} + C^*_{t3} + ... + C^*_{tn}$ players. And suppose that we take these weights to indicate the proportion of 'high d' and 'low d' types, with the corresponding ES outcome determining the mix that proceeds to the next period, $t + 1$. That is, the presumed proportion of low versus high d memes in the population will be the total of their respective levels of consumption, divided by the total of all consumption.

This assumption warrants discussion since there are several interpretations we can give to it, as well as alternative assumptions. First, if we again revert to a more economic perspective, this assumption in effect views units of consumption as units of participation in the market and a measure of the proportion of types. If individuals who are characterized as having low *ds* consume out of proportion to their numbers and if this proportion exceeds the ES value of p, it will appear that there is an 'overpopulation' of low d memes, in which case some low d types will 'convert' to become high *d*s. Alternatively, suppose individuals take relative rates of consumption as a measure of performance: If they see a type enjoying greater rates of consumption than they are, they will attempt to mimic that performance by adjusting their type in the direction of the better performing one. This adjustment, if it is incremental, will cease only when the overall proportion of apparent high and low d types equals the ES proportion. Two stopping rules here, though, are possible: in one, which better matches the first interpretation, the system will converge (if it converges at all) to where the proportion of consumption among low d types matches p. In the second, the system is directed to equating the actual proportion of low d types to equal p. We explore both rules in subsequent chapters, but for the present we will rely primarily on the first.

There are now only three possibilities: (1) 'All high d' is ES, and all low ds become high ds; (2) 'All low d' is ES, and all high ds become low ds, and; (3) a mixture is ES, in which case the population moves in the direction of the corresponding ES proportions.[2] For case 1, unless low ds receive a supplemental income, they disappear. For case 2, all high ds will disappear and the return on their investment will be given to the transformed low ds. And unless there is a supplemental income, the process will terminate with everyone consuming all resources in the next period. Finally, for case 3, either some low ds must become high ds or some high ds must become low ds (with the resources of transformed players being carried with them) so as to satisfy the proportions dictated by the ES value of p.

In our actual analysis we allow for the possibility that adjustments occur incrementally – that not all high or low ds transform themselves immediately if the ES p dictates change. Indeed, this adjustment rate proves later to be a critical parameter of our analysis. To illustrate, then, the process envisaged by our analysis and with Figure 4.1 in mind:

- Suppose, for the period in question, we have ph and $pl = 1 - ph$ as the relative proportion of high and low d types respectively in a population of n individuals.
- If K_h denotes the resources controlled by all high ds in that period, and K_l the resources of low ds, the share of resources consumed by high ds is $ch = rK_h$, whereas that consumed by low ds is

$$cl = K_l \text{ if } 1 - d_{low} - rd_{low} > 0$$

or

$$rK_l \text{ if } 1 - d_{low} - rd_{low} \leq 0.$$

For purposes of this discussion suppose low ds consume K_l and invest nothing. Thus, it appears as if the population contains $n\, K_l/(rK_h+K_l)$ low ds.

- Proceeding to the game G, let the ES outcome dictate a mix between high and low ds of $(1-p)$ and p respectively, $0 < p < 1$. However, the proportion of low ds actually represented 'in the market' on the basis of rates of con-

---

[2] In the analysis that follows, we consider only polymorphic populations since the consequence of one for which the ES outcome is all low or all high ds is obvious.

sumption will correspond precisely to this proportion only in equilibrium. So suppose, to illustrate, that there are too few low *ds*, i.e., let $n\,K_l/(rK_h+K_l) < np$.

- A proportion of high *ds* must now be transformed to low *ds* in order to move the overall ratio of types in the direction of the ES *p*. If *cl* and *ch* denote the total consumption of low and high *ds* respectively and if $cl/(cl+ch) < p$, let $(cl + C)/(cl+ch) = p$, so that *C* is the total of consumption that must be equated with low *ds* in order to match the *p*. Hence, $C = p(cl+ch) - cl$, and the proportion of high *d*'s that must be converted is *C*/*ch*.
- However, rather than assume that this transformation occurs in one step, assume an *incremental adjustment rate* of *ar* ($0 < ar < 1$) so that the proportion of those needed to be transformed is *ar*. The new proportion of low *d*'s, then, is

$$pl' = pl + arCph/ch = pl + arC(1-pl)/ch$$

- Low *ds* do not earn anything since they invest 0, whereas high *ds* have earned $(1+r)(1-r)K_h$. But this return must be shared with low *ds* in the next round since some high *ds* will have been transformed.
- Low *ds* will enter the next period with *nkpl'*, their new share of the total supplemental income, plus the resources of the transformed high *ds*. Since high *ds* earned $(1+r)(1-r)K_h$, the resources of the low *ds* entering the next period will equal this supplemental income plus their share of the high *ds* return on investments, or

$$nkpl' + [arCph/ch](1+r)(1-r)K_h$$

and high *ds* will have

$$[1 - arCph/ch](1+r)(1-r)K_h + nk(1-pl').^{3}$$

- All persons make new investment/consumption decisions and the process repeats itself.

[3] Not all low *ds* are equal. Some have only their supplemental income, while others have this plus the return on investment they earned when high *ds*. However, owing to the linearity of our model, we can sum these resources and analyze the situation as if there is a single consumer with a weight equal to the total of its consumption.

Figure 4.1: Simulation summary

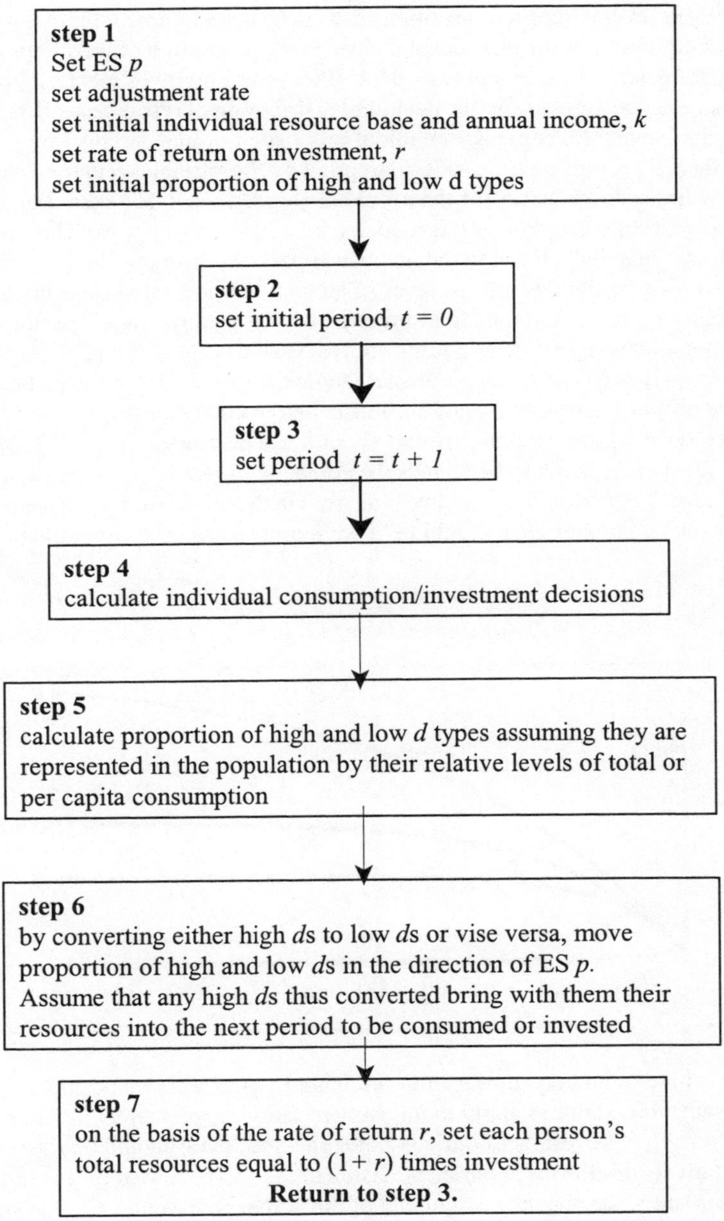

4.2 Analysis

There are several questions awaiting answers now, including: if there is a stable population, what is the relationship between the proportion of, say, high *d*s and the parameters of the model (i.e., the market return on investments, *r*; the rate of incremental adjustment, *ar,* the initial distribution of resources across types; the supplemental income, *k;* the initial proportion of high versus low *d* types, and the ES proportion of low *ds*, *p,* implied by the variables *u* and *v*)? Also, if we return to the scenarios of the previous chapter, what, for example, are the social costs of corruption as it is modeled here in terms of people who discount the future heavily? Owing to the number of parameters in our model, we forgo closed form analytic results in favor of a computer generated assessment, and to begin, Figure 4.2 graphs total invested capital against 'time' (periods) and illustrates convergence to an equilibrium for various values of the ES proportion of low *d* types (*p*) when supplemental income, *k*, equals 0.2, *r* equals 0.05, the initial proportion of low *d*'s is zero, and an incremental adjustment rate of 0.10. There is, of course, nothing special about these parameter values. We set *r* = 0.05 for no reason other than that is the number commonly employed as a 'normal' rate of return on financial investments. On the other hand, we have no idea what such a number is or should be if we impute a return on investment when

Figure 4.2: Total investment vs. t: r=.05, k=.2, ar=.1

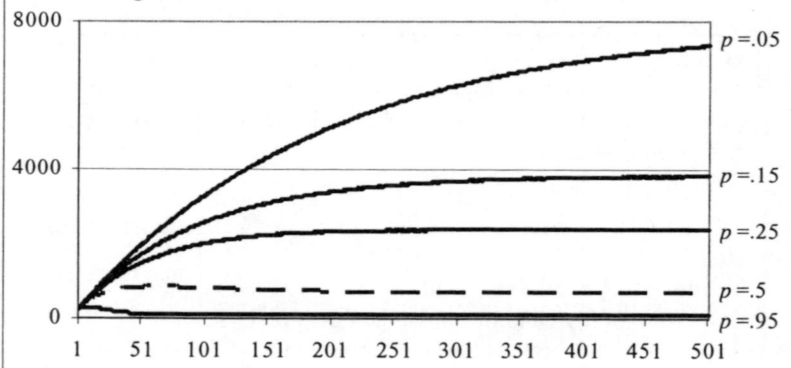

what is invested is, say, time spent exercising so as to secure a healthier future. The same uncertainties apply to adjustment rates, supplemental incomes, and even *p* itself. The results of our computer analysis, then, should be interpreted qualitatively despite the apparent precision of the numbers we report as outputs. It is not, then, the absolute magnitude of the numbers in Figure 4.2 that should interest us, but rather the confirmation it provides of the wholly expected result that as the ES proportion of low *d*'s decreases in a population, the greater will be the long term total level of investment. The increase in investment we see at a diminishing rate in this figure for all but the highest values of *p* occurs simply because those who value the future are slowly building their capital from one period to the next. Another way to view this data is to look at the equilibrium value of investment plotted against *p* as in Figure 4.3. Once again we see that as ES *p* increases, total investment decreases. However, note that the rate of decline is especially pronounced when ES *p* is low. That is, an initial increase in the evolutionary stationary proportion of those who value the future slightly and, thereby, consume everything, has a disproportionate impact on total investment. If one accepts, then, our argument linking valuations on the future with corruption, the implication of Figure 4.3 is that the initial stages of the expansion of corruption in a society are especially critical to that society's overall level of invested capital.

Figure 4.3: Total investment vs. p: r=0.05, k=0.2, ar=0.1

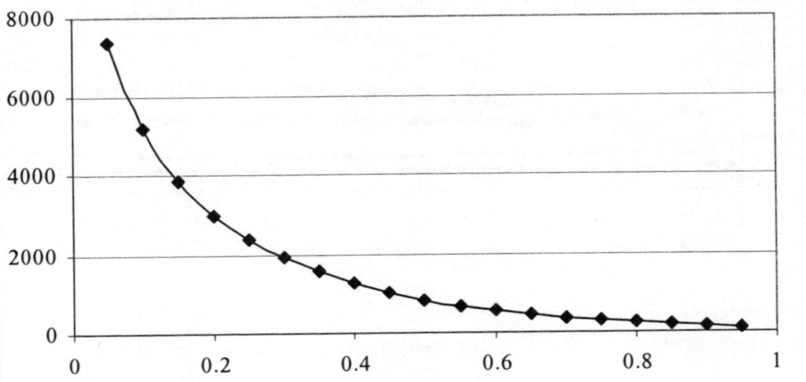

Although this result may seem intuitive, *p*s impact on total consumption is less obvious, since there are countervailing forces in operation here. Although high *d* types invest, we might speculate that they invest so much that although total investment increases, consumption declines. Indeed, letting investment be our sole criterion of welfare ignores the fact that a society which invests everything is a society that consumes nothing. Figure 4.4, then, graphs total consumption against time for the same adjustment rate (0.10) and values of *p* as in Figures 4.2 and 4.3, and reveals that consumption does in fact increase as *p* declines. There are two things to be noted about this figure. First, note the oscillations in consumption for lower values of *p*. These oscillations, though, have a ready explanation; namely, an adjustment rate that allows for an over compensation of the proportions of high and low *d*s. These oscillations, then, are as much an artifact of the incremental mechanism that is a part of our model as anything else and doubtlessly could be eliminated with alternative algorithms. It is the second pattern of this figure, then, that is of greater interest. As Figure 4.5 shows in detail, initial periods are marked by greater consumption when *p* is high. It is only later, when the investments of high *d*s allow them to consume more, that societies with fewer low *d* types in equilibrium consume more overall. Here, then, it is tempting to draw a parallel to the excesses of consumption reported in, say, Russia among 'new Russians,' whereby a burst of consumption for luxury goods, often at inflated prices seems to characterize anyone who finds a way to benefit from the initial emergence of a market economy.

Figure 4.4: Total consumption vs. t: r = 0.05, k = 0.2

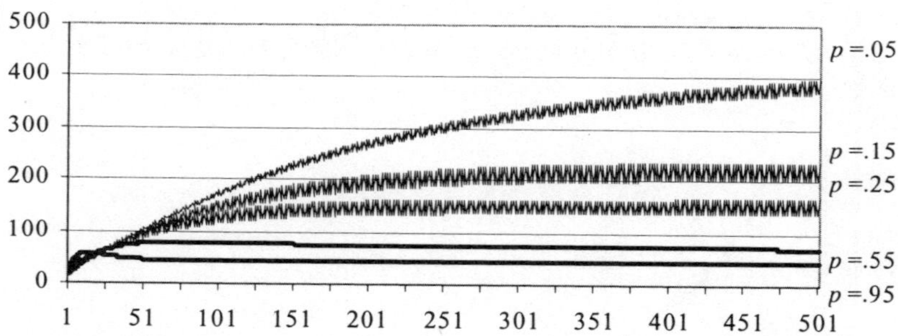

Figure 4.5: Total consumption vs. t: r= 05, k = .2, ar= .1

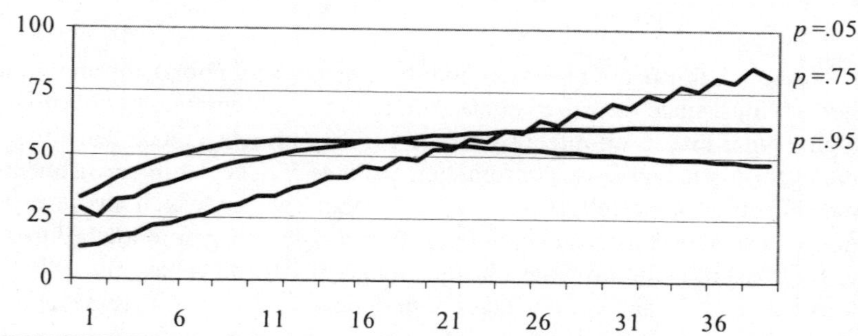

Our next two figures, 4.6 and 4.7, show the impact of *r* on both total investment and consumption. Here our intuition is less secure since variation in *r* yields two contradictory consequences: as *r* increases, the return on investment increases, but any increase in *r* also implies an increase in the rate of consumption among high *d* types. One might suspect, then, that the impact of *r* on overall investment and consumption will depend on parameter magnitudes. And indeed, this is what we find. At least for the functional forms in our analysis, as Figure 4.6 shows, overall investment does not increase as *r* increases, but rather decreases. Interestingly, however, overall long-term consumption also decreases as *r* increases, although not quite in a simple way. Figure 4.7 reveals that although the *equilibrium steady-state* value of consumption is greater when consumption is 'suppressed' by low values of *r*, consumption is greatest for high values of *r* in the early periods. With low values of *r*, of course, high *d* types are

Figure 4.6: Total investment vs t; k = .2, p = .5

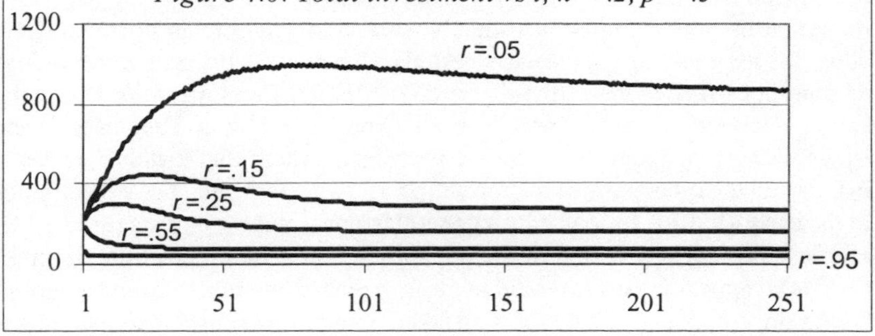

Figure 4.7: Total consumption vs. t, k = 0.2, p = 0.5

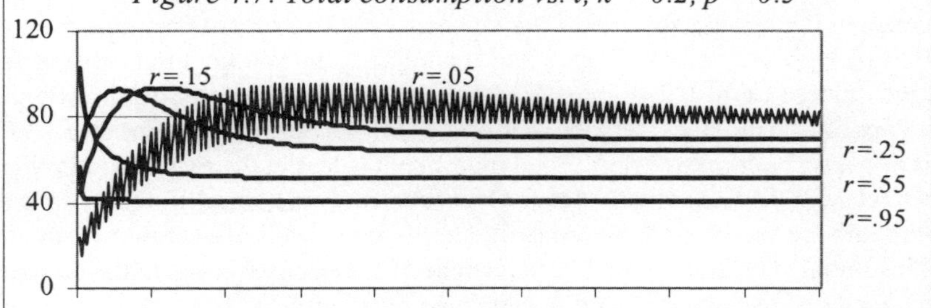

encouraged to invest, and over time this allows them to overtake 'societies' that encourage consumption with high values of *r*. Again we are reminded of the seemingly excessive consumption and absence of investment that characterized behavior in Russia, especially in the years immediately following the dissolution of the Soviet Union despite the rates of return offered by private banks and the government that often exceeded 300 percent. These rates, of course, were the consequence of Russia's uncertain history with respect to public and private finance and a corresponding lack of faith in the government and private sector; but, as in Figure 4.7, these rates resulted in neither high savings (except for capital that moved offshore) nor significant domestic investment.

Figure 4.7 reveals several additional interesting patterns, although we must again caution the reader that these patterns are doubtlessly specific to the functional forms of the analysis. First, note the oscillations in consumption that occur when $r = 0.05$. What is happening here is that, given an adjustment rate of 0.10, the proportion of high versus low *d*s is being over- then under-compensated for. Put simply, the step size is too large for a smooth convergence to the equilibrium. Second, for "intermediate" rates of return, consumption initially rises, then falls. This initial burst of consumption derives from a rate of conversion of high *d*s to low *d*s (in order to match an ES p of 0.5, keeping in mind that the starting point is a society with no low *d*s) that is slow enough to allow for some savings in the early periods. However, as the proportion of high *d*s continues its decline in the direction of p, total investment, as Figure 4.6 shows, declines with the consequence that overall consumption begins to diminish as well. Here again, then, we see perhaps a pattern that matches events in Russia and the other parts of the former USSR, whereby a burst of consumption in the early 1990s is followed by a general malaise and even crisis (as in 1998) after a sufficient proportion of the population adjust their behaviors to mimic those who appear to be successful in the new post-communist market economy.

Again, considerable care must be taken in imputing substantive meaning to these patterns. Not only can we not preclude the possibility that they are specific to the details of our formulation, but they might also be specific to the values of other parameters. Thus, to see how these other parameters might affect things, consider Figures 4.8, 4.9a and 4.9b which assess the impact of the initial proportion of high versus low d types and the adjustment rate *ar*. First, setting the adjustment rate at 0.2, Figure 4.8 portrays the impact of choosing different values for the initial percentage of low *d*s in the population, p_{init}. The effect here is not altogether unexpected: Total investment is less in the early periods when we set $p_{init} = 0.95$ as compared to 0. Over time, however, this difference 'washes out' and the two alternatives converge to the same level of overall investment. This initial condition at least appears to be of little consequence to the ultimate character of the society that emerges in equilibrium.

Of course, we must then ask whether this invariance depends on the adjustment rate *ar*, and that is the issue Figures 4.9a and 4.9b address. First, as Figure 4.9a alone shows after we set the initial proportion of low *d*s equal to 0, the impact of *ar* is neither intuitive nor inconsequential. Specifically, the lower the value of *ar*, the greater is the overall steady state level of savings, although for

Figure 4.8: Total Investment vs. t: k=.2, p=.5, r=.05

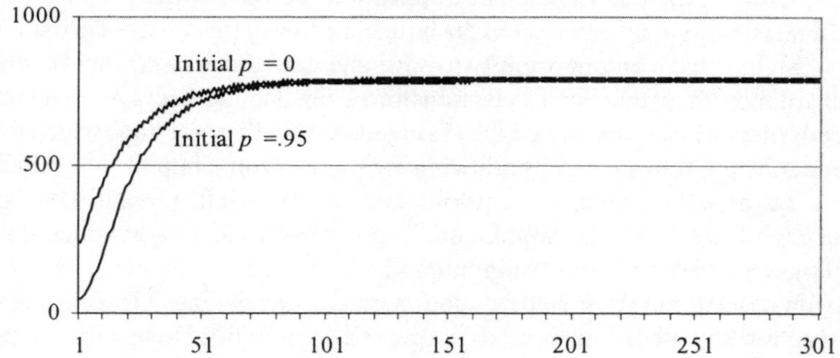

Figure 4.9a: Total invest. for alt. values of ar; p_{init} =0

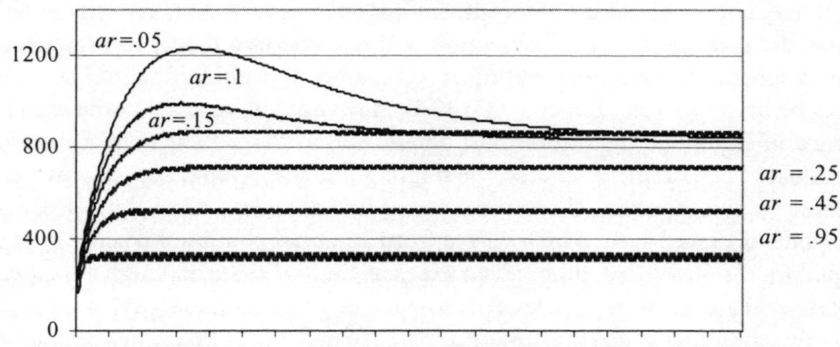

Figure 4.9b: Total invest. for alt. values of ar; p_{init} =.95

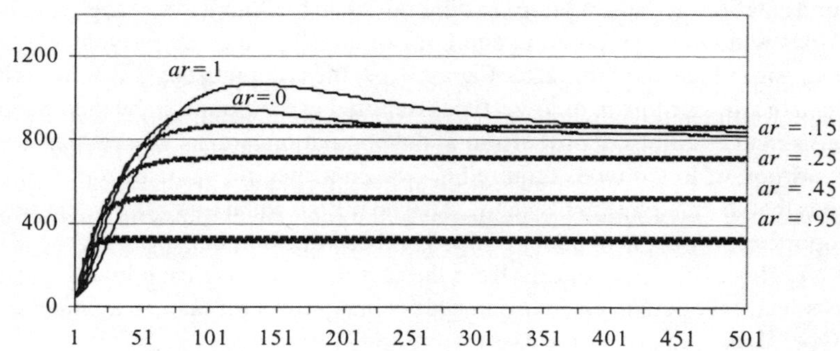

'very low' values (i.e., $ar = 0.05$, 0 .1 and 0.15) those levels are nearly the same. The explanation for this dependence appears to be this: If the proportion of individuals who give little weight to the future is allowed to increase only slowly from 0, high *d*s have an opportunity to save and earn a return on investments. The share of high *d*s that need to be transformed in accordance with *p*, then, is reduced, since when transformed they bring with them their savings, which then become consumption — thereby allowing the share of con-sumption attributable to low *d*'s to approach *p* without requiring a corresponding increase in the actual proportion of low *d*'s in the population. Thus, a lower incremental rate allows for a larger share of the population to remain high *d* types.

Again it is tempting to speculate about substantive meaning. Returning again to corruption, the preceding patterns suggest that a 'solution' to low investment and increasing corruption is not attempting to stamp out this form of market exchange, but to slow its rate of increase, which is consistent with the evolutionary argument that ' when the environment changes quickly, the evolving population may lag ... behind the environment it is "attempting" to track. In the latter case, the process of natural selection will not improve the fit of organism to environment. The environment must be reasonably stable, if natural selection is to be an improver' (Sober 1984: 174). Potential investors who have confidence in and value the future must be allowed to make their investments and dominate the market. Of course, we might guess that a rapid adjustment rate is best if there is an initial excessive share of low *d*s (relative to ES *p*). However, the data in Figure 4.9b, which derive from an initial condition where p_{init} is set equal to 95, shows that although an excessive initial share and high adjustment rates yield a slightly greater level of social capital in the initial periods, eventually the investments that high *d*s are allowed to make over time when adjustment rates are low increases social capital above these numbers. So that once again a slower rate of adjustment is best – at least here in the long run.

Returning to the assumption that the proportion of types is taken to be their representation in consumption, in general the actual numbers of high and low *d* types will not correspond, in equilibrium, to ES *p* since those types will not be consuming at the same rate. Figure 4.10, then, depicts the relationship between the proportion, *p*, of low *d* types required in consumption by the ESS and their actual equilibrium proportion in the population, and as we see, the actual proportion of low *d* types required to sustain a specific value of *p* in fact exceeds that dictated by the ES value. That is, if ES *p* equals, say, 0.30, the actual proportion holding low values of *d* will exceed this number (from Figure 4.10, 55%). This difference follows from the fact that, because each low *d* type invests less, that person also consumes less in equilibrium than an average high *d* type. Hence, to satisfy the expression

$$[\textit{total low d consumption}]/[\textit{total consumption}] = p$$

we must have more low *d* types than this proportion. Substantively, this result suggests that any increase in ES *p* will result in a disproportionate increase in the actual number of people who discount the future greatly. Or, with respect to our inferences about corruption, any increase in the ES proportion of people participating in corrupt activities will yield a disproportionate increase in the actual number of people participating in those activities.

Figure 4.10: Actual prop.of low d's vs p: r=.05, k=.2, ar = .1

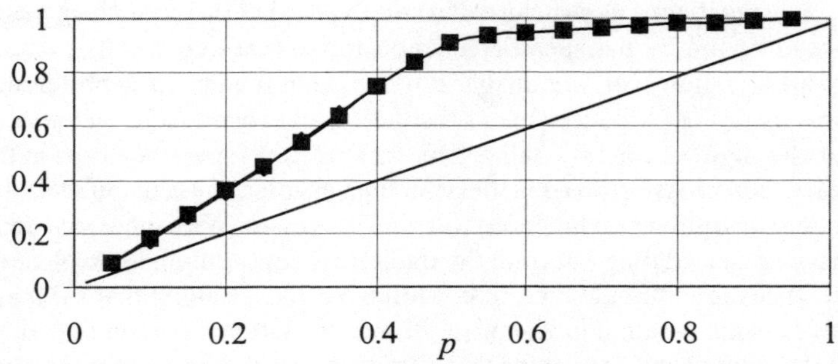

4.3 Initial Conclusions

We would not argue that the way we model endogenous discount rates is the uniquely best approach. Certainly it is worthwhile to consider ways of making *r*, for instance, endogenous. And although we can associate the parameter *p* with various substantive scenarios, we have not yet linked those parameters to specific political-economic or social features of society. Nevertheless, if, for instance, we focus specifically on the issue of corruption, our model is a departure from the approaches offered by those formal analyses that look at corruption as the product of some system of supply and demand relationships. By focusing on private discount rates and arguing that there is a logical connection between these rates and people's propensity to engage in corrupt acts, we are able to offer a number of hypotheses that have heretofore escaped notice. We see, for example, that the rate at which people shift between holding high versus low time discounts may be critical to the overall level of welfare (as measured by invested capital as well as by aggregate consumption; see Figures 4.9a and 4.9b) – and, by implication, between welfare and corruption. Second, although we find that high levels of corruption in the form of a population that for the most part discounts the future heavily might initially yield higher levels of consumption so as to give the impression that corruption 'greases the wheels' of an otherwise inefficient bureaucracy, Figure 4.5 suggests that this benefit is temporary. Third, any factor that causes the equilibrium value of p, the proportion of the population that consumes rather than invests, to increase above zero will have, as Figure 4.3 shows, an initially disproportionate impact on social capital. Fourth, although it might seem that high rates of return on investments will induce savings, the effect in equilibrium is, in general, the opposite – a hypothesis that is not contradicted by patterns of domestic investment and inflated interest rates in the former Soviet Union.

Of course, citing empirical parallels to our model without any in-depth empirical application is a tenuous enterprise since that model, like any other, is limited by the specifics of its structure. Put simply, it is not theoretically general but is instead bound by the specific function forms that define its structure. More importantly, however, our analysis to this point renders time preference endogenous by doing little more than substituting an exogenously set value of ES p. And although we can rationalize p by various fitness parameters as in the evolutionary scenarios portrayed in the preceding chapter, those parameters are most certainly themselves endogenous to social processes. We cannot say, then, that we have advanced much beyond the traditional conceptualization of fixed time preferences found in the literature. Moreover, by assuming that ES p applies to all persons within a group or to all memes within a person seems, in spirit at least, equivalent to assuming that all persons share the same time preference. To make the requisite advance promised by the title of this volume we need to embed the analysis of this chapter in a more encompassing model that allows individuals and subpopulations to influence each other in a more explicitly endogenous way so that ES p (or the fitness parameters that dictate its value) becomes endogenous. It is that task to which we now turn.

5 A 3-Node Network: Part 1

Men nearly always follow the tracks made by others and proceed in their affairs by imitation (Machiavelli, The Prince)

Social learning can be more efficient than reliance on direct experience because learning is then distributed over a population who search in parallel for a solution. The higher an agents cumulative payoff, the greater the probability that the agent will influence its neighbors and the lower the probability that the agent will be influenced (Macy and Sato 2002: 7216)

5.1 Introduction

The previous chapter embeds an analysis of discount rate change in a simple linear multi-period investment model, and although it is tempting to suggest substantive implications, our model to this point is limited by the fact that it takes the evolutionary stable value of time discount types as fixed. Thus, it suffers from the same critique we level against those who take discount rates themselves as exogenous and unchanging. It may be that we look at a dynamic adjustment of private discounts and its consequences for investment and consumption and that we can take a specific scenario and restate the parameter p as a function of the payoffs to some evolutionary game. But the distribution of types to which private discounts converge is exogenously set. There is no learning *per se*, but only 'memes' which take consumption as the measure of proportions and adjust accordingly. That is, the dynamic adjustment that we describe thus far corresponds to a simple fixed algorithm whereby the proportion of high versus low discount types is adjusted to match an exogenously set target. In this chapter, then, we allow for imitation to the extent that we allow the ES parameter itself to change in a manner that seems to accord with notions of learning.

Our analysis here, which embeds the previous model in a simple 3-node network in anticipation of a more general n-node analysis, is divided onto two parts. Recall that in the previous chapter we assume that the relative rates of consumption measure the proportion of high versus low discount types. We make this assumption in large part because consumption is the sole observable behavior that individuals can use as their signal about the proportion of types, but it is not immune to challenge since it is also reasonable to assume that an unstated evolutionary process adjusts memes or personal time preferences on the basis of the actual number of high and low *d* types. To impose this assumption in the previous analysis in effect decouples the types: high and low *d*s make their investment-consumption decisions while proportions are adjusted in the direction of an exogenous and fixed *p* regardless of consumption rates. However, once we embed the analysis in a network, a second mechanism of adjustment – one that provides a better representation of learning or mimicking – becomes available. Specifically, we can let a node's consumption be taken simply as a measure of its relative performance and assume that nodes mimic those who perform best. That is, we can suppose that society is described as a network of (generally intersecting) subgroups whereby ES *p* within one subgroup adjusts as a function of what people (or the memes in them) observe in other subgroups to which they are 'connected'. Because of the availability of this second mechanism of adjustment, we divide our analysis into two parts. In the first, which this chapter details, we take consumption both as the measure of proportions and as a measure of performance, whereas in the second part, which follows in the next chapter, we take the actual numerical proportion as the proportion within each node and use consumption rates solely as the measure of relative performance. However, before we formalize these ideas, we review some of the research on networks, especially as it pertains to learning and the dynamic adjustment of discount rates.

5.2 Networks, Learning and Trust

It is evident that social networks are an important part of everyday life and the sociologists abstract conceptualization of them is an important analytic tool for understanding social processes generally. We also know that many things contribute to individual success, including education, employment, or being born into a wealthy or 'politically connected' family. And in this context it is not uncommon for people to ask: What is more important in achieving personal goals: *what* we know or *who* we know? An equally important question is: to what extent is what we know (or at least what we believe to be the case) a function of who we know?

Much of the literature on networks can be interpreted as an attempt to answer such questions. Or, restated: What are the precise mechanisms whereby interpersonal connections influence beliefs, preferences and, ultimately, actions, and to what extent do beliefs, preferences and actions influence social connections? One answer is that although we are, as human beings, complex creatures, we also have a simple (patterned) range of reactions to circumstances. Among other things (and aside from valuing information for its own, non-instrumental, sake such as when we memorize baseball batting averages), we try to minimize the information we must process to complete tasks, in which case we are likely to rely on readily available and relatively 'inexpensive' sources of information (for the application of this idea to our understanding of electoral processes in light of a seemingly uninformed electorate, see McKelvey and Ordeshook 1990). A networks informational role in this context is perhaps no better illustrated than in the sociological study of labor markets. From the perspective of neoclassical economics, individuals simply 'meet' in the market to conduct economic exchange and the only factors that drive their interaction are the price and quantity of goods and peoples tastes for one good relative to another. This assumption is a convenient tool for extensive modeling, but the results do not fully reflect reality. Granovetter (1995), for example, makes the point that a labor market is far from a perfect example of some simple balance between supply and demand. Many people find jobs through personal contacts, in which case the dynamics and equilibria of the corresponding market for labor is a byproduct of other social processes (see also Fernandez et al. 2000 and Petersen et al. 2000). In fact, in addition to demonstrating the importance of networks in a specific substantive context, research here provides us with two general constructs: strong' and 'weak' ties. strong' ties are with well-known contacts, typically close friends and family members with whom people usually spend a considerable amount of time; 'weak ties' are the acquaintances and job-related contacts that are less intimate and presumably carry less weight in ones choices. The utility of these two constructs in understanding flows of information, which we need not detail here, suggests that networks are themselves complex and require an array of concepts to describe. But this much seems certain: to the extent that a model of social processes – political, economic or otherwise – ignores these constructs, it circumvents a factor that most likely is critical to any adequate understanding of those processes.

A variety of other constructs exist to model network complexity. For example, Uzzi (1996) introduces the notion of embeddedness to emphasize transfer and joint problem-solving. Podolny et al. (1996) link the idea of networks as sources of information to the idea of network density. Podolny (2001) introduces the analytical distinction between two types of market uncertainty: egocentric, which refers to a focal actors uncertainty regarding the best way to convert a set of inputs to an output; and altercentric, which denotes the uncertainty confronted by an actors exchange partners regarding the quality of the output brought to the market. In this conceptualization, networks provide information about the 'quality' of market actors in a way that is critical to the encouragement of social cooperation.[1]

---

[1] An especially good illustration of networks facilitating cooperation (joint action) is Goulds (1995) study of the Paris Commune in 1848. One critical factor of the uprising was 'preexisting patterns of social interaction in the capitals residential districts', with spatially defined communities as important units of political activity' (pp.150 – 51). The importance of residential social networks in this case is that it facilitated

Networks also help formalize the notion of culture. Aside from our discussion of corruption in Russia and Ukraine, well-articulated examples can be found elsewhere. Wank (1996; see also Guthrie 1998), for example, notes that networks commonly referred to as *guanxi* connections (patron-client ties), as a rational response to dealing with the (often uncertain) actions of government, are important descriptors of Chinese society and culture.[2] The concept of *guanxi*, however, and its role in a contemporary capitalistic society should at least cause us to question the universal value of networks as facilitators of cooperation to the extent that they can also become a source of higher opportunity costs and rigidity in response to environmental change. If network relationships become the sole source of economic exchange and information, actors can lose access to potentially profitable opportunities outside of the network. Podolny (2001) notes the detrimental results of high density production networks, Uzzi (1996) emphasizes that 'too much embeddedness' brings more liability than opportunity, and Brinton and Tang (2002) discuss the efficiency of the job placement system of Japan (consisting of the close connections between high schools and firms) under conditions of economic crisis, and how instead of facilitating meritocracy, it contributes to a deepening inequality.

Although there are any number of approaches to understanding the role of networks, there is a consensus that they play one critical role in particular, namely as channels for learning. Much of what people know about the world is based not only on personal experience, but also on the experiences of those with whom they interact. And here we can imagine two forms of interaction – immediate experiences (e.g. everyday market transactions) and what they observe among 'more distant' acquaintances. We can think, then, of people existing in a two level network – immediate acquaintances (i.e., 'differential associations' as described in social learning theory (see Akers 1985)) and indirect relationships. Both exert an influence on attitudes, beliefs and perceptions, and by extension, on the survivability of alternative memes. More specifically, if it is true that cooperation can be sustained in social but otherwise anarchic processes only if the shadow of the future is sufficiently great so that the gains of a short term defection from cooperation are offset by potential long-term losses that result from lost opportunities and sanctions, then for such incentives to

transcending class boundaries, thus enabling a higher mobilization rate.

2 Another example, is the Japanese *keiretsu*, or clustered corporations. Member corporations are linked through the many business activities, including loans, interlocking boards of directors, joint projects and so on. Members of a *keiretsu* prefer to trade with one another rather than with outsiders, which provides a form of joint economic insurance against economic difficulty but at the same time can thwart innovation and rapid market adjustment.

operate, at least one of two things is required: either (repetition) interactions between people need to occur with the expectation that they will be repeated so that an injured party can administer punishments in the future; or (reputation) if they are not repeated between specific pairs, each individual in question needs to know something of the history of cooperation of his/her potential partners and feel secure that that history can be communicated to others so that others can administer sanctions if necessary. This is the essence of Bendor and Swistaks (2001) treatment of norms. The issue that arises when we attempt to establish an equilibrium, evolutionary or otherwise, wherein one person punishes another because that other person has defected from an agreement with yet a third person when punishments are costly, is that punishments are much like a public good. Bendor and Swistak's solution though is to make such punishments, in effect, private by introducing the notion of a meta-norm whereby 'player 1, who failed to punish player 2, who failed to punish player 3, and so on, must also be sanctioned' (p. 1514). To circumvent what might seem an unreasonable assumption about information, their analysis proceeds under the assumption that all persons are classified as either friend or foe, and persons remain in the category of friend only if they 'cooperated with all other friends and defected against all foes ...' (p. 1515). And here, of course, networks are critical in order to associate people with the proper category. As Bendor and Swistak note, 'to know a partners standing in the community tomorrow one only need keep track of that partners status today and, if the partner is currently a friend, his current actions. Social mechanisms like gossip greatly reduce the burden of the second requirement – news about defections among friends spread amazingly quickly in some very large and complex networks' (p. 1515).

The idea of learning through network connections is, of course, not new, but to the extent that networks encourage a form of informational stability that promotes cooperation by allowing for a high probability of finding ones exchange partners at the same place to 'punish' or to 'reward' or transmitting information so that others can reward or punish, networks are the key vehicle whereby expectations labeled trust develop (see Cohen et al. 2001). Sociologists, of course, regard trust as a 'central element in economic and social exchange ... a form of social 'glue' that can be described as consisting of three variants: (1) family or kinship based trust; (2) trust based on reputational mechanisms as present in repeated interactions ... and (3) extended trust, understood as the *ex ante* willingness to cooperate with anonymous others' (Raiser, 1999: 16). The concept of trust, then, is intimately connected with the notion of reputation.

As a side note, here we see once again the problem of enforcing cooperation in corrupt political systems. If secrecy is to be sustained, then to repeat the argument we offered earlier, cooperation requires a more atomistic and less anonymous form of exchange than the forms of market cooperation that correspond to, say, international investments or corporate finance. Extended trust is difficult to sustain – corrupt officials simply do not have the luxury of establishing an extensively advertised brand label. This is not to say that all reputational components of cooperation in markets necessarily reside in the shadow of a formal legal system. But reputation–based enforcement, combined with the absence of legal backup, necessitates closer monitoring by all parties and, correspondingly, a more atomistic structure to the networks that might describe patterns of cooperation. This argument parallels Yamagishi and Yamagishi's (1994) discussion of trust and assurance in social processes, although there is room here for some confusion of terminology.[3] Briefly, they propose a distinction 'between trust as a cognitive bias in the evaluation of incomplete information…and assurance as a perception of the incentive structure that leads the interaction partner to act cooperatively' (p. 129). In fact, our description of the basis of enforcement of 'corrupt cooperation' largely corresponds to their notion of assurance whereby monitoring and the social sanctions that monitoring allow impacts the incentive structure of participants in such a way as to establish conditions for cooperative exchanges. That is, 'trust' in the context of corruption derives from the explicit incentives that monitoring and subgame perfect sanctioning allow. Thus, as with the analysis of differences in Japanese versus American youth, the sanctions available in the enforcement of corrupt forms of cooperation necessarily apply best to closely knit networks of interpersonal relations (see also Yamagishi and Cook 1993, Yamagishi et al. 1998).

More generally, students of social psychology, exchange theory and sociology have been intensively interested in the subject of trust, because it is looked upon as an essential element of cooperation, exchange and the general functioning of society. For example, Gibbons (2001) associates trust with long-term self-interest in repeated games, linking it to the shadow of the future in a way that parallels Axelrods analyses of the emergence of cooperation. Similarly, Buskens and Weesie (2000), in their analysis of car dealerships, find that 'learning is more important than control for the explanation of a buyers trust in a dealer' (p. 249), where 'learning' corresponds to accumulation of past inter-actions with a dealer as opposed to some future opportunity to punish a dealer by not buying from him later. Raub and Weesie (1996) develop the idea of reputations that

---

[3] The confusion here pertains to the words 'trust' and 'assurance'. In their analysis, assurance refers to a situation in which there are two competing equilibria, one of which is preferred by both persons (players) to the other. Trust, on the other hand, refers to a system of beliefs and expectations that allows for successive punishments and rewards. This, however, is a false distinction. The repeated Prisoners' Dilemma, for example, in which cooperation is assured if beliefs and perceptions are such that people can reward cooperative behavior and punish defections owing to relatively ongoing and closed networks (Yamagishis 2001 model of trust) is in fact an assurance game in the sense that not cooperating remains an equilibrium even though both persons prefer the equilibrium in which both players cooperate.

'depend on the 'embeddedness' of interactions in structures or networks of social situations' (p. 626), and conclude that 'efficiency (through the use of reputation) is more easily attained as a result of individually rational behavior in perfectly embedded systems' (p. 626) as opposed to atomized societies.[4] And at a more general level, consider this argument: 'We hypothesize that embedded relations are not just a provincial alternative to participation in the open market but are the classroom where the skills are developed for navigating outside the neighborhood, as argued by Stolle [2001: 305]. "The important question is how the trust that we obviously build for people we know well can be extended and used for the development of generalized trust, or trust for people we do not know well."' (Macy and Sato 2002: 7214). However, even if we accept the idea that the particular importance of social networks with respect to the development of trust is that they allow actors to signal information about themselves even to those 'weakly' connected to them and thereby render non-repeated social interactions among members of a network equivalent to repeated ones, the question remains as to what is being communicated – what precisely are the components of information being transmitted when we say that 'people learn to trust each other'?

In answering this question it is useful to look more closely at the concept of trust in the context of the rationalist paradigm. Presumably the existence of trust causes people to behave or act differently from when it is absent – to cooperate in one instance and not cooperate in others. The expression of trust is action: 'trust is conceptually different from cooperation ... [it is] a decision whether to play the Prisoners Dilemma game or to exit' (Macy and Sato 2002: 7215). However, if we look at the elements of choice that constitute a description of, say, Prisoners' Dilemmas in their various forms (e.g., single play, repeated play, as part of a game with informational uncertainty, and so on) we do not see 'trust' *per se* as a primitive construct, but rather outcomes, beliefs about the nature of the interaction (the set of available strategies or actions and the mapping between actions and outcomes), preferences over outcomes, and beliefs about the preferences of others. It is these things that we find in the rationalists abstract representation of a two-person 'game' used to understand the likelihood of cooperative versus non-cooperative action. The development of trust, then, is necessarily a process whereby these constructs take on values so that cooperation becomes more or less likely. And, as we already know, an essential compo-

[4] The idea of social network and its connection to the notion of trust is also reflected in the work of Coleman (1990), who equates trust and social capital since trust reduces transaction costs. Similarly, Putnam (2001), in his discussion of American communities, attributes the development of trust to norms of reciprocity and networks of civic engagement, and Knight (2001) stresses the within-network enforcement of compliance, which induces trustworthiness and generalized trust. See also DiMaggio and Louch 1998, and Insead 2000 and Cook and Hardin 2001.

nent of the construct 'preference' is the time discount people apply to the future. Thus, quite directly,

> *arguments about the importance of the interaction between trust and social networks are simultaneously arguments about understanding the interaction of social networks and individual time discounts that encourage cooperation.*

Put differently, to the extent that individual time discounts are the endogenous product of a social process,

> *we cannot have a complete understanding of the dynamics of time discounts absent understanding how networks impact this object of study. At the same time, we cannot have a complete understanding of the interaction of networks and trust and the role of each in facilitating cooperation until and unless we understand the interaction of networks and time discounts.*

5.3 Networks, Crime, Addiction and Corruption

What often makes it difficult to describe the complexity of social systems and conceptualize them in abstract terms is that a great deal of the inter-connectivity – the networks – that render society a society' is informal or unconscious. We rarely see people drawing diagrams of their core network, and even fewer still could calculate the 'power' or strength of connection' of some relationship in comparison to others. Consequently, the social scientist, in order to study the role of networks and their influence on behavior, must not only invent terms to characterize things (e.g., power, strength of relations, embeddedness, clique, etc.), but he or she must also find ways to measure and represent these concepts, along with their theoretically formulated implications, so that empirically testable hypotheses can be formed and assessed. We appreciate that networks are not universally explored in the social sciences and that sometimes it is analytically more tractable to assume that the relevant social group is homogeneous. However, in the cases at least of corruption, gang formation and addiction, the preceding discussion of trust, reputation and learning suggests that a formal consideration of network structures can introduce a level of analytic complexity which brings theorizing considerably closer to reality. But it is not merely complexity that motivates introducing networks here. It is, as we conclude in the previous section, the fact that private discount rates are not fixed parameters of a person's character and that these parameters are themselves the product of human interaction and experience in contexts in which networks appear to be critical determinants of how people value the future relative to the present.

To expand on what we have just said, when we look at society we do not see a mass of identical individuals. People differ by gender, age, income, race, etc. But they also differ by experience, taste, belief, access to opportunities, status, accomplishment, and so on. People are, thereby, 'laboratories of social experimentation' for others to whom they are socially connected such that people 'connected' to each other are able to observe each others behavior, success, and so on and copy that behavior if it appears advantageous to do so. This notion of imitation is not new. Echoing Machiavelli's words that introduce this chapter, as Bikhchandani et al. (1998: 152) observe, 'social observers have long recognized imitation as important in human society.' The logic of mimicking, moreover, is straightforward. Cognitive abilities are limited, and very often, instead of creating their own solution to the problem, another solution, already proved to be successful, is borrowed from others. The solution need not be perfect as long as it yields satisfactory' outcomes and saves time and effort. It is for this reason that we often find members of a close-knit group more homogeneous across a variety of dimensions than two randomly selected people from two different groups. That does not preclude other groups from observation and copying, but these copying abilities can be limited if one group is a closed community without free and easy access.

The notions of copying and imitation move us to the domain of social learning, and here we recall Macy and Sato's (2002) description of their model of adaptive learning: social learning is based on adaptation to direct and vicarious experience: 'Social learning can be more efficient than reliance on direct experience because learning is then distributed over a population who search in parallel for a solution. The higher an agent's cumulative payoff, the greater the probability that the agent will influence its neighbors and the lower the probability that the agent will be influenced' (p. 7216). Macy and Sato attempt to formalize what we know substantively (and somewhat imprecisely) about the social processes that describe the various scenarios offered earlier concerning addiction, crime and corruption. First, with respect to social deviance generally, '[s]ocial learning theory...views miscreant behavior as a learned response to a configuration of social reinforcements and definitions provided by individuals with whom one associates' (Winfree et al. 1993: 102). And with respect to drug and alcohol addiction in particular,

> the probability of abstinence decreases and the frequency of use increases when there is greater exposure to using rather than to abstinent models, when there is more association with using than with abstinent peers and adults, when use is differentially reinforced (more rewards, fewer punishers) over abstinence, and when there are more positive or neutralizing than negative definitions of use ... users are more likely than abstainers to associate with peers who are also users and this relationship remains whether friends' use is measured by or independently of the individuals perception of friends' use' (Akers et al. 1979: 639).

In addition, 'friends provide social reinforcement or punishment for abstinence or use, provide normative definitions of use and abstinence, and, to a lesser extent, serve as admired models to imitate ... [as a result] the adolescents use drugs and alcohol to the extent that the behavior has been differentially reinforced through association in primary groups and defined as more desirable than, or at least as justified as, refraining from use' (Akers et al. 1979: 647).

Akers and his colleagues anticipate the conclusions of decades of subsequent research into the relationship among peer networks, addiction and social learning. For example, '[i]t is during pre-teen and adolescent years that nearly all alcohol and drug use begins, although health consequences of practices such as smoking and the use of marijuana may not be apparent for some years afterwards. The *peer group* is now widely regarded as an important social environment capable of influencing both the *initiation of substance use and the maintenance* of patterns involved' (Cotterell 1996: 129, emphasis added); and '[w]hen drugs are actually used, it is almost always in a *peer context*. Peers initiate the youth into drugs. Peers help provide drugs. Peers talk with each other about drugs and model drug-using behaviors for each other and in doing so shape attitudes about drugs and drug-using behaviors' (Oetting and Beauvais 1987: 137 emphasis added).

Cotterell (1996) sees this process of social learning as wholly general: '[i]nformally structured face-to-face groups ... are emotionally supportive structures, affirming the individual members as somebodies with distinctive personalities. The central characteristic of these groups is an *attachment relation* ... the most important motivational factor in the formation of peer groups is the desire for a *personal, affective response* from a few other individuals. These groups will continue to exist because they supply positive emotional experiences to their members, through *acceptance and recognition* of the individual as a contributor to what the group stands for, and by providing in return the *sense of belonging and solidarity* with the group which confirms the group identity for each individual member' (p. 44 emphasis added). And echoing many of the constructs of social learning theory with respect to imitation and so on, 'for many decisions like finishing high school, attending college, joining a gang, taking drugs, or engaging in premarital sex, youth possess limited information about the potential outcomes of their choices ... youth look to role models and their social networks for advice [and] to the extent that social networks are drawn from the neighborhoods in which youth live, the *information and choices* of youth will depend critically on the *choices made by and the outcomes* of others in their neighborhood' (Heavner and Lochner 2002: 1, emphasis added).

What applies to addiction should apply to gangs and crime, and following largely in the footsteps of Sutherland (1939), 'who is usually credited as being the intellectual ancestor of the differential association school, which emphasizes the importance of social interactions' (Glaeser et al. 1996: 510), Winfree et al. (1994) note that 'gang members reward certain behaviors in their peers and punish others, employing goals and processes that are undistinguishable from those described by Akers ... The concept of differential associations – typically operationalized as the proportion of ones best friends that engages in some illicit act – has a natural linkage to gang research. The greater the proportion of ones best friends who are youth-gang members, the greater the social forces impelling the youth to become a gang member' (pp. 149 – 50). Winfree et al (1994) prescriptions for combating crime and delinquency reenforces this connection between networks, learning and payoffs: 'negative social punishers exhibited an inverse and independent relationship to gang membership. Consequently, getting out the specific message that gangs create problems for youths with their parents and with the law may make gang membership less appealing. It may be possible to use these same forces to '*unlearn*' gang motivations and definitions, or prevent the process from occurring' (p. 170, emphasis added). And in a related study, Winfree et al (1993) note that 'differential peer associations played the largest role in explaining cessation of drug use, particularly that of alcohol, marijuana, and stronger drugs [p. 103] ... current users reported having more drug using peers and higher levels of personal approval [p. 113] ... and those who have never used drugs tended to describe learning environments that encouraged abstention. They associated with non-using peers and were exposed to definitions that are disapproving of drugs ... Abstainers were exposed to non-using peers with anti-drug orientations; current users generally had drug-using friends who approved of drug use [p. 120] ... peer influences may be sufficient to constrain drug abstainers or to initiate and sustain users, ...parents and other influences may be required to effect the change in status from use to resistance that occurs for former users, a process that ... conforms to the principles of social learning theory ... [and] it is conceivable that former users may have come to the conclusion that drugs and education do not mix by *observing and interacting* with college-bound youth; consequently, if they, too had aspirations for higher education, they may have decided to quit using illicit drugs' (pp. 118 – 19, emphasis added).

Thus, contemporary research regarding crime (delinquency) and addiction both emphasize the role of networks (peer groups), adaptive imitation, and learning. The role of networks and learning is no less evident with respect to corruption. Returning to the idea of corruption as a form of cooperation, we can see here the essential components of repetition, reputation and trust, as well as the relevance of Rose-Ackermans idea that corrupt officials need networks to promote their services' to potential clients. As we noted earlier, a corrupt individual cannot easily advertise their character as might, say, the manufacturer of baby food or toothpaste and instead needs to rely on word-of-mouth advertisement. Thus, people who pay bribes as well those who accept them must either create new social networks or use preexisting ones for that purpose. For a more specific instance of the connection between networks and corruption, consider Russia and the former USSR. Although one might expect the role of networks, notably 'informal' ones, to diminish as society becomes more urbanized and developed industrially (and therefore in Émile Durkheim's terms, more anomic), Russia, because of the years of totalitarian oppression, made family and close friends crucial for survival: 'Russians have developed extensive social networks with high levels of political capacity... which are important means for the diffusion of ideas' (Gibson, 2001: 51). Networks were not only a means of cultural and spiritual survival – 'groups of friends who met in the kitchens of their apartments and led endless conversations about the meaning of creation, art and politics' (Greenfeld, 1992: 23) – they also provided a partial solution to recurring deficits, and allowed members access to goods and services in short supply and ways around formal bureaucratic procedures. These networks even had a semi-formal label: 'blat'. Similar types of networks could be found in East Germany before 1990, and as Volker and Flap (2001) note, 'provision networks' were characterized by weak ties with a lesser level of trust than that of very close, 'niche' networks, which consisted of close relatives and friends. Emphasizing the role of trust here, 'blat' networks differed, then, from family relations in that they not only concerned specific services, they also 'included ... people who one knows only indirectly by recommendation' (Ledeneva, 1998: 117). Networks of this type, moreover, were complicated formations, where one could distinguish between 'horizontal' and 'vertical' structures – the former 'composed of people of similar status, known as 'people of the circle', and the latter 'composed of people of different social strata interested in each others connections and linked by kin, personal contacts, or most often, intermediaries known as 'useful people'' (Ledeneva, 1998: 121).[5]

The importance of networks in the former USSR is underscored by Remington (1981: 804): '[t]he Soviet case is intriguing, because, despite a modern, elaborate media system, which is used regularly ... by fully 80 percent of the population, studies show that strikingly high proportions of the population – as many as half or more – also regularly obtain information from informal social contacts such as neighbors, relatives, and co-workers'. This is in sharp contrast to the United States, where 'only a very small proportion cite conversation with other people as one of the sources from which they obtain most of their news' (Remington, 1981: 806). The years of totalitarian rule and overall oppression made Russians distrustful in general ('Russians in 1998 expressed more distrust than trust – 59.0 percent said one cannot be too cautious, while only 30.5 percent claimed that other Russians can be trusted' (Gibson, 2001: 61)), and made them rely on closer circles of kin and friends ('trust in acquaintances is relatively high (nearly 60 percent), members of the social networks are trusted even more, and most people estimate that others trust them' (Gibson, 2001: 61)). It

[5] The group described above as 'useful people' is close to our notion of the corrupted official searching for clients through network connections.

comes as no surprise, then, that 'Russian political leaders are quite widely distrusted. But when it comes to ordinary Russians, levels of trust among people who know each other are high and roughly double the levels among strangers' (Gibson, 2001: 61). It is interesting to see how Russians themselves describe their everyday life and the role of networks in it: 'Everything here depends on personal relationship with authorities: if you are a relative or an acquaintance, then you will get a nice, fertile piece of land, if you are on hard terms or totally strange to the village, you might receive far distant plots with lots of ravines' (Klyamkin and Timofeev 2000: 320) notes the 21-year-old son of a Russian farmer. And with respect to Russian business partnerships generally: 'It happens that the farmer won't sell his produce to somebody he does not know, instead he sells it to somebody he was selling it to before even though the new buyer would offer a higher price. The farmer will sell to 'his' buyer, because this action will preserve good a relationship through the years; while a new buyer might not come back next year and the produce thus will be wasted' (Klyamkin and Timofeev 2000: 324). Even getting credit from bank is not easy: 'My dad managed to receive credit, but he made an agreement through personal connections, because nobody in the bank will lend the money to an unknown person' (Klyamkin and Timofeev 2000: 330).

These networks were in the past crucial for the survival of the average Russian. Now, when Russia and other states of the former USSR experience seemingly cataclysmic economic and social change, we can see why networks continue to be an important parameter for understanding social processes. An individual, losing the ground from under his feet as official ideology is suddenly discarded and formal political-economic relations change, has only friends and connections to 'hang onto.' Indeed, regardless of era – Tsarist or Soviet – Russians have learned to award a minimum level of trust to state and state-sanctioned institutions, and to relate primarily to friends and acquaintances for support and information. In this context, it is unreasonable to suppose that these networks do not continue to play an important role not merely in patterns of cooperation generally, but in those patterns we might label corrupt in particular – in teaching people what forms of cooperation are likely to be sustained, what actions are likely to be taken by people outside their networks, what actions are likely to be sanctioned and not sanctioned, the preferences and likely actions of state bureaucracies, and, in Bendor and Swistaks (2001) terms, the identities of friends and foes. Indeed, we might even hypothesize that to the extent that corruption requires special modes of monitoring, we can begin to understand the apparent pervasiveness of corruption there. Preexisting modes of cooperation provide a ready supply of networks both among the population generally and among ex-Communist party officials in particular that can be readily adapted to corrupt forms of cooperation and that, via the usual processes of social learning, allowed people to adapt to changing circumstances in a way that leads to widespread forms of corruption infecting even 'normal' economic market exchange.

5.4 Investment in a Triad Network: Initial Structure

To translate the preceding ideas into a model that allows us to study the social determinants of time preferences, imagine that the members of subgroups within a network discount the future greatly while members of other subgroups place a higher value on the future. There are any number of questions we can now ask. Will people in one group mimic other groups if they appear to be more successful? How will such mimicking impact overall economic performance? Will the size of a subgroup be a critical parameter in determining the direction and velocity of change? Such questions open the door to the possibility of two mechanisms of social change, one of which is not present in the model offered in Chapter 4. First there is the within-group change that we do model whereby the perceived proportion of high and low d types (as measured by rates of consumption) incrementally adjust in the direction of some exogenously set ES proportion. But second, there is the cross-group change, which occurs when one subgroup either directly or indirectly influences the overall equilibrium of other groups to whom it is 'connected'. It is the nature of this 'connectedness' and the process of both modes of evolutionary change to which we now turn.

We cannot proceed immediately to consider all issues and concepts associated with social networks and individual discount rates, but as a first step in that direction, instead of assuming a homogeneous society, we suppose in this chapter and the next that society consists of three groups (nodes). Each group, in turn, is assumed to consist of a mixed population of two types: those with low and those with high valuations of the future. Before proceeding, however, we want to emphasize again the flexibility of our conceptualization in that groups can be viewed here as individuals, where the 'population' of low and high d types, or as individual decision makers, each of whom corresponds to a collection of memes. However, throughout our discussion, we refer to the nodes of networks as groups for linguistic convenience.

The groups themselves are distinguished initially by the proportion of high and low types in the population that would exist in equilibrium as determined by an evolutionary model in which the proportion of high and low types is adjusted at an incremental rate within each consumption period in the direction of the ES p associated with each group. This is the convergence process explored in Chapter 2 as a function of the parameters of the ESS games, rates of change and relative magnitudes of supplemental incomes. The focus of this chapter, in contrast, is change in the ESS values themselves. To illustrate what we mean, consider the polymorphic solution to our first corruption scenario (Table 3.6), which, if we set $d_{low} = 0$ for expository purposes, equals

$$p = \frac{(X - C) - A(1 - d_{high})}{R(1 - d_{high}) - C}$$

Suppose a person updates the values of these parameters as a function of what he or she observes within their network. A person (or, more conveniently, a persons memes) might, for instance, witness instances of corruption being especially profitable, leading to an increase in the estimate of A and a decrease p; or especially profitable patterns of non-corrupt cooperation might be observed, thereby raising X and increasing p. We will remain agnostic as to what precisely is being 'learned' in the dynamic process we describe except to allow p to change as a function of either the relative per-capita 'performance' of subgroups or the overall relative performance of subgroups.

The type of network connections explored in this chapter, now, are of an especially simple sort. Our previous analysis assumes, in effect, that a group is wholly isolated and that all individuals (or memes) within it are *connected* to all others in the sense that overall levels of consumption are observable by all. Here we assume that, in addition to being cognizant of its own pattern of consumption, a group (network node) observes the pattern or consumption of every node to which it is connected.

However, earlier we also assumed that the ES value of p is exogenously fixed, whereas here we allow p to change incrementally as a function of the per capita or total consumption of each group to which the group in question is connected. Specifically, suppose group Y is connected to both other groups and that per capita consumption (welfare) is highest in group X. Then we let group Y incrementally adjust its ES p in the direction of the ES p of the seemingly 'more successful' group. This is admittedly a simple formulation of learning via mimicking. Nevertheless, just as it is best to teach children to behave in a certain way by ones own example, we incorporate this idea in our network where the members of one group 'learn' through the observation of connected groups. When members of one group see the members of another group being more successful (i.e. their consumption being higher) they mimic the character of the most successful group by way of mimicking their ES p value.

Before continuing with any other description of our network simulation we need to elaborate further on the nature of the connection between groups, and here we consider two distinct types of connections.

The *first* type derives from the one allowed in our earlier model. It involves adjusting a groups proportion of low *d*s in the direction of its ES p proportion, except that now the adjustment within a group depends on what it observes in sum among all the groups to which it is connected. To explain this process in more detail one can offer the following example: imagine two connected groups and let the ES p of one group be 0.5 and the other 0.9. Suppose that the apparent proportion (based on relative consumption as in the previous chapter) of low *d*s in the first group be 0.4 and in the other 0.8. Since these two groups are connected, what the members of each group observe is the proportion 0.6 (the average between 0.4 and 0.8, assuming that the two groups are of equal size). Therefore the first group will incrementally decrease the number of low *d*s in it whereas the second group will increase that proportion. In this way the connection of one group to another influences the adjustment of the other group and vice versa.

The *second* type of connection involves changing the ES *p* within a group. In this case if a group observes (is connected to) another performing better (in terms of per capita or total consumption), it changes its ES *p* in the direction of the more successful node. Taking our previous example, imagine that the group with the greater ES *p* has higher consumption per capita and thereby gives the appearance of better performance. Its connected counterpart thereby incrementally increases its ES *p* from 0.5 in the direction of 0.9 (at the rate of ES adjustment, which means if the rate of ES *p* adjustment is 1, the ES proportion changes instantaneously to 0.9; if the adjustment rate is zero, it does not change at all). This change is equivalent to letting one or more of the parameters of an implicit evolutionary model such as the ones we offer in Chapter 2 change in the direction of the mimicked group. That is, this incremental mimicking of ES *p*s is equivalent to revising the payoffs of an evolutionary game on the basis of what one observes about the relative performance of the groups to which one is connected.

Insofar as other parameters are concerned, in addition to ES *p* values, the three nodes or groups in the network we are about to explore can also all differ by the various parameters of the model (size, supplemental income, initial proportion of low *d* types and incremental rate of adjustment). Adjustments in individual discount rates, moreover, can be a function of per capita consumption (in which case, relative size of the groups is irrelevant) or total observed consumption within each group. Each pair of groups, in turn, can be connected or not to form a network. This notion of connection in the triad, then, gives rise to four distinct graphs. However, even these four possibilities give rise to a larger number of alternatives to the extent that the character of the groups assigned to the nodes may differ. In the simulations that follow we simplify our analysis and reduce the number of parameter configurations that we must consider by assuming that groups 1 and 2 are twins and that group 3 is distinct from the other two in terms or size and/or its initial ES *p*. This, then, gives rise to six alternative network structures as portrayed in Figure 5.1 and as summarized in Table 5.1.

Finally, apart from looking at how the structure of the network can influence per capita consumption, there are two other dimensions that can play a decisive role in determining the outcome of a simulation. The first is whether the adjustment of the ES proportions is based on per capita or total group consumption. Admittedly, we are unsure as to which alternative corresponds to the better assumption about how individuals measure the relative performance of groups. When deciding whether or not to pattern ones behavior after the seeming affluence of 'new Russians' and their often absurdly ostentatious consumption, it is generally the behavior of individuals that we observe as opposed to the

Figure 5.1: Network Structures

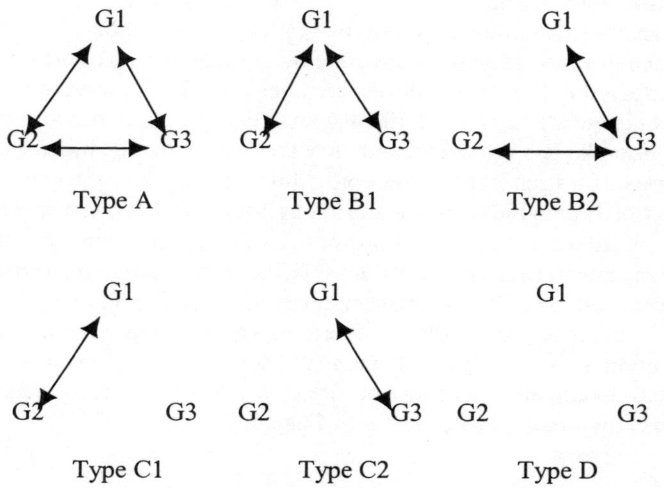

Table 5.1: Six Network Structures

	A	B1	B2	C1	C2	D
groups 1 and 2	1	1	0	1	0	0
groups 1 and 3	1	1	1	0	1	0
groups 2 and 3	1	0	1	0	0	0

total consumption of this class. On the other hand, Moscow shops awash in luxury goods sporting price tags that might make the purveyor of comparable items along Rodeo Drive blush, serve as a measure of the overall capacity of this class to consume. Or, if we interpret each node of a network as a single individual populated by a collection of memes, then total consumption might be the more relevant measure, whereas if a node is a collection of individual decision makers, each characterized by a single meme, then perhaps per capita consumption provides the better model. Hence, we explore both alternatives (note, incidentally, that the size of groups matters only when total consumption is the measure of performance). The second dimension is rooted in whether we allow a group's ES p to change and, if so, the rate of allowable change. Just as the rate of adjustment within a group is revealed in the preceding chapter to be a relevant parameter, we cannot suppose that the rate of adjustment across groups is less important.

The analysis that follows requires two caveats. First, although we attempt to assess the relevance of network structure by examining each of the six alternatives portrayed in Figure 5.1, those structures are taken as exogenously fixed input 'parameters'. It is not until Chapter 7 that we allow network structure itself to be endogenous. And second, note that the preceding discussion makes no allowance for strength of connection'. Moreover, in this chapter and the next, a connection is both bidirectional and strong': if group 1 is connected to group 2, then 2 is connected to 1. And if group 1 takes group 2 into account when making any internal adjustment, it does so without any discounting factor. There are no connections of intermediate strength whereby a group might 'look to' another with some probability or where, in assessing the overall average of consumption across groups, it deferentially weights the groups to which it is connected. Again, allowance for network connections that are asymmetric and also somehow weak is postponed until Chapter 7.

5.5 Analysis

General Parameter Values: Our simulations begin by letting the ES proportion of low *d*s in the three groups be 0.05, 0.05, and 0.95 with respective sizes of 100, 100, and 50. The reason for this is that we want to see what influence a smaller, but 'bad' group (a group with a high ES proportion of low *d*s and, therefore, with a minimal propensity to invest) has on two larger, 'good' groups. This particular starting point arises from the suggestion that 'loud' or highly visible groups of, say, corrupt individuals, gang members or addicts, despite their initial small numbers, can exert an inordinate influence on a population (note again that if a group observes others on the basis of per capita consumption, as opposed to total consumption, the size of all groups is obviously irrelevant). In examining this scenario, we can, as in earlier chapters, iterate on different parameters, which include the initial proportion of low *d*s, supplemental income per period per person, return on investment, the rate of a groups adjustment in its internal proportion of high versus low *d*s, initial ES *p,* and, finally, the rate at which a group adjusts its ES *p*. Once we choose one parameter on which to iterate through a range of likely values, the remaining parameters stay fixed, and unless otherwise specified, the following values are used for those parameters we treat largely as constants in the analysis that follows: Incremental income, *k*, equals 0.2; the internal return on investments, *r*, is set to 0.05; the initial proportion of low *d*s in each group, p_{init}, equals 0.05, and the internal rate with which a group adjusts its proportion of high and low *d*s, *ar*, is set at 0.025. Table 5.2, then, summarizes the parameter values employed in the simulations that follow unless stated differently.

Table 5.2: General Parameter Values

parameter	group 1	group 2	group3
n	100	100	50
ES *p*	.05	0.05	0.95
k	0.2	0.2	0.2
r	0.05	0.05	0.05
p_{init}	0.05	0.05	0.05
ar	0.025	0.025	0.025
ES *p* adjustment	0.025	0.025	0.025
basis of evaluation	per capita	per capita	per capita

Rate of ES p Adjustment: Our first simulation results, now, assume that all adjustments within groups occur on the basis of per capita consumption. Hence, the size of groups is irrelevant. We begin with Figures 5.2a–c, which compare the final equilibrium value of per capita consumption against the rate of adjustment in ES *p*s – the rate at which a group adjusts its ES proportion of individuals (or memes) with high and low *d*s – for network structures A versus D, B1 versus B2, and C1 versus C2, respectively. The data here are reported in triples: the first, fourth, seventh and so on data points in each chart correspond to the per capita consumption of group 1 for adjustment rates of 0, 0.025, 0.05, and so on respectively; the second, fifth, eight and so on data points correspond similarly to group 2, and the third, sixth, ninth, and so on data points correspond to group 3's per capita consumption at the assigned adjustment rates.

There are several things, now, to note about these figures. First, if groups 1 and 2 are not connected to 3 (network D in Figure 5.2a and C1 in Figure 5.2c), they converge to a uniform (and relatively high) level of per capita consumption regardless of the rate of ES *p* adjustment. This invariance with adjustment rate derives in the case of network D from the fact that with a wholly disconnected network there can be no adjustment in ES *p* whereas, with network C1, groups 1 and 2 are identical and remain so throughout the simulation so that neither group sees' another with any greater level of performance. In the case of these two network structures, then, our model is equivalent to the one we offer in the preceding chapter (except that now we can examine three groups simultaneously).

The comparisons offered in Figures 5.2b and 5.2c with respect to network types B and C are perhaps more interesting. First, note the uniformly better overall performance of all groups in network B1 as compared to B2 – the network for which group 1 is the 'focal point' of a hierarchical structure as opposed to when 3 is the 'focal point'. This might seem an intuitively plausible, even

Figure 5.2a: per capita consumption vs. ap

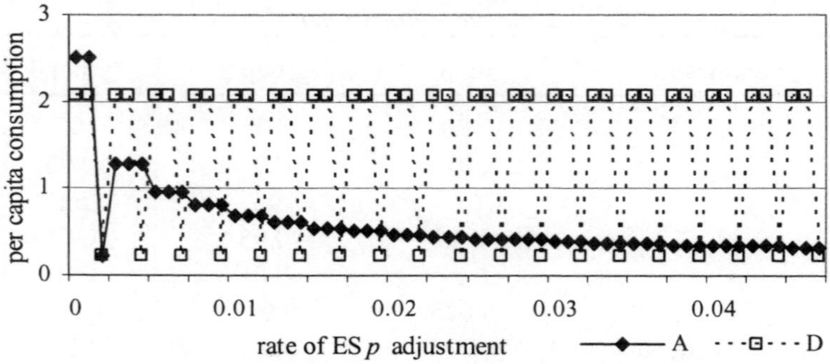

Figure 5.2b: Asymmetric networks B

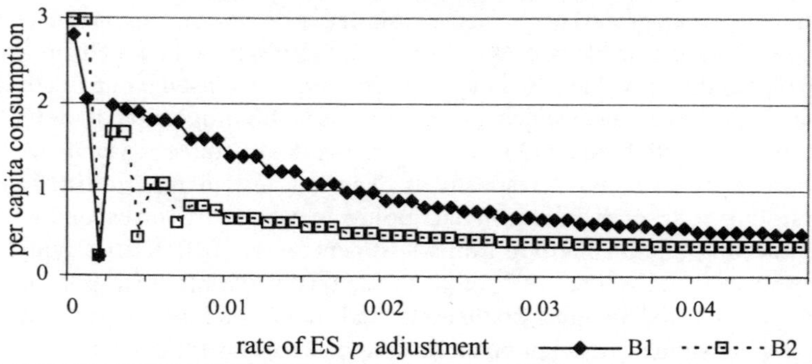

Figure 5.2c: Asymmetric networks C

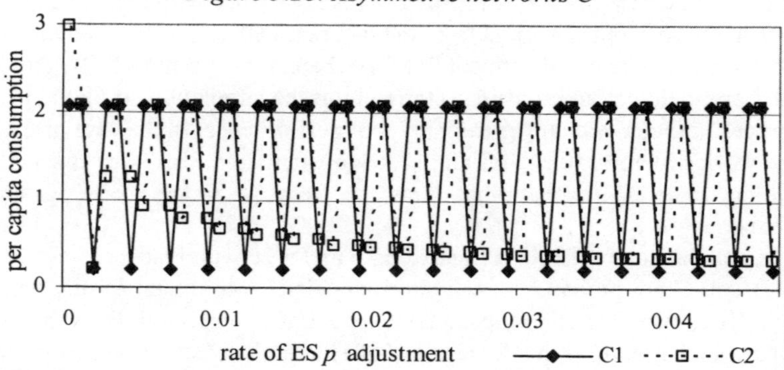

expected, result since 1 and 2 in B2 can only mimic the group with the highest ES proportion of high discounters, group 3, or themselves, whereas in B1, 2's behavior is not directly and immediately influenced by 3, which eventually allows 2 to act as a counter-weight to 3's influence on 1. This is, in effect, largely what occurs. For example, for the parameter values we use to generate Figure 5.2b and a rate of ES *p* adjustment equal to 0.025, node 2 in network B1 mimics 1 throughout the process of convergence whereas in B2 it mimics group 3 instead until period 26 before its shifts to mimicking itself. For these first 25 periods in B2, then, 2 moves its ES *p* in the direction of 3's considerably higher value whereas in B1, although group 1 mimics 3 for 25 periods before mimicking itself, by mimicking 1, 2 tracks a considerably lower value of ES *p* and thereby makes far smaller upwards adjustments in its ES *p*. This permits 2 to build up a level of capital that allows it eventually to increase its long term consumption to the point that it becomes an attractive node for 1 (and itself) to mimic. Indeed, although in B1 group 1 mimics 3 for the first 25 periods before it switches to itself, by period 37, 2's consumption has increased sufficiently that 1 turns to mimicking it.

This interplay between 1 and 2 is not without consequences for 3. Although 3's pattern of mimicking is the same in both B1 and B2 (itself for the first 25 periods and 1 thereafter), because 1 shifts to tracking 2's ES *p* after period 37 in B1, group 1 in B1 is a more productive target for 3 to 'chase' than it is in B2. Thus, 3's ES *p* ultimately converges to a lower value and its per capita consumption converges to a higher value in B1 than it does in B2.

Figure 5.2b also portrays a pattern that holds generally across a range of parameter values. Specifically, unless they are wholly isolated, group 3's influence on groups 1 and 2 and its corresponding ability to 'drag down' the long-term per capita consumption of groups 1 and 2 increases as the rate of ES *p* adjustment increases. Thus, just as we found earlier that an increased rate of adjustment within a group with respect to adjusting the proportion of high and low *d* types in the direction of *p* is generally detrimental to overall social welfare, we find the same is true with respect to this additional adjustment rate.

Our next comparison concerns network structures C1 and C2. As expected, when neither group 1 nor 2 are connected to 3 (C1), the system performs as if no node is connected: because 1 and 2 are identical in all respects, neither influences the other. However, if 1 alone is connected to 3, group 2 continues to converge to the same level of consumption as before, but groups 1and 3 influence each other, with 3 pulling 1 down and 1 pulling 3 up. And again, group 3s influence grows as the rate of ES *p* adjustment increases – as group 1 is given less time to build up its investments for future consumption.

Finally, consider structures C2 and B2 (which differ in that 2 is connected to 3 in B2 but not C2) and the performance of groups 1 and 3. Although the progression of data differs here somewhat, there is no order of magnitude difference in consumption levels attained (for example, per capita consumption converges approximately to 0.5 in both cases at an ES adjustment rate of 0.0175 in Figures 5.2b and 5.2c). Thus, in B2, 2's influence on 3 with moderately high rates of ES *p* adjustment is not sufficient to impact 3's influence on 1, which again illustrates that if groups who invest initially are not allowed time to build those investments early in the process of adjustment and convergence, the influence of groups that save little will predominate. Stated more generally,

> *if the rate of mimicry is 'sufficiently high', and adjustments are based on per capita consumption, a small cadre of visible individuals can change the behavior of the mass of the population and it is only when that rate of adjustment in ES p is low that the mass can either exert a positive influence on some smaller cadre or resist the influence of that cadre.*

This conclusion can be presented even more forcefully, and in a way that illustrates the interaction of parameters by noting first that if we compare networks A and B2 in the preceding figures, we might conclude that isolating nodes 1 and 2 from each other (B2) has an advantage over connecting them (A) when both are otherwise connected to 3. Although differences are slight, at least for low rates of ES *p* adjustment, groups 1 and 2 achieve a higher level of per capita consumption in B2 than in A. This fact is shown even more clearly in Figure 5.3a where we raise the return on investments, *r*, from 0.05 to 0.10. But now consider Figures 5.3b and 5.3c whereby *r* is increased to 0.20 and 0.30 respectively. There are three patterns to note in this progression of figures. *First*, as *r* increases, the overall level of per capita consumption achieved decreases. Although all three rates of return yield approximately the same level of per capita consumption at the highest rates of ES *p* change (approximately 0.05), for lower rates of change a lower return on investment is clearly advantageous to all three groups. *Second*, as in Figure 5.2b, group 3's per capita consumption is maximized for all three values of *r* here for some intermediate value of rate of ES *p* change. When $r = 0.05$, that maximum occurs when the rate of ES *p* change equals 0.01, when $r = 0.1$, that maximum occurs at 0.015, when $r = 0.2$ it occurs at 0.03 and when $r = 0.3$, it occurs at 0.0425. Thus, the greater the return on investments, the greater is the rate of change in ES *p* that maximizes 3's welfare as measured by per capita consumption. *Finally*, although groups 1 and 2 clearly achieve a higher level of per capita consumption in network B2 than in A when the rate of ES *p* change is not 'too high', that advantage slowly disappears as the return on investments, *r*, increases. On the other hand, the range of values of rate of ES *p* change for which B2 holds this advantage over A increases as *r* increases. When $r = 0.10$, groups 1 and 2 are essentially indifferent between A and B2 for rates of ES *p* change at or above 0.0175, when $r = 0.20$, indifference begins at 0.0325 and when $r = 0.30$, at 0.045.

We hesitate to attempt to draw substantive inferences from these numbers and patterns, but at least this much is clear:

Figure 5.3a: network A vs. B2, r = 0.1

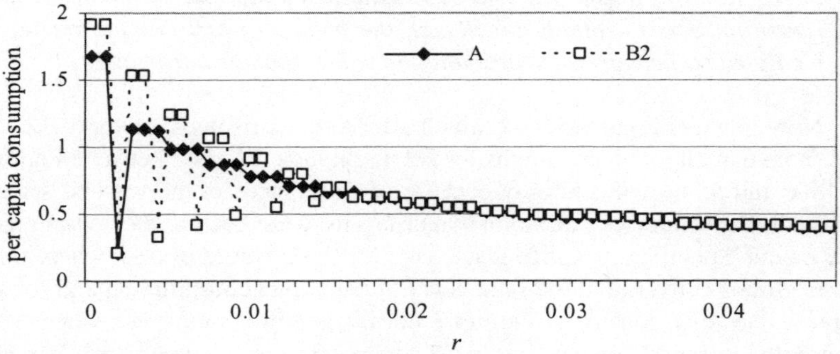

Figure 5.3b: network A vs. B2, r = 0.2

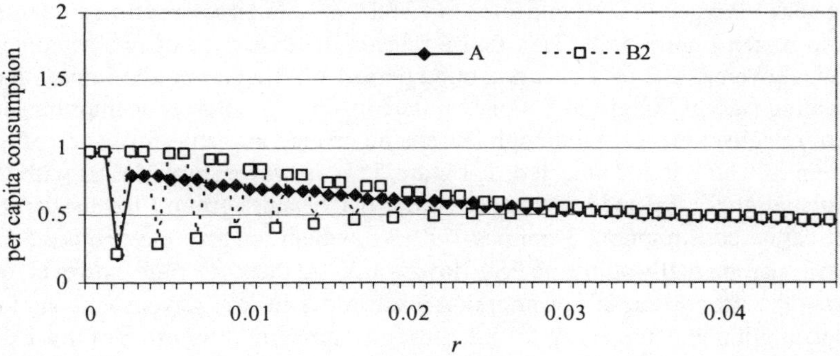

Figure 5.3c: network A vs. B2, r = 0.3

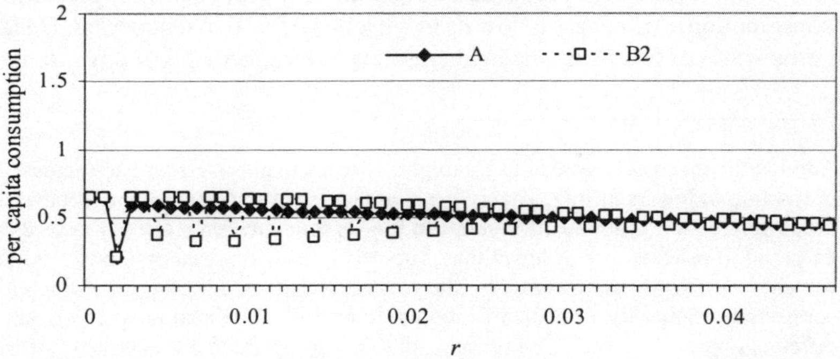

the relative performance of one network versus another is mediated by the interaction of parameters – in this instance by the rate of mimickry and return on investments. Equivalently, the impact of any one parameter is mediated by network structure and the value of other parameters.

Now consider Figures 5.4a–c which offer a comparison of networks B2 and C2 (note that although the graphs for B2 might look different between Figures 5.4b–c and 5.3b–c, this is only because of differences in the vertical scale). This comparison raises a question that brings us to the heart of the operation of our model. Specifically, notice that except for those values of ES p wherein all three groups converge to the same level of per capita consumption and for all three values of r considered, group 3 does worse in network B2, where it is connected to 1and 2, than it does in C2 where it is connected only to 1. It might seem that a connection to two groups that invest heavily would have a more salutary impact on 3 than being connected to only one such group. To trace the source of these differences, let us begin with Figure 5.5a, which focuses on network C2 wherein 3 is connected to 1 and 2 is isolated after setting $r = 0.2$ so as to match Figure 5.4b. This figure reveals the existence of two seemingly critical events. The *first* occurs around period 1424 wherein after a period of seeming stability in group 3's per capita consumption, that consumption suddenly begins to rise even though there is no change in the performance of the group to which it is connected, 1. Figure 5.5b, however, provides us with the explanation for this sudden change. Notice that because group 1 has the greater per capita consumption, 3 mimics 1's ES p, which requires (see Figure 5.5a) that 3 incrementally adjust its ES p downwards. At the same time, however, 3's *apl* – the proportion of consumption 3 attributes to low d types (the sum of consumption in it and group 1) and, thus, the 'apparent proportion of low ds' – is constant since it and 1s consumption patterns are constant. The source of this constancy is this: Up to period 1424, group 3's ES p (approximately 0.09118 in period 1424) is greater than its *apl* (approximately 0.09112). This could trigger an increase in the proportion of low ds in 3, except that the proportion of consumption attributable to low ds in 3 itself is 0.9036 in that period. Hence, the proportion of low ds in 3 is held constant.[6] More formally, if $apl < p$, then

---

[6] Admittedly, an equally plausible assumption here is to allow group 3 to increase its proportion of low ds further. That is, we could let the group 'ignore' the fact that the proportion of consumption attributable to low ds solely within it already exceeds its ES p and to instead raise or lower that proportion under the assumption that it must somehow attempt to compensate for the proportions across all groups to which it is connected. Rather than pursue this model, however, in the next chapter we take a different approach, namely, we presume that groups operate on the basis of the true proportions within it and that network connections only impact the mimicking of ES ps.

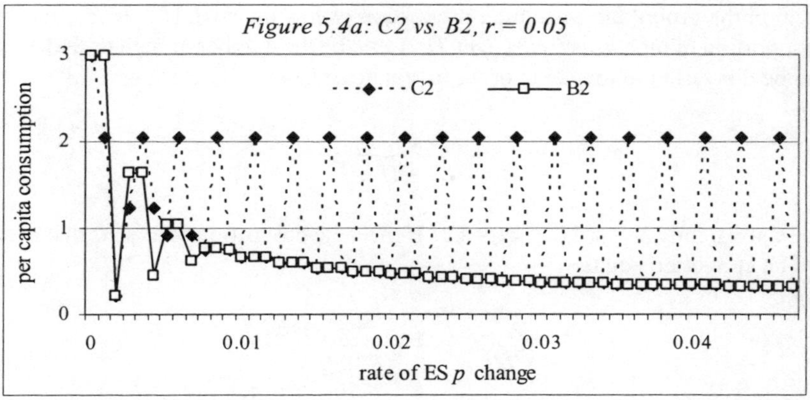
Figure 5.4a: C2 vs. B2, r.= 0.05
C2
B2
per capita consumption
3
2
1
0
0
0.01
0.02
0.03
0.04
rate of ES *p* change

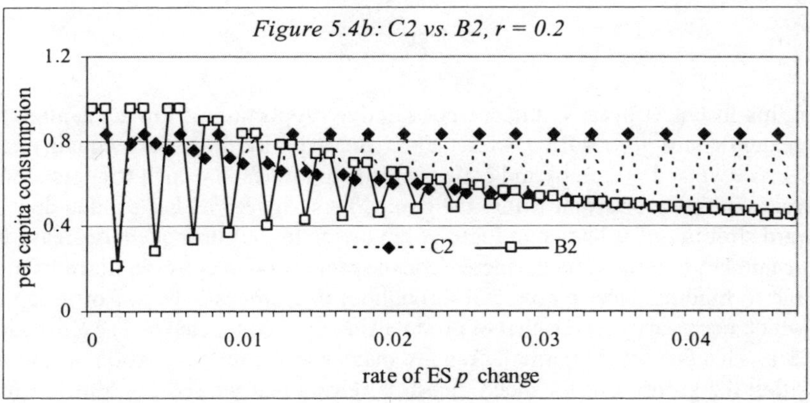
Figure 5.4b: C2 vs. B2, r = 0.2
C2
B2
per capita consumption
1.2
0.8
0.4
0
0
0.01
0.02
0.03
0.04
rate of ES *p* change

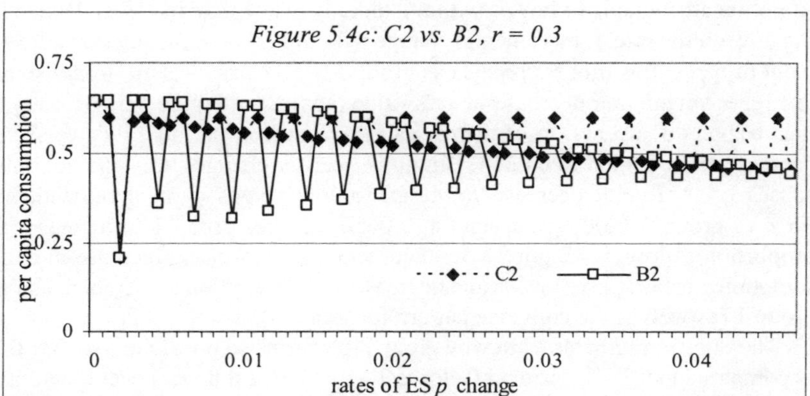
Figure 5.4c: C2 vs. B2, r = 0.3
C2
B2
per capita consumption
0.75
0.5
0.25
0
0
0.01
0.02
0.03
0.04
rates of ES *p* change

within the group, the presumed proportion of low ds is $CL/(CL+CH)$ and the proportion of high *ds* is $CH/(CL+CH)$. Let X be the number of high *ds* that need to be converted to low *ds* in order to equate *apl* and p. That is

$$\frac{X + CL}{CL + CH} = p$$

or equivalently $X = p(CL+CH) - CL$. Hence, the proportion of high *ds* that need to be converted equals

$$\frac{X}{CH} = \frac{pCH - (1-p)CL}{CH}$$

But note that $X/CH < 0$ if

$$p < \frac{CL}{CL + CH}$$

In this instance, because the current relative proportion of low *ds* is already greater that the groups ES p, we set $X = 0$. But in period 1425 the inequality *apl* $< p$ is reversed – thereby indicating too many low *ds* – which triggers a decrease in that proportion within the group. This trigger yields a sudden downward shift in *apl* as shown in Figure 5.5b owing to the subsequent decrease in the number of low *ds* and the increase in the proportion of consumption attributable to high *ds*. Notice now that throughout this process, the *apl* of group 1 (which necessarily equals that of group 3 since 1 is connected to 3) exceeds its ES p. This fact would normally require that we lower the proportion of low *ds* within the group. But for much the same reason that we set $X = 0$ earlier for group 3, group 1 does not decrease it proportion (i.e, the proportion of consumption attributable to low *ds* within it already is less than its ES p). Depending now on the rate of incremental change in both ES p and the internal adjustment of types, this process repeats as group 3's ES p 'catches up' to *apl* so as to trigger yet another decrease in *apl*s value (again, see Figure 5.5b). Group 3 thus triggers a succession of decreases in *apl* until in period 2790 we see a decrease from 0.0665 to 0.0654. But unlike earlier changes, this one not only causes 3, as before, to decrease the number of low *d* types within it, but with the ES p of group 1 holding constant at 0.0656, induces group 1 to increase its proportion of low *ds*. Figure 5.5c, then, shows the resulting increase in consumption attributable to low *ds* and decrease in consumption among high *ds* for group 1 (as well as the converse pattern for group 3).

The question remains as to why group 3 performs so poorly in network B2 as compared to C2. The critical factor is 3's *apl*. In C2 it hovers near 0.09 until 3's ES p decreases sufficiently to trigger a decrease in *apl* (see Figure 5.5b).

Figure 5.5a: network C2, groups 1 & 3, r = 0.2

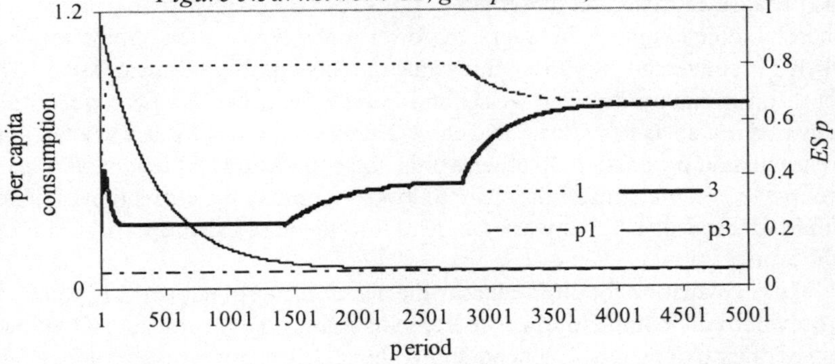

Figure 5.5b: group 3, network C2, ES p and apl, r = 0.2

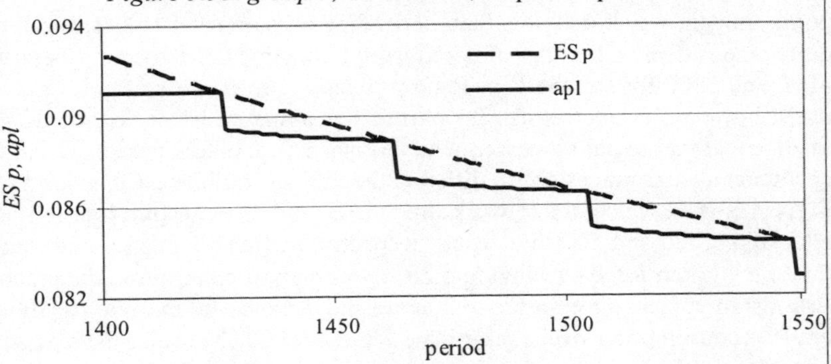

Figure 5.5c: CH and CL, network C2, r = 0.2

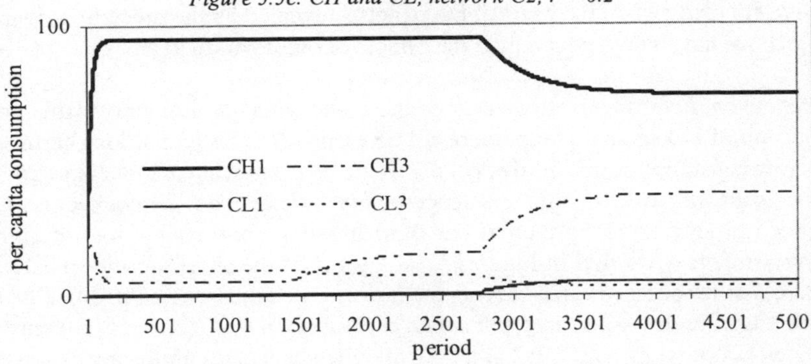

But in B2, because 2s consumption pattern is now added to 3s calculation of *apl* and because 2 invests and subsequently consumes the product of that investment (thereby indicating to 3 that there are fewer low *d* types in the population than when 3 is connected only to 1), that value quickly equilibrates near 0.053. The net result is that unlike what we see in Figure 5.5b, group 3's ES *p* never falls below its *apl* so as to indicate that there are too many low *d*s and trigger a decrease in their proportion. In other words, the connection to 2 as well as 1 leads group 3 to 'underestimate' the share of low *d*s within it, the correction of which would otherwise, as in network C2, lead it to increase the proportion of types who would invest.

This explanation for differences in the performance of networks is, of course, only partially complete since we have not considered the impact and interaction of other parameters – most notably *r*, the return on investments, and *ap*, the rate at which individual nodes adjust their proportions of high and low *d* based on the disparity between *apl* and their current ES *p* values. Nevertheless, hopefully the preceding discussion provides some insight into the model. At this point, though, we should note that, depending on parameter values, it can require a considerable number of iterations for convergence to occur – between 4000 and 5000 for the charts in Figures 5.2a–c. The process of convergence itself, however, is interesting and warrants examination since we should not assume that real social processes achieve equilibrium before parameter values encounter shocks that establish different theoretical equilibria. Consider, then, Figures 5.6a–c, which, for network structure B1, graph per capita consumption after 500, 1000 and 5000 iterations, respectively. (hence, Figure 5.6c reproduces the graph for B1 in Figure 5.2b). Note in particular, now, the gradual 'lifting' of consumption in group 3 as we move from 500 to 1000 iterations, until 3's consumption level equals 1 and 2's after 5000. Note also the existence of an 'optimal *ar*' from group 3's perspective that decreases as we progress towards 5000 iterations and that, unlike what we see in Figure 5.2b, disappears upon full convergence. Admittedly, however, it is premature to assert any definitive conclusions on the basis of the patterns revealed by the preceding figures since we have not yet examined the effects of other parameters.

Return on Investments: If we next consider the initial evolutionary stable proportion of low *d*s in a group, there are no surprises to be found: long-term per capita consumption uniformly declines as the ES proportion of low *d*s in a group increases, regardless of the existence or non-existence of the connection between groups. Looking then at our third iteration parameter – the return on investments – we find in Figures 5.7a–c some of the same non-linear effects here that we observe when varying the rate of ES *p* adjustment. Specifically, as the rate of return *r* increases, per capita consumption first increases (but only in networks of types A, B1, B2 and C2), and then decreases uniformly across all networks. That is, if the rate of return is 'too low', investment capital builds too

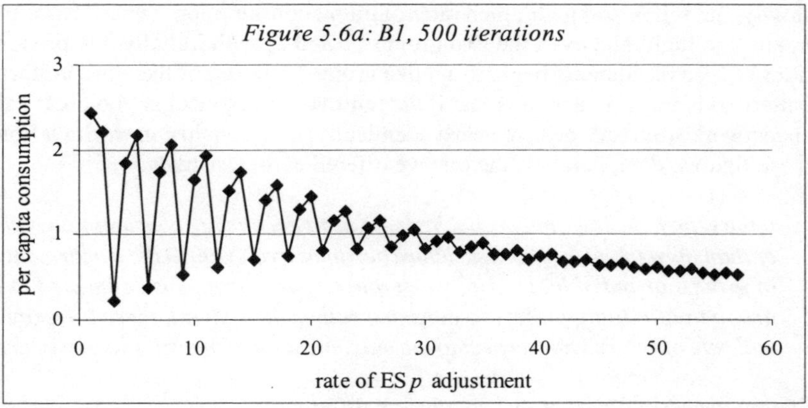
Figure 5.6a: B1, 500 iterations
per capita consumption
3
2
1
0
0
10
20
30
40
50
60
rate of ES p adjustment

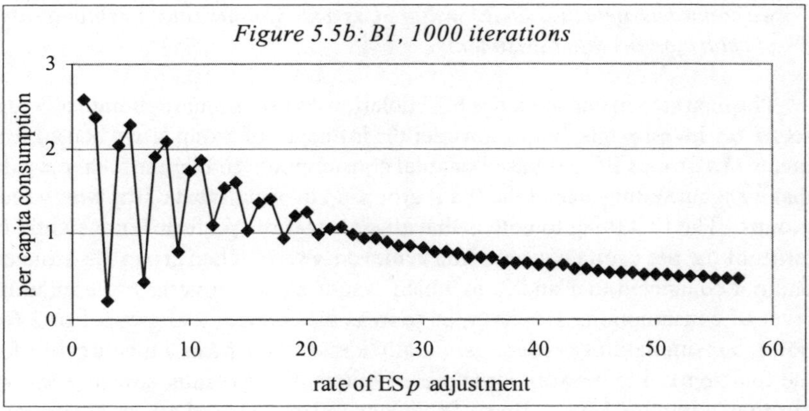
Figure 5.5b: B1, 1000 iterations
per capita consumption
3
2
1
0
0
10
20
30
40
50
60
rate of ES p adjustment

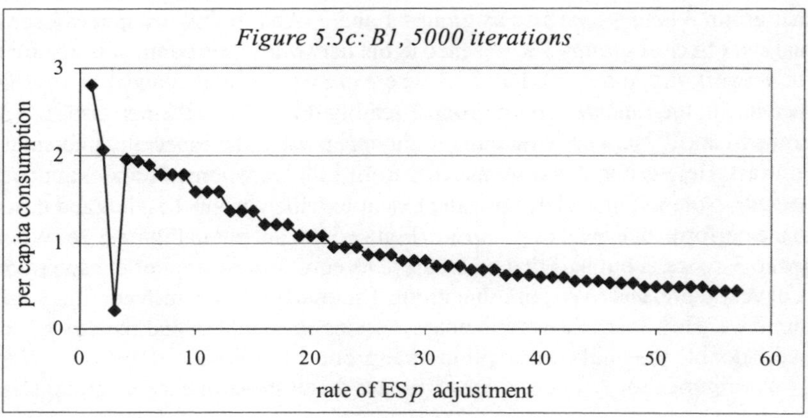
Figure 5.5c: B1, 5000 iterations
per capita consumption
3
2
1
0
0
10
20
30
40
50
60
rate of ES p adjustment

slowly; but if it is 'too high', then rates of initial consumption, regardless of ES *p*, are 'too high' and even the two groups (1 and 2) with initially low proportions of high discounters begin to act like group 3. Indeed, if there is a uniform pattern to Figures 5.7a–c, it is that if the return on investment is 'too great', all six network structures behave nearly identically. The conclusion we draw from these figures, then, parallels the one we offered earlier, namely:

> *if the return on investments is ' high' so as to encourage consumption rather than investment, network structure becomes irrelevant to the performance of groups or individuals. And when one or more groups is connected to a group that initially prefers to consume rather than invest, there is an optimal rate of return with respect to the maximization of overall social welfare.*

Stating the second part of this conclusion differently,

> *we cannot assume that social welfare has a monotonic relationship to rates of return on private investments.*

The next step in our sequence of simulations is to continue to iterate on *r,* the return on investments, but to consider the influence of group size when adjustments in a groups ES *p* is based on total consumption. In Figures 5.8a–c we do that by again setting size of the 'bad' group (3) to be half that of the two 'good' groups. The first thing to note is that all six networks perform equivalently in terms of the per capita consumption achieved, except when group 3 is isolated and not connected to 1 and 2, in which case it alone converges to a minimal level of consumption. However, as soon as we connect 3 to group 1 or 2 (or both), our simulations produce essentially identical charts as a function of *r* for the four remaining network structures. Contrast these results now to what we observe in Figures 5.7a–c. Because adjustments were based on per capita consumption there, the results portrayed in those figures are equivalent to assuming that group 3 is the same size as groups 1 and 2. And in this instance we see a marked effect of groups 3's presence in the network. This contrast is apparent for network structures A, B1 and B2, where in a wholly understandable way, the increase in the relative size of group 3 readily drags down the performance of groups 1 and 2. Network structure C2, though, presents an especially interesting contrast. Here group 2 is disconnected from 1 and 3, whereas 1 and 3 are connected. Note in particular, now, the extent to which group 1 is dragged down to the performance level of group 3 in Figure 5.3c, but not in Figure 5.8c. When group 3's size is but half that of group 1, its equilibrium level of consumption is never so great as to require that group 1 mimic 3's ES *p*; instead, it is 3 that mimics 1. However, once we increase 3's size to equal 1's and thereby effectively double the total consumption of 3, group 1 mimics 3.

We return shortly to consider in greater detail the influence of group size,

Figure 5.7a: Iteration on rate of return on investments

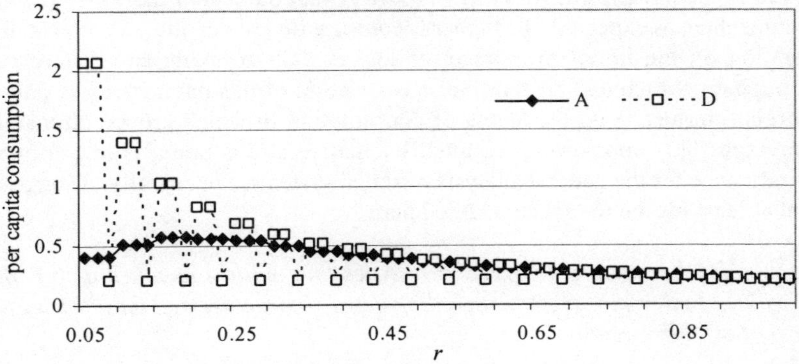

Figure 5.7b: Asymmetric networks B

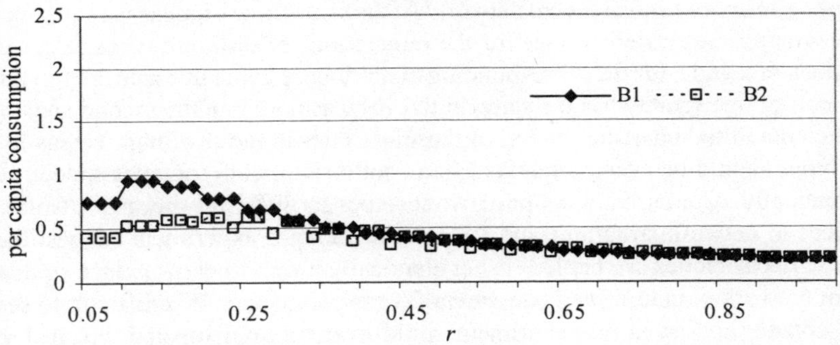

Figure 5.7c: Asymmetric networks C

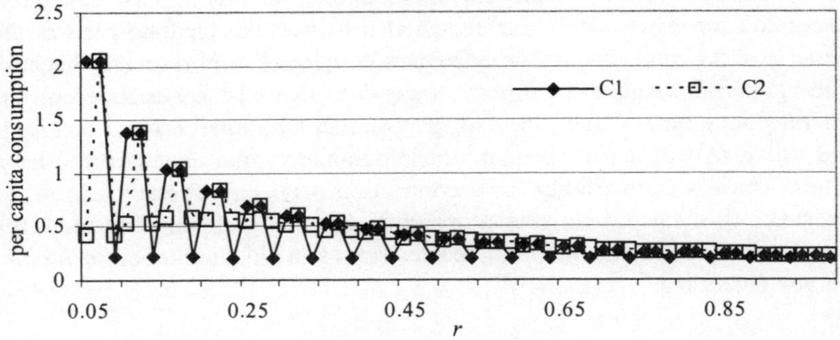

but first insofar as other parameters are concerned, we can summarize a large sequence of simulations by simply reporting that the higher the supplemental income then, as expected, the higher is consumption per capita. In contrast, the iteration on the initial proportion of low *d*s fails to show any differences throughout, which tells us that the starting point of this parameter has only a marginal influence on the levels of consumption to which groups ultimately converge. This conclusion, admittedly tentative at this point, is of profound significance for the general analysis of social systems. Specifically, it suggests that at least for the model considered here,

> *it is not the starting distribution of types that matters as much as it is the value of other parameters, including network structure and those variables that establish initial ES ps.*

This conclusion allows us to imagine any number of thought experiments that admittedly smack of the always dangerous enterprise of social engineering, Suppose, for example, that we populate two societies from scratch, each with a pre-configured and identical network. At the same time, suppose we can somehow induce equivalent values for the interactions of different types (e.g., the values of *u* and *v* for the corresponding evolutionary game in Table 3.1) so that node *j* of one society has the same initial ES *p* as node *j* in the second society. Then the initial characterization of these societies in terms of high versus low *d* types should be nearly irrelevant – in equilibrium, both societies should act identically. Alternatively, suppose two societies act differently because of differences in network structures and ES *p*s. Our thought experiment suggest that these two societies can be 'led' to act identically if we somehow induce equivalent network structures and equivalent fitness parameters. It is difficult to see, of course, how such an experiment could ever be implemented, but it does highlight the importance of network structure and those things that determine initial evolutionary values.

The interesting dimension to such thought experiments is that they do suggest that the problems of official corruption and criminality in those day to day economic exchanges that characterize states such as the separate parts of the former Soviet Union are not trapped in their unsavory equilibrium only because of the past and whatever beliefs and incentives that past has induced on the general population. Rather, they suggest that an additional critical feature is their initial network structures and those parameters that dictate initial ES *p* values. Thus, we can readily see a connection between our analysis and the observations of various commentators when discussing the ongoing centrality in the political-economic life in these societies of an old guard of *apparatchiks* and *nomenclatura*.

The Internal Adjustment Rate, ar: Once again, of course, any such conclusion

Figure 5.8a: Iteration on r, total consumption

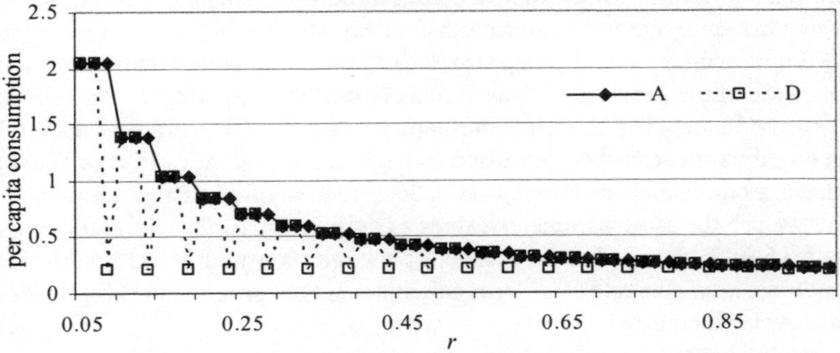

Figure 5.8b: Asymmetric networks B

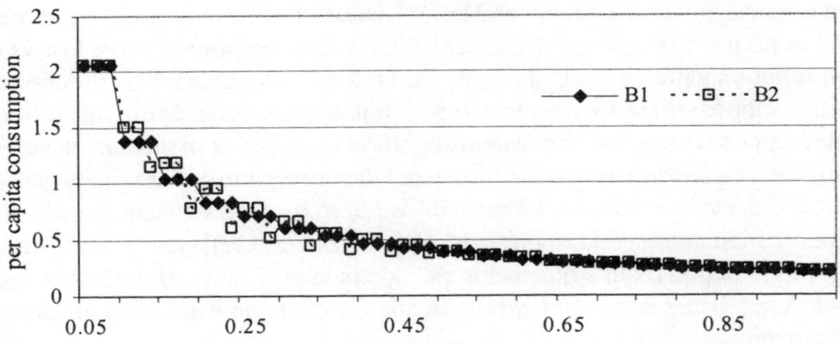

Figure 5.8c: Asymmetric networks C

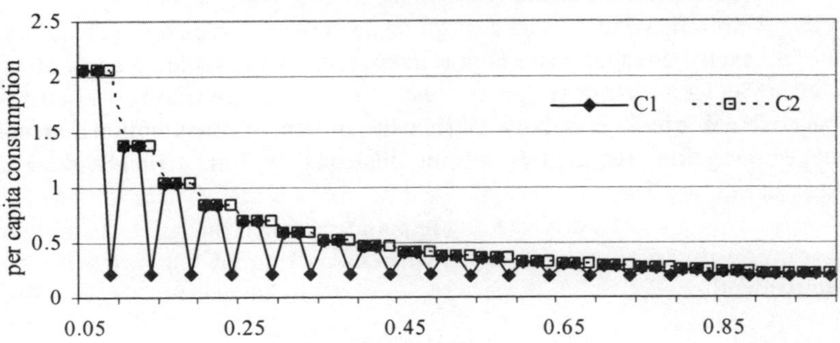

and speculations need to be treated with care since we have thus far examined but a small proportion of alternative configurations of parameter values. Indeed, if we return once again to requiring that all adjustments be based on per capita consumption rates (so that group size doesn't matter or, equivalently, so that all groups are of the same size), there is one parameter whose impact is somewhat unexpected except in hindsight – the iteration on size of the within-group incremental adjustment in the proportion of high and low d types (the rate, ar, at which a group adjusts its population in the direction of its current ES p value). Here we use the same parameter values as outlined in Table 5.1and looking at Figures 5.9a–c, we see that the change in per capita consumption is anything but monotonic as ar, the rate of internal adjustment in the proportion of high versus low d types, increases.

The full nature of this instability is revealed in Figure 5.10, which graphs the actual sequence of per capita consumption for group 1 in network B1 over the first 500 iterations after setting $ar = 0.25$ and the rate of ES p adjustment equal to 0.025. To trace the source of these oscillations let CL denote the total consumption of low ds in a group, and let CH denote the total consumption of high ds. Let ES $p = 0.05$ and $PC = CL/(CL+CH)$ denote the proportion of low ds in consumption within a group. Imagine PC is slowly declining in the direction of p (i.e., suppose there are too many low ds initially in consumption) but at some point suppose PC over-compensates slightly and declines a small amount below p so that we must adjust now by decreasing the proportion of high ds. In particular, if the current situation is described as $[CL, CH]$, then we want to make the new situation correspond to $[p(CL+CH), (1-p)(CL+CH)]$ – or at least move it in this direction if our adjustments are incremental. That is, of the total amount currently being consumed, which equals $CL+CH$, we want, after adjusting, the amount

$$p(CL+CH)$$

to be attributable to low ds and

$$(1-p)(CL+CH)$$

to be attributable to high ds. Now let K be the amount of consumption that must be taken away from the high ds to bring this equality about. That is, we want to set

$$CH - K = (1-p)(CL+CH)$$

or equivalently,

$$K = CH - (1-p)(CL+CH)$$

Figure 5.9a: Iteration on rate of incremental adjustment, ar

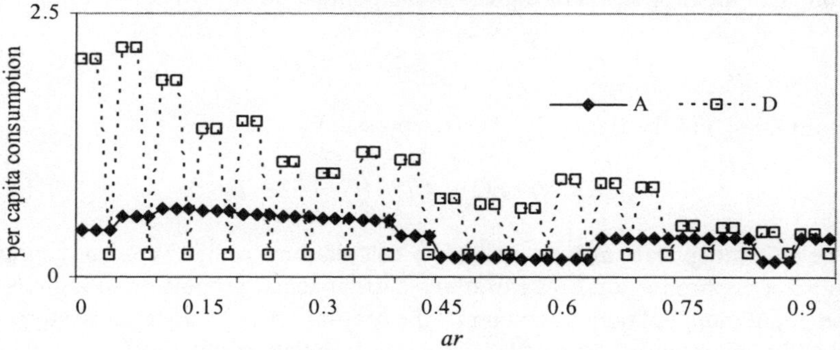

Figure 5.9b: Asymmetric networks B

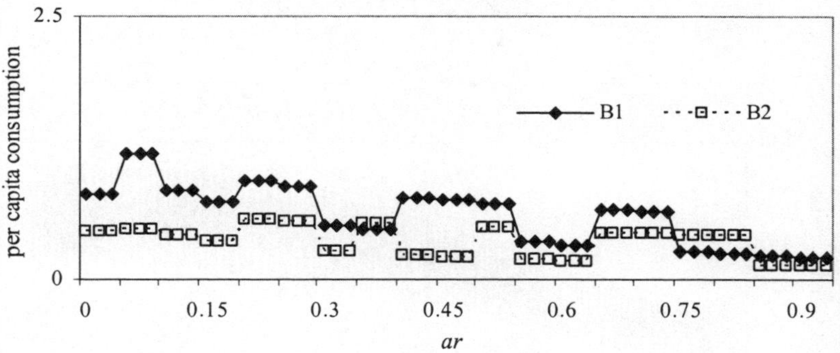

Figure 5.9c: Asymmetric networks C

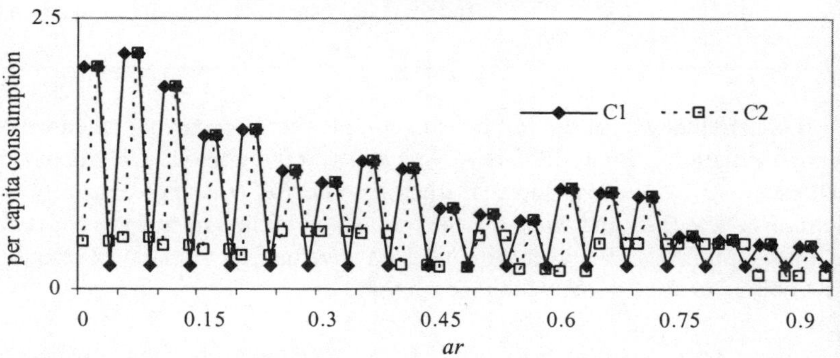

In other words, the proportion of high *d*s that need to be converted (if we assume that the proportion of high *d*s in the population is $CH/(CH+CL)$) is

$$K/CH = 1 - (1 - p)(CL+CH)/CH$$

Multiplying out the right side of this expression gives,

$$K/CH = p(1 + CL/CH) - CL/CH \qquad (5.1)$$

The adjustment rate *ar* now enters the calculations in that we do not simply subtract the proportion K/CH from the current actual proportion of high *d*s in the population, but rather we subtract the proportion arK/CH. That is, suppose *pl* is the *actual* proportion of low *d*s in the population and $1-pl$ is the *actual* proportion of high *d*s (as opposed to the proportions *in consumption*). Then we set the new actual proportion of low *d*s to be

$$pl_{new} = pl_{old} + arK/CH(1-pl_{old}).$$

Figure 5.10: group 1, network B1, ar = 0.25

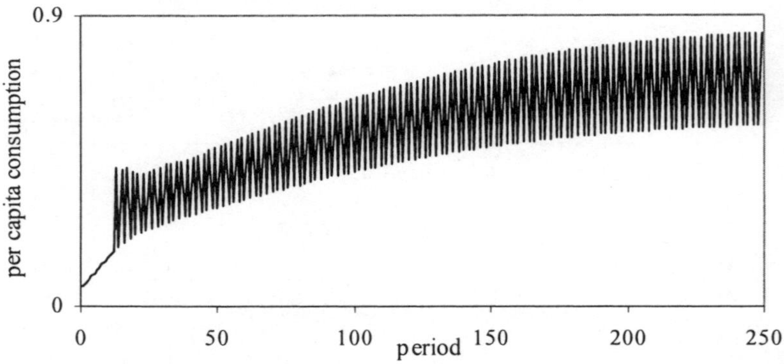

It is here that we get the full explanation of what is occurring when we observe oscillations. First, if *CL* is very small (near 0 relative to *CH*), then from expression (5.1), K/CH is approximately *p* and arK/CH is approximately *arp*. So if *ar* is, say, 0.25 and *p* is 0.05, the proportion of high *d*s that are converted to low *d*s is 0.0125. But when a high *d* is transformed, all of its saved resources now belong to the low *d*s, which means that:

- the proportion (*ar* times *p*) 0.0125 of saved resources immediately become the 'property' of low *d*s, all of which is consumed in the next period;

- the proportion of low *ds*, as measured by consumption, immediately soars;
- the system begins once again to decrease the proportion of low *ds*.

This process can be seen in the numbers in Table 5.3,which are taken from an actual run after setting *ar* equal to 0.125 in order to best illustrate the process. In period 3 the proportion of low *ds* suddenly slips below 0.05, the ES value of *p*. At that point *CL*, the consumption of low *ds* increases from 0.52 to 47.97 as they consume all the resources of the transferred high *ds*. But the consumption of high *ds* does not drop by much, because we only transform approximately 0.25x.05 = 0.0125 of them (actually 0.01185). That is, consumption of high *ds* decreases only slightly. Note in the column denoted *PL*, the actual proportion of low *ds* in the group in question increases only slightly – from approximately 0.026 to 0.037 whereas *apl*, the proportion in consumption when looking at the group in question (group 2 and the group it is connected to, 3), shows the proportion 0.1956.

Table 5.3: Actual Run of the Program

period	*PC*	*CL*	*CH*	*PL*	*apl*
1	0.050139	0.52	198.68	0.025815	0.002592
2	0.050025	0.52	199.16	0.025815	0.002586
3	0.049911	47.97	197.26	0.037393	0.195600
4	0.226567	0.61	197.74	0.030434	0.003069

The oscillations portrayed in Figure 5.10, then, are wholly understandable – they are a function of the fact that when high *ds* (persons or memes) become low *ds* they carry with them what they saved, which they then immediately consume and thereby give the appearance of an excessive population of low *ds*. This, in turn, explains the apparent incoherence of Figures 5.9a–c. Because both the period of oscillations (the number of periods it takes for a group to 'recover') and their magnitude will change with *ar*, the per capita consumption graphed there, which corresponds to what happens only in the 5000th period, need not exhibit a clear pattern. Look, for example, at Figure 5.9b and the levels of consumption for network B2 that correspond to the 5th through 8th values for *ar* (or equivalently, Figure 5.9c and network C2). Here we see a sudden rise in the consumption of groups 1 and 2 that increases for four successive values of *ar*. Clearly, then, for these values of *ar*, the 5000th iteration corresponds to a peak in the oscillations, the magnitude of which is increasing slightly as *ar* increases.

It is evident, then, that this apparent 'instability' is a consequence of two features of our model – an incremental adjustment mechanism that employs a fixed step size and, more critically, the assumption that relative rates of consumption are the measure of proportions when making internal group adjustments. It is partially for this reason, then, that we consider an alternative to this assumption in the next chapter. However, we should ask if there are empirical counterparts to the 'instability' that Figures 5.9a–c portray and here we can only wonder whether the political fortunes of Russia's oligarches and their new-Russian minions in the last decade of the 20th century match such a cycle. Initially seen as the purveyors of the new market economy to a decayed Soviet society, their actions were widely admired and even imitated. Soon enough, however, in the mid-1990s, their actions were seen as excessive and even resented, with countless jokes portraying this new class as silly and stupid except in those instances where the very richest of them were viewed as dangerous. With the succession of the former KGB functionary Putin as president, a concerted media campaign transformed them into social outcasts. Many of the most economically successful ones were then required to flee Russia to escape politically motivated prosecutions by a leadership that seems intent on resurrecting a strong Czarist-style state in which control of the economy, media and Russias superficial democracy all lie in the hands of a new oligarchy ensconced in the Kremlin. It is, admittedly, wildly speculative to say that such events correspond to anything like the oscillations in our model. Nevertheless, our model alerts us to the possibility that

> *social networks will operate in 'more understandable' ways only to the extent that the changes within them are not 'too abrupt'; but if change is rapid, with the various components of a network reacting and 'over-reacting' to each other, an impression of incoherence may prevail.*

This proposition might seem self-evident, but again we want to emphasize that it is not predicated on any incoherence in the iteration of any parameter of the model. Rather, whatever incoherence we find has, underlying it, a coherent pattern. Figure 5.10 is not 'incoherent' or in any way stochastic but instead simply points to the possibility that outcomes (in this case, levels of per capita consumption) need not converge to a specific value but can, in 'equilibrium' oscillate over a wide range of values. And even if there is not a clear analogy to these oscillations to be found in real social dynamics, they nevertheless offer a note of caution about assuming that social systems converge in simple ways to singular outcomes.

Group Size: Our final series of simulations concerns the influence of group size within a network. Naturally, the larger a group (node), the greater its influence provided that adjustments are based on some measure of total as opposed to per capita consumption. And indeed, this is precisely what we see when we iterate on the rate of change in ES *p* and allow the groups in a network to adjust their ES *p*s in accordance with total consumption. Figures 5.11a–c set group 3's size equal to one half that of groups 1 and 2, and just as we compared Figures 5.8a–c with Figures 5.6a–c when iterating on rate or return of investments, we can compare these new figures to Figures 5.2a-c, which in effect compare networks when n_3 = 100 versus 50 (and $n_1 = n_2$ = 100). The impact of decreasing the relative size of group 3 from equality (Figures 5.2a–c) by half (Figures 5.11a–c) is dramatic: all groups in network A now converge to a uniformly higher level of per capita consumption for even low rates of ES *p* adjustment; the same is true for network B1 and C2. Group 3 performs poorly only when it is wholly disconnected from 1 and 2 (network configurations D and C1) or when, as with network B2, the rate of adjustment is low (although even there it performs better when small, Figure 5.11b versus 5.2b).

The pattern portrayed for network B2 as compared to B1 in Figure 5.11b is perhaps especially interesting, since here for low rates we have a situation in which groups 1 and 2 converge to higher levels of per capita consumption when 3 rather than 1 is the 'focus' of the network. This pattern is the opposite of what we report earlier when discussing Figure 5.2b as well as Figure 5.7b The explanation for this change in pattern lies in the calculation of each groups *apl* – the share of consumption each attributes to low *d*s. Briefly, although *apl*s fluctuate, in general they satisfy the following inequalities:

in B1		in B2
apl_1	<	apl_1
apl_2	<	apl_2
apl_3	>	apl_3

The logic of these inequalities is as follows: in network B1, node 1's calculation of *apl* is based on the consumption of all three nodes whereas in network B2, node 2 (which is an initially 'high investing' node) is excluded from 1's calculation. Hence 1 will attribute a greater share of consumption to low *d*s in B2. Similarly, node 2's calculation of its *apl* will wholly exclude 3 in B1, but will include only 3 plus itself in B2. The inequality for 3, on the other hand, is reversed since it begins with a high ES *p* (0.95). It sees only itself and 1 in B1 but 2 as well in B2 and will thereby calculate a lower *apl* in B2 since in that network a dominant share of the consumption it sees as soon as investments accrue their returns will be attributed to the consumed savings of the high *d*s in 1 and 2 (see Figure 5.12a). In addition, owing to the size of groups 1 and 2 relative to 3, in both B1 and B2, node 3 mimics 1 (since, if adjustment rates are

Figure 5.11a: ES p, total consumption, n=(100, 100, 50)

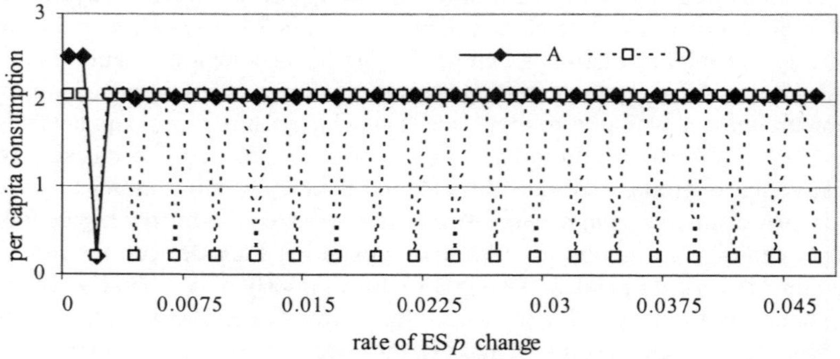

Figure 5.11b: ES p, total consumption, n=(100, 100, 50)

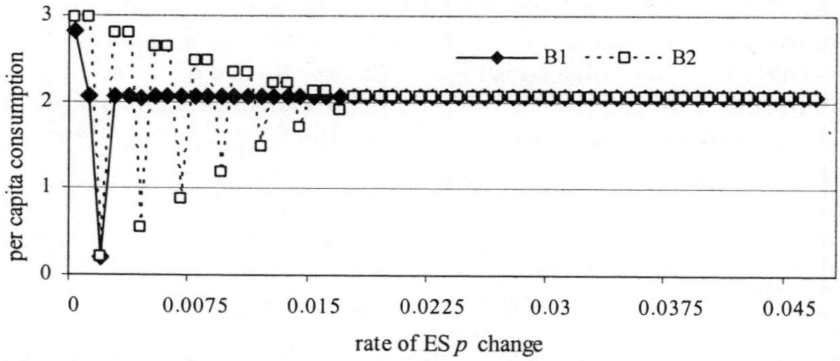

Figure 5.11c: ES p, total consumption, n=(100, 100, 50)

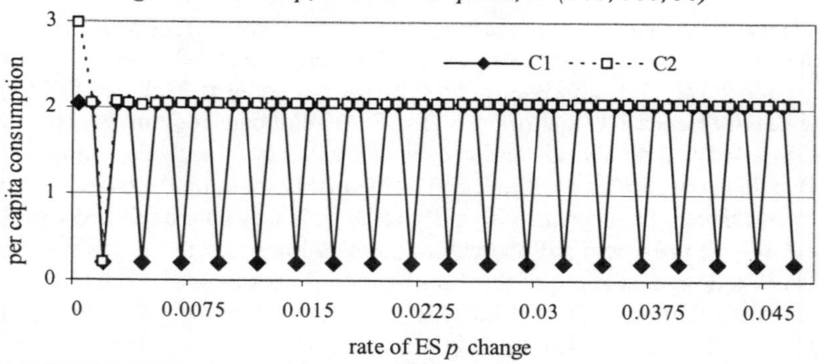

Figure 5.12a: Group 3s apl

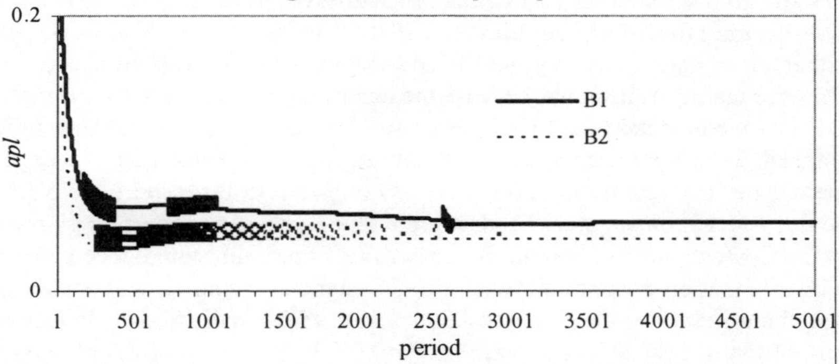

Figure 5.12b: Proportion of low ds in group 3

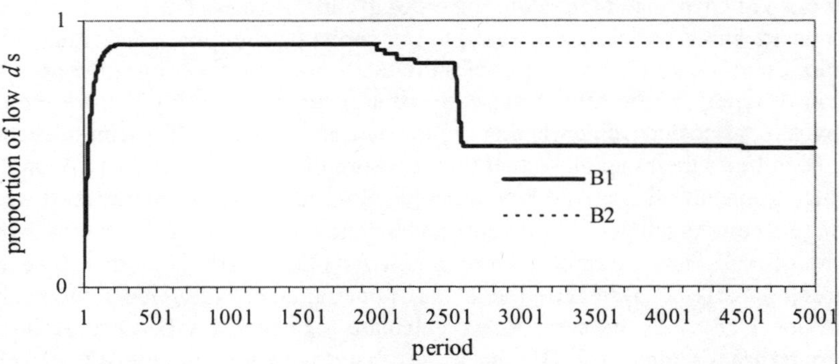

Figure 5.12c: Per capita consumption in group 3

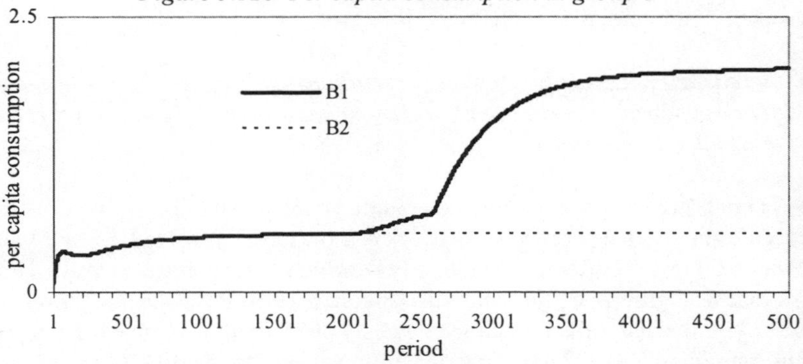

sufficiently low, investments will build to allow significant future consumption in these groups) whereas 1 and 2 mimic themselves or each other. As node 3's ES *p* declines from 0.95 in the direction of 0.05, the node, in both networks, will either adjust its proportion of low *d*s upwards or hold that proportion constant. But, as Figure 5.12b shows, because the inequality $p < apl_3$ is satisfied first in B1, this incremental buildup of low *d*s will halt and reverse itself first in B1 (indeed, 3's ES *p* never decreases below its *asp* in network B2). Hence, 3's greater per capita consumption in B1 as compared to B2 (see Figure 5.12c). Nodes 1 and 2, on the other hand, begin with ES *p*s and a proportion of low *d*s of 0.05, which, given 3's initially greater rate of consumption (since it begins with a higher proportion of low *d*s who consume everything), will occasion a calculation of $apl > 0.05$. These two nodes then, will actually initially reduce the proportion of low *d*s within them below 0.05. But because of the previously describe inequalities, this incremental reduction will, for these two nodes, be reversed first in B. Hence, their better performance in network B2.

Of course, all that we have said here depends critically on parameters such as rates of change and the relative sizes of groups. Relative sizes, when evaluations are based on total as opposed to per capita consumption, determine who mimics whom and how groups (of individuals or memes) estimate the proportion of types. At the same time, rates of adjustment interplay with return on investments to determine whether groups such as 1 and 2 will be given sufficient time to build their capital so that their consumption is weighted heavily on the calculations of others. And how quickly ES *p*s adjust helps determine at what stage a node is triggered to reverse course and either increase or decrease the mix of types. One additional fact warrants emphasis here. Note that if we set group 3's size equal to zero, the levels of per capita consumption achieved by groups 1 and 2 is the same as we calculate for network type D wherein all groups are disconnected. But this means that if network structure B2 pertains, then as we increase node 3's relative size from 0 to 100, the long-term level of consumption achieved by nodes 1 and 2 for low rates of ES *p* adjustment will rise (from 2 in Figure 5.11a to between 2 and 3 in Figure 5.11b and then fall, to less than 2 in Figure 5.2b). That is,

> *there is not necessarily a monotone relationship between the performance of one group in a network and the size or 'importance' of some other group to which it is connected.*

There is at least one additional qualitative conclusion to be drawn from the preceding analysis. Specifically, network A is the most connected, types B1 and B2 next, C1 and C2 after that, and finally D, which is not connected at all. Note now that if all groups are the same size (Figure 5.2a–c) B1 is the most 'productive' when the rate of ES *p* change equals 0.0025 (with a summed per capita consumption of 5.8), C2 next (at 4.58), followed by D and C1 (4.33 each), then

A (3.81) and finally B2 (3.77). And when group 3 is half the size of groups 1 and 2, this order is nearly reversed with B2 the most productive (at 5.87), A, B1 and C2 next (all three at 5.12), and C1 and D both (at 4.20) the least productive. All such comparisons depend, of course, on rates of adjustment, returns on investment and the relative sizes of groups. Less-connected networks can lead nodes to both over- and under-invest so as to produce rates of consumption in excess of or below what would be observed in less complete structures. And although we see no simple generalization that can describe the relationship between connectedness and overall long-term productivity (as measured by per capita consumption), at least this much is clear:

there need not be a monotone relationship between overall social welfare and the connectedness of networks.

To complete our analysis of size, Figures 5.13a–c portray the intermediate case of letting the relative size of node 3 be 80 (as compared to 100 for nodes 1 and 2). As expected, we see patterns of per capita consumption that lie between those we find in Figures 5.2a–c and Figures 5.11a–c. Specifically, in network A the per capita consumption of nodes 1, 2 and 3 all fall somewhere between what they enjoy when $n_3 = 100$ and 50. For example, if the rate of adjustment in ES p equals 0.20, per capita consumption is approximately 0.33 when $n_3 = 100$ (Figure 5.2a), 2.0 when $n_3 = 50$ (Figure 5.11a), and 0.80 when $n_3 = 80$. Similarly, for network C2, if the rate of adjustment is again at 0.20, per capita consumption is approximately 0.40 when group 3 is the same size as groups 1 and 2, 2.0 when it is half their size, and approximately 0.75 when $n_3 = 80$. Figures 5.14a–d, in turn, combine Figures 5.2a–c, 5.11a–c and 5.13a–c for four of the network types (A, B1, B2 and C2). These figures serve to reenforce the otherwise intuitive result that

the larger a group that has little initial incentive to save, the lower is the overall level of consumption of all groups to which it is directly or indirectly connected.

Setting the rate of change in ES p equal to 0.02, Figure 5.15 summarizes much of what we have learned about the relationship between the relative size and the level of consumption achieved by group 1and shows some of the sensitivity of that relationship to the structure of the network considered. The results here follow a logical progression. Node 1 does best when it but not 3 are connected to node 2 (network B1). In this way node 3's potential negative influence cannot be felt on 2 directly and thereby gives 1 an alternative 'reference point'. Far worse from 1's perspective is when the network is complete (A) or when 1 is isolated from 2 (networks C2 and B2). Once again, though, we emphasize that these conclusions depend critically on the assumption that nodes take total

Figure 5.13a: n = (100, 100, 80), total consumption

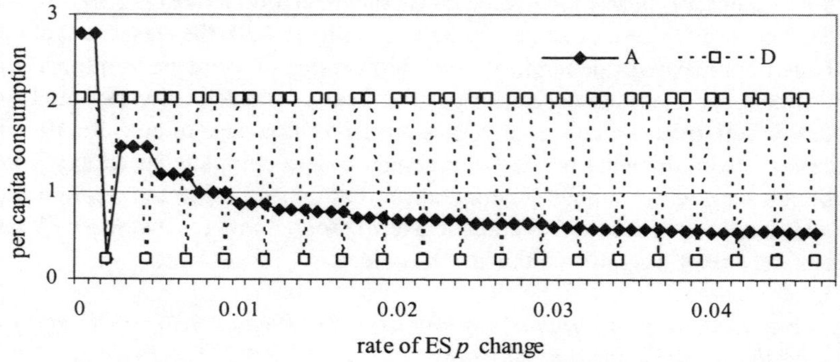

Figure 5.13b: asymmetric networks B

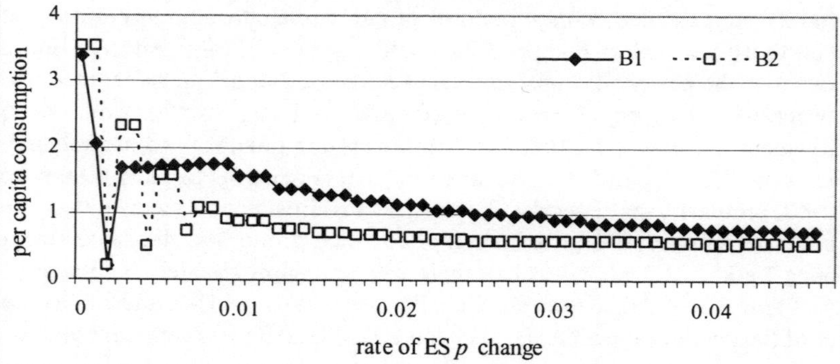

Figure 5.13c:asymmetric networks C

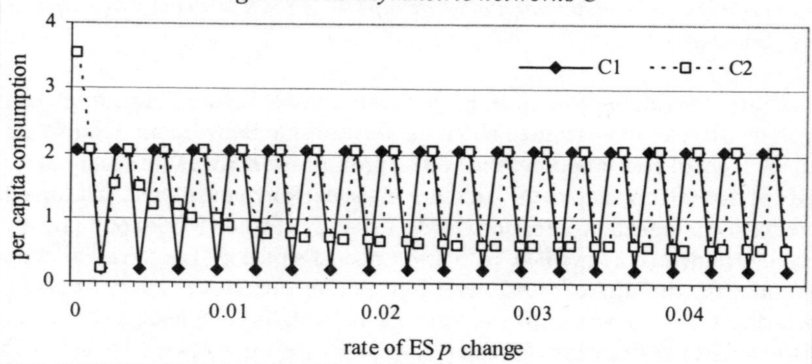

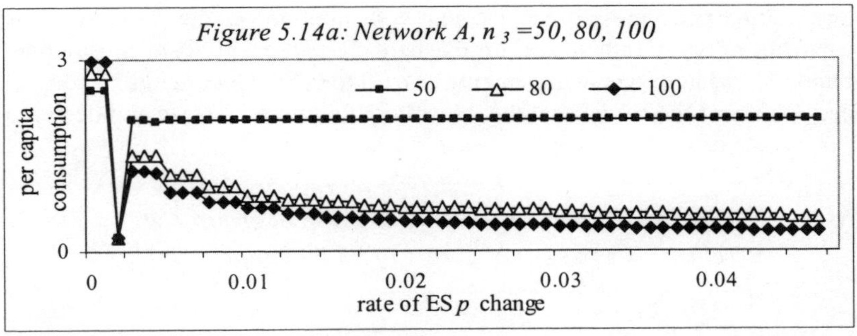
Figure 5.14a: Network A, n 3 =50, 80, 100
3
0
per capita consumption
50
80
100
0
0.01
0.02
0.03
0.04
rate of ES p change

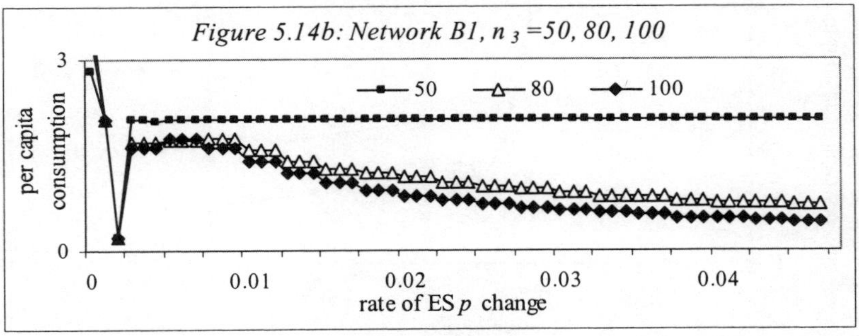
Figure 5.14b: Network B1, n 3 =50, 80, 100
3
0
per capita consumption
50
80
100
0
0.01
0.02
0.03
0.04
rate of ES p change

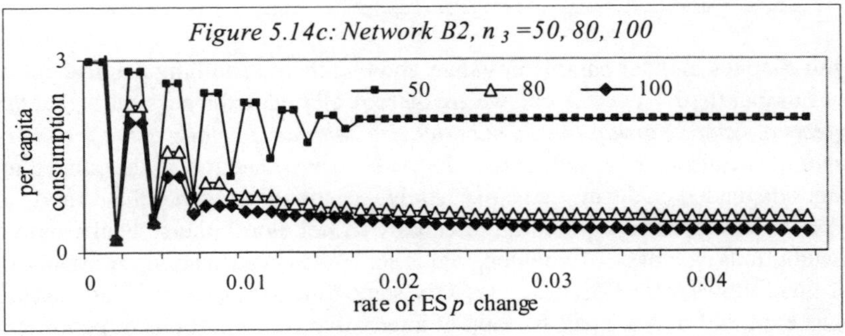
Figure 5.14c: Network B2, n 3 =50, 80, 100
3
0
per capita consumption
50
80
100
0
0.01
0.02
0.03
0.04
rate of ES p change

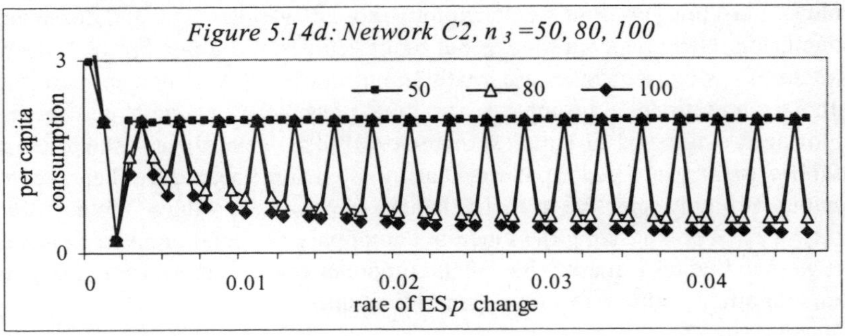
Figure 5.14d: Network C2, n 3 =50, 80, 100
3
0
per capita consumption
50
80
100
0
0.01
0.02
0.03
0.04
rate of ES p change

consumption (as opposed to per capita consumption) as the measure of the proportion of types. If there is a surprise here it is the sensitivity of consumption to node 3's relative size in the vicinity of 0.70 to 0.75. Once again, though, we need to remind the reader that this sensitivity will both increase and decrease

Figure 5.15: per capita consumption of group 1 vs. relative size of group 3, rate of ES p change = 0.02

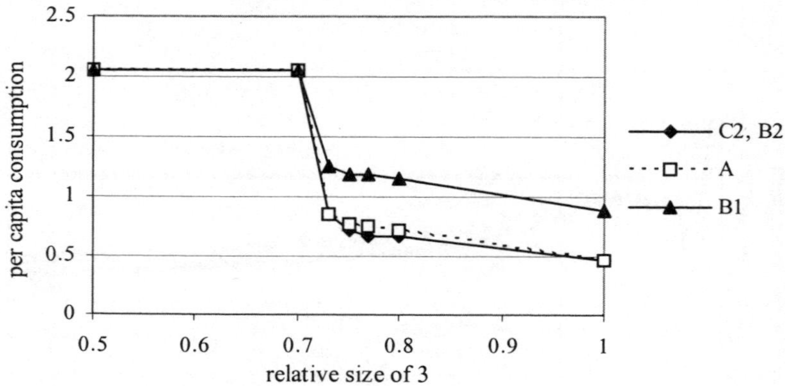

with changes in other parameter values and that there is nothing magical about the number 0.70. Nevertheless, we are thereby alerted to the possibility that *the effects of relative group (network node) size can be especially acute for intermediate values of size.* Of course, the patterns we describe in the preceding pages depended on the mechanism whereby groups adjusted their ES *ps* and the types within them. The potential of equally (if not more) plausible alternative assumptions cannot be discounted, which serves as a justification for the analysis that follows in the next chapter. At the same time, we also appreciate that we have explored only a small fraction of alternative configurations of parameter values. The configurations we do explore, though, yield one general dominant conclusion. Namely, even though out model consists of a number of 'linear' structures, its consequences are anything but linear. We implement here the simplest investment/consumption structure possible along with adjustment algorithms that are of the simplest incremental sort. Nevertheless, the product of those assumptions yields patterns that are often non-linear and either discontinuous or highly sensitive to small changes in parameter values. We note this all by way of emphasizing the inherent complexity of social processes even if we were to find (as is most unlikely) its component parts (e.g., learning, adaptation, cognition) subject to simple generalization.

6 A 3-Node Network: Part 2

6.1 Introduction

The previous chapter assumes that network nodes adjust their proportions of high and low *d* types so as to bring the proportions they 'see' in line with their ES *ps*. To formally state this assumption, suppose PL_i and PH_i are the current proportions of low and high *d*s respectively in node *i* (with $PL_i + PH_i = 1$) and let $c_{ji} = c_{ij}$ be the strength of connection of node *i* to *j* (where strength here is either 0 or 1). Next, let CH_i and CL_i denote the consumption attributable to high and low *d* types, respectively, in *i*. Finally, let p_i be the ES *p* of *i*. The total consumption that *i* 'sees' attributable to high and low *d* types, then, equals

$$CH_i^T = c_{ii}CH_i + c_{ij}CH_j + c_{ik}CH_k$$

$$CL_i^T = c_{ii}CL_i + c_{ij}CL_j + c_{ik}CL_k$$

and the overall proportion of consumption *i* might attribute to low *d* types is $P' = CL_i^T/(CL_i^T + CH_i^T)$. In this scheme a group takes account of all nodes (including itself) to which it is connected. As will generally be the case, P' will not equal p_i so suppose the group attempts to adjust its current proportion of types in order to move p_i in the direction of P'. For example, suppose P' is less than p_i so that it seems necessary to increase the proportion of low *d*s within it. To match p_i identically, the proportion of its population it must convert is $p_i - P'$. Here, however, we introduce the parameter *ar* to formulate a model in which adjustments occur incrementally, with the actual proportion transformed being $ar(p_i - P')$. And if $(1 + r)I$ is the total return on investment that high *d* types would otherwise carry into the next period, we assume that the newly created low *d*s bring with them their share of this sum, $ar(p_i - P')(1 + r)I/PH_i$.

As we note earlier, the preceding assumption might correspond to a situation where individuals judge the proportion of types in a population either by their visibility in the market or by the number of services directed at those with high rates of consumption. Minimally, this assumption appears to require a market analogy for its applicability. It is also true that it suffers from the critique of leaving unanswered the question 'How do people know that a unit of consumption is being made by a low versus a high *d*?'A no less plausible assumption, then, is to suppose, as before, that a node 'looks to' others when assessing which is performing best so that it can mimic that node's ES *p*, but to let its assessment of whether to increase or decrease the proportion of types be based solely on the actual proportion within it and its current ES *p*. In this instance, then, the actual mechanism of internal adjustment is unspecified and thereby conforms more to the 'invisible hand' presumed to operate in those evolutionary models that do not rely strictly on biological adaptation.

One can imagine, nevertheless, a variety of social processes to which such an assumption might correspond. For example, we can think of a node as a single individual described by a collection of high and low *d* memes. The experience of it and other individuals within its view (i.e., individuals to whom it is connected) dictates the parameters of an ES model that, in turn, determines its ES *p*. However, we can also suppose that the proportion of meme types within the individual rises and falls as a function of personal experiences when making a decision to which that meme type applies (a decision that depends on time preferences). Note, moreover, that if we assume people sample from their population of available memes randomly, the rates of 'consumption' and 'investment' in our analysis can be viewed as 'expected levels'. It is the analysis of this somewhat less market-dependent model to which we now turn.

6.2 Analysis

Rate of ES p Adjustment: Our first series of simulations, in Figures 6.1a–c, parallel those in Figures 5.2a–c, which iterate across ES *p* adjustment rates. There are two things to note here. First, the general pattern is essentially the same as in Chapter 5: Nodes 1 and 2, logically enough, do best when not connected to 3 whereas as before, 3 does best when, in network B1, it is connected to 1 but not 2 and 1 is connected to 2 and 3. That is, 3's performance as measured by per capita consumption is best when 1 stands at the top of a hierarchical network. Although 3 drags down the ultimate performance of 1 and 2, once again, because it is connected to 2, 1 'resists' this influence to some extent and, ultimately, bring 3 up to a level of consumption it fails to achieve otherwise. Regardless of adjustment model, then, we see in bold relief the indirect influence of one node (#2) on another (#3) through an intermediary (#1) so that we are once again reminded that *we cannot understand the interaction of groups or pairs of individuals absent an understanding of the complete social context in which they act.*

The second point of comparison concerns ultimate levels of consumption, since those levels are uniformly greater with this second adjustment alternative. This might seem odd since, if we consider network D wherein no node is connected to any other, the initial proportion of low *d*s, 0.05, is precisely the ES *p* of nodes 1 and 2 and, therefore, doesn't change under the current assumption of internal adjustment. But with our earlier assumption, because high *d*s save $1 - r$ of their resources and because it takes a while before their consumption

Figure 6.1a: Per capita consumption vs. ap

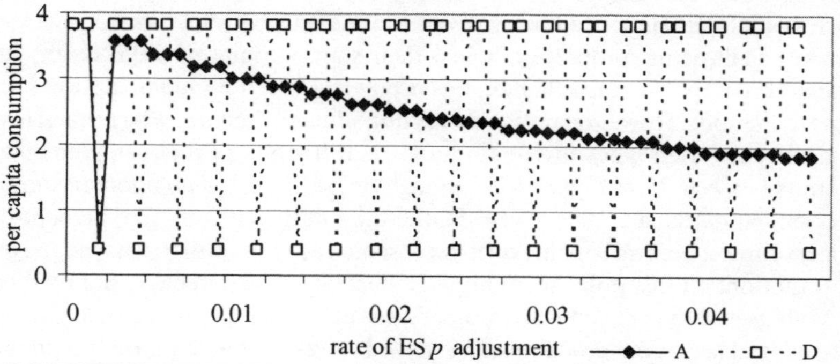

Figure 6.1b: Asymmetric networks B

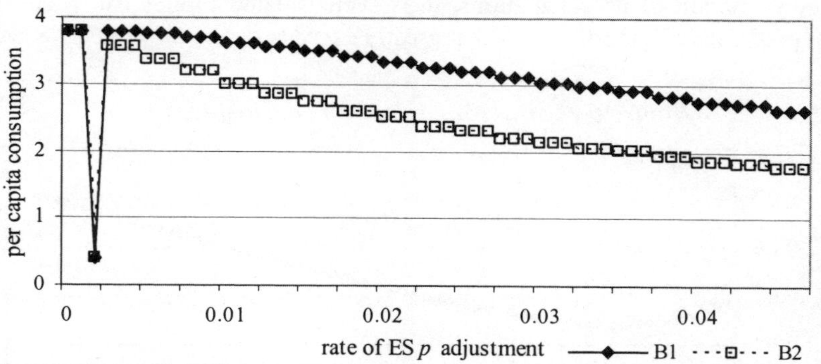

Figure 6.1c: Asymmetric networks C

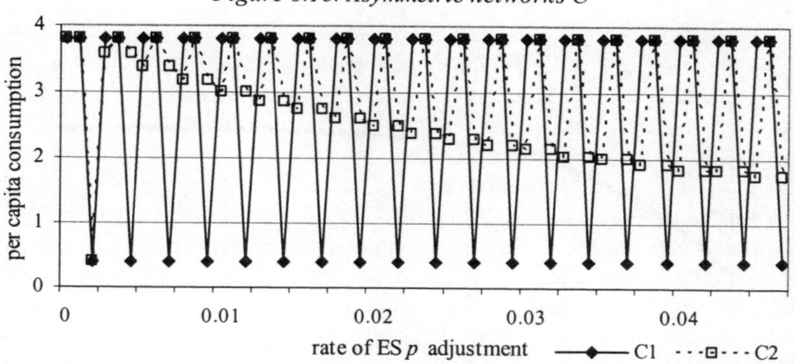

matches that of a low *d* as investments accrue interest, if the proportion of types is measured by consumption, it will initially appear within a group that there are too few high *d*s. Hence, some share of low *d*s will be transformed, thereby reducing their actual proportion below their starting point of 0.05. Using the assumption of the previous chapter, this is precisely what Figure 6.2a shows for the early periods. However, unlike those individuals or memes who remain low *d*s, these converted types invest all but a small fraction of their supplemental income ($1-r = 0.95$ in Figure 6.2a) along with their high *d* counterparts until invested capital is such that, even though they only consume five percent of their resources, high *d*s become over-represented in the population as judged by consumption. At this point some high *d*s must be converted back, but by now the savings and thus the consumption of all remaining high *d*s are such that an increasing share of the group must be made low *d*s to equate its ES *p* with the proportion of consumption attributable to low *d*s. This process continues until, as seen in Figure 6.2b, their true proportion exceeds fifty percent.

If the algorithm that dictates adjustments in proportions uses relative rates of consumption in its calculations, then even isolated groups can become 'trapped' with a share of low *d*s that greatly exceeds their ES *p*. But if, as we

Figure 6.2a: PL1, adj. based on consumption

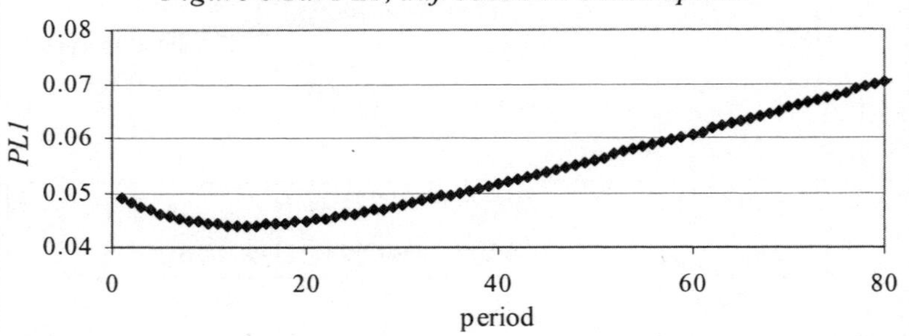

Figure 6.2b: PL1, adj. based on consumption

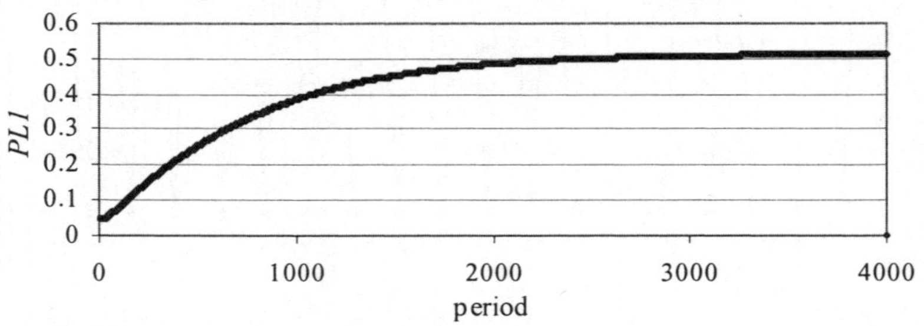

assume in this chapter, adjustments are based on true proportions, such 'trapping' cannot occur. At least in these simulations, then, the 'invisible hand' works best and the per capita consumption levels reported in Figure 6.1a-c exceed those in Figures 5.2a–c. However, perhaps the starkest contrast between these two sets of figures concerns the impact of group 3 generally on groups to which it is connected. Note for network A in Figures 5.2a, for B1 and B2 in Figure 5.2b, and for network C2 in Figure 5.2c that with a sufficiently high rate of ES *p* adjustment, group 3 succeeds in dragging down any group to which it is connected to nearly the same level of consumption it achieves when it is wholly isolated. But in Figures 6.1a–c group 3's impact is not nearly so great, and even for the highest rates of adjustment, the groups to which it is connected do not have their equilibrium levels of consumption reduced by much more than half. More specifically, in Figures 5.2a–c, a connection to group 3 can reduce per capita consumption by a factor of nearly seven, from 2 to approximately 0.30 (an 85% reduction), whereas in Figures 6.1a–c, per capita consumption falls by a factor of a little more than 2, from 4 to approximately 1.75. Thus,

> *although adjusting the proportion of types with the assumption that current proportions are measured by rates of consumption across nodes affords some nodes the opportunity to reduce the proportion of low ds, the impact is primarily to pull down otherwise high performance nodes. On the other hand, decoupling this influence by letting nodes adjust proportions in accordance with true ES p values (the 'invisible hand') leads to higher levels of consumption among otherwise high performance nodes.*

Both adjustment mechanisms are similar, however, in one respect. Consider Figures 6.3a–b, which graph per capita consumption for network B1 after 500, 1000 and 5000 iterations. Comparing these figures with Figure 5.5a–c we again see a pattern in which 3's welfare rises then falls as a function of the rate of ES *p* adjustment. However, if we use these figures to gauge the rate of convergence, we also see that convergence is more rapid with an 'invisible hand' directing internal adjustments. For example, in Chapter 5 with 500 iterations, node 3's welfare is maximized when the rate of ES *p* adjustment equals approximately 0.50 whereas here it is maximized when that rate is 0.35. With 1000 iterations, the difference is slight and not easily seen in Figures 6.3b and 5.5b: nevertheless, group 3's maximum does occur a bit sooner in the current model as opposed to the one employed in Chapter 5. And clearly, Figure 6.3b better approximates 6.3c than does 5.5b approximate 5.5c. That is, convergence here is more nearly complete after 1000 iterations than with the internal adjustment assumption employed in the previous chapter.

> *If we are so bold, then, as to suppose that a model in which people assess the proportion of types in a population by rates of consumption applies best*

Figure 6.3a: B1, 500 iterations

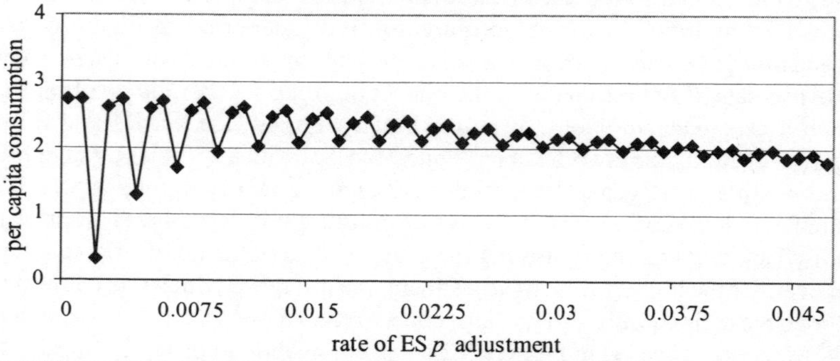

Figure 6.3b: B1, 1000 iterations

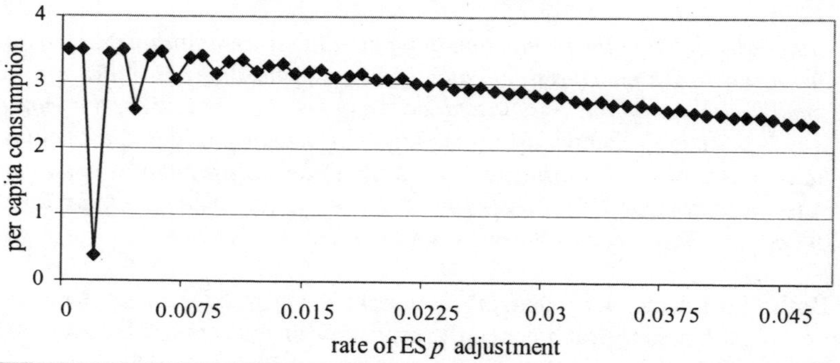

Figure 6.3c: B1, 5000 iterations

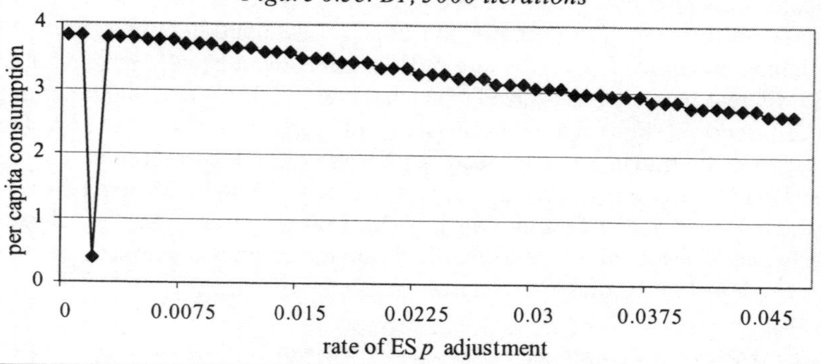

to market-like processes in which the nodes of a network are collectivities the model considered here applies best when nodes are conceptualized as single individuals, each 'populated' by high and low d memes, then we can say that networks of individuals adapt more quickly than do networks of collectivities.

This might seem a contradictory statement since, networks of collectivities are, ultimately, networks of people. But the issue here is the unit of analysis and whether we are concerned with the transformation of memes within people or the transformation of people themselves, each consisting of a single meme. The assumption here – and it is, after all, only an assumption or rather the interpretation of an assumption – is that individuals can be misled in their estimate of proportions to the extent that they base their estimates on the actions of others whereas memes, operating and competing within a persons head, will evolve more nearly in line with traditional conceptualizations of evolutionary processes. Memes, presumably, are not complex decision-making entities that gather data, weight alternatives and formulate estimates of consequences, but rather correspond to simple behavioral heuristics that survive or are replaced on the basis of relative performance. Thus, an algorithm, such as the one employed in Chapter 5, that assumes a process of estimation based on observable patterns seems more appropriate for modeling networks whose nodes consist of collections of people who use observable acts to estimate circumstances whereas one that presumes a simple (invisible hand) evolutionary heuristic seems more appropriate for modeling networks whose nodes consist of single individuals and whose memes adjust in accordance with objective experiences.

Adjustment Rate ar: Before we consider the two critical parameters of group size and return on investments, recall in Chapter 5 that consumption follows anything but a coherent pattern when we consider alternative rates of internal adjustment in the proportion of high and low *d*s, *ar* (see Figures 5.9a–c). Our explanation for that 'incoherence' rested primarily on the over-compensation that is encouraged when groups base their estimate of proportions on consumption rates. If that explanation is correct, coherence should return in the current model, and, in fact, this is what we see in Figures 6.4a–c. Although there is some slight over- and under-compensation at the highest rates of adjustment, we see nothing that even approximates the incoherence reported in the previous chapter. But there is an additional pattern here that does not arise in our earlier analysis: per capita consumption *increases* rather than decreases as *ar* increases.

The explanation for this difference is that, absent a 'misreading' of the proportion of types allowed by an adjustment based on the summed consumption across connected nodes, nodes converge more directly here to their ES *p* values even though those values can adjust as groups mimic their best-performing counterparts. Nodes 1 and 2, of course, achieve their highest performance when

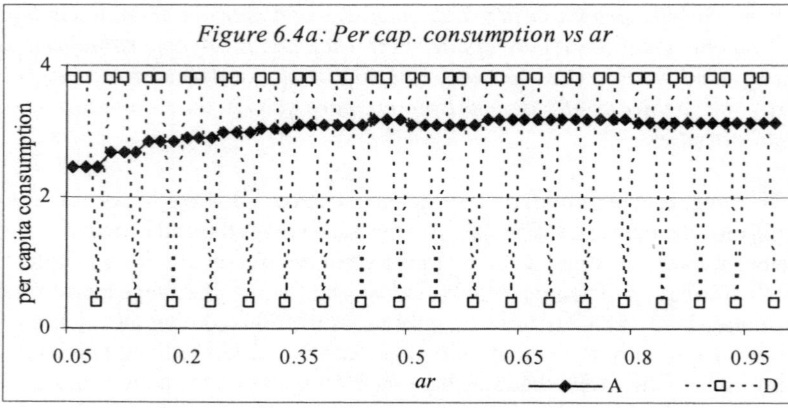
Figure 6.4a: Per cap. consumption vs ar
4
2
0
per capita consumption
0.05
0.2
0.35
0.5
0.65
0.8
0.95
ar
A
D

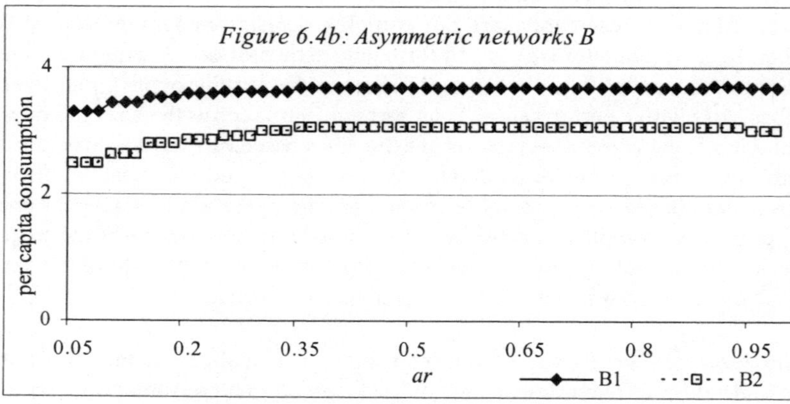
Figure 6.4b: Asymmetric networks B
4
2
0
per capita consumption
0.05
0.2
0.35
0.5
0.65
0.8
0.95
ar
B1
B2

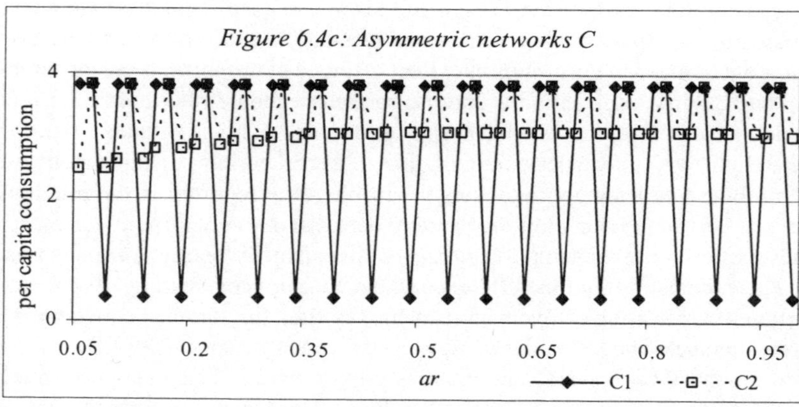
Figure 6.4c: Asymmetric networks C
4
2
0
per capita consumption
0.05
0.2
0.35
0.5
0.65
0.8
0.95
ar
C1
C2

they are not connected to 3 (1 and 2 in network D, and 2 in C2) whereas 3 performs worst when it is wholly isolated (networks D and C1).When there is no possibility of being influenced by node 3, nodes 1 and 2 make no adjustment in ES *p* or internal proportions, and high *d*s simply save and ultimately consume at optimal long term levels. Node 3, on the other hand, begins with an "under-supply" of low *d*s and we might expect that its equilibrium level of consumption will be affected by the internal rate of adjustment in types. However, as Figure 6.5 reveals, although the level of consumption is indeed impacted in the early periods by the value of *ar*, per capita consumption con-verges rapidly even when *ar* is set at the lowest value, 0.05.

Figure 6.5: Node 3's per cap. consumption for alt. ar

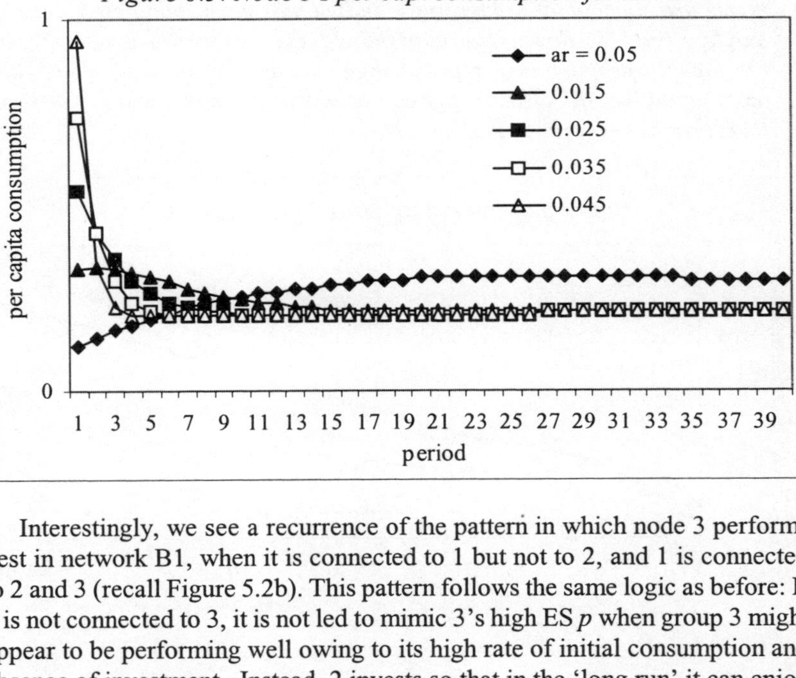

Interestingly, we see a recurrence of the pattern in which node 3 performs best in network B1, when it is connected to 1 but not to 2, and 1 is connected to 2 and 3 (recall Figure 5.2b). This pattern follows the same logic as before: If 2 is not connected to 3, it is not led to mimic 3's high ES *p* when group 3 might appear to be performing well owing to its high rate of initial consumption and absence of investment. Instead, 2 invests so that in the 'long run' it can enjoy a high rate of consumption. And although group 1 initially mimics 3and incr-ementally moves its value of ES *p* in the direction of 3's, once group 2's con-sumption increases sufficiently, 1switches to mimic 2. Group 3 then mimics 1 instead of itself as 1's consumption grows owing to its investments. The conse-quences of these switches in mimicking are shown in Figure 6.6, where we let $ar = 0.05$ and graph per capita consumption for the first 150 iterations. Note 3's initially greater consumption owing to 1and 2's choice of investment over con-sumption. Node 2's consumption, moreover, initially lags that of node 1 since

1 (but not 2) attempts in the early periods to mimic 3. However, by period 25, 2's consumption surpasses that of 3, at which point 1 begins to mimic it. By period 23 group 1's consumption surpasses 3's, at which point 3 mimics 1 rather than vise versa. And by period 40 group 3, because it is now mimicking 1's ES *p*, has increased the proportion of high *d*s sufficiently that, as a group, it has begun to save and increase its per capita consumption until, as Figure 6.4b, the consumption of all three groups equalize. Despite this similarity, we can offer a general conclusion with wording that parallels our previous conclusion:

> *if a model in which people assess the proportion of types by rates of consumption applies best to market-like processes wherein network nodes are collections of people whereas if the model considered here applies when nodes are conceptualized as single individuals each 'populated' by high and low d memes, then networks of individuals are less sensitive to the rate at which the proportion of types change than are networks of collectivities and that unlike collectivities, per capita consumption increases as this rate increases in networks of individuals.*

Figure 6.6: Per capita consumption, ar=.05

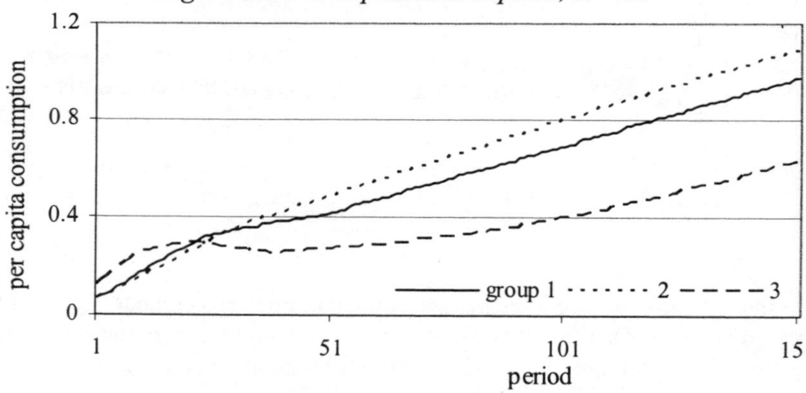

Return on Investment: We concluded in Chapter 5 that we cannot suppose that social welfare has any simple monotone relationship to the return on investment, largely because that rate has two somewhat contradictory consequences: (1) higher values of *r* encourage consumption among high *d* types; but (2) those same values also yield greater returns to investment and, thus, greater consumption in the future. As a consequence in the previous chapter, then, we found that if our objective is to maximize social welfare as measured by per capita consumption, there is an optimal rate of return that falls somewhere between 0 and

1 (see Figures 5.7a–c). As Figures 6.7a–c reveal, however, this finding does not appear to generalize to our 'invisible hand' model of internal adjustment of types.

This might seem a confusing result since our argument for an optimal *r* in the open interval (0,1) does not depend on how groups internally measure the proportion of high and low *d*s. Rather, it depends on the fact that increasing *r* yields a greater return on investments in the long run, but also encourages consumption among high *d* types in the short run. But we can use this fact to emphasize that the absolute values of our parameters have little meaning. Only relative values matter. That is, a return of 5 percent might seem reasonable when discussing investment opportunities in the real world, but here the number 0.05 has meaning only insofar as it is smaller than, say, 0.1 but larger than 0.01. To say that per capita consumption is maximized at, say, 0.2 does not mean that a society in which rates of financial return equal 0.05 or even 0.95 is necessarily performing suboptimally. With this in mind, consider Figure 6.8, which for network structures B1 and B2 graphs per capita consumption for values of *r* smaller than those considered in Figures 6.7b; namely, between 0.0025 and 0.045. And indeed, here we see once again that per capita consumption, and by inference, social welfare, is maximized at some intermediate value of *r* (approximately 0.02 for both network structures). We can, in fact, offer this general conclusion:

> *regardless of which model of within-group adjustment in the proportion of types is considered, there is a rate of return on investment, r, between 0 and 1 that maximizes per capita consumption. However, the value of r that maximizes consumption under the assumption of an 'invisible hand' (on the basis of objective ES p values) is generally less than the value that maximizes consumption under the assumption of adjustments based on relative rates of consumption.*

Group Size: Since in this model, like the one we considered earlier, the relative size of nodes enters into ones assessment as to who is performing best and, therefore, who to mimic when adjusting ones ES *p*, size will play much the same role here as it did earlier. However, we have already seen how parameter values can interact in complex ways and we should not suppose, a priori, that they will do so in precisely the same way under the assumption of an 'invisible hand' as they do when internal adjustments in the proportion of types are based on observed proportions in consumption. Hence, paralleling the treatment given to 3-node networks in the previous chapter, consider Figure 6.9, which for network structures B1 and B2 reproduces Figure 6.7b, except that now we allow nodes to evaluate the relative performance of other nodes on the basis of total as opposed to per capita consumption. In this case node 3 is recognized as being half the size of 1and 2 and the pattern to be noted is that networks B1 and B2 perform identically. It does not matter whether 3 is connected to 1alone or to 1 and

Figure 6.7a: Per cap. Consumption vs r

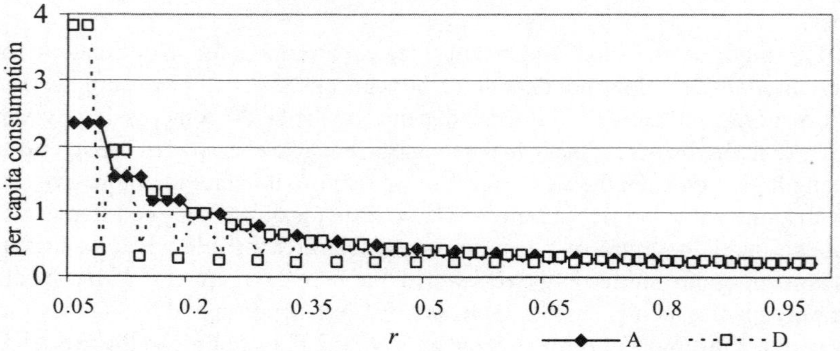

Figure 6.7b: Asymmetric networks B

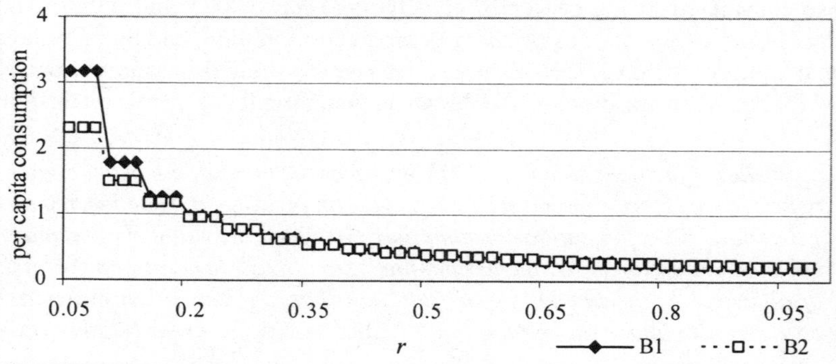

Figure 6.7c: Asymmetric networks C

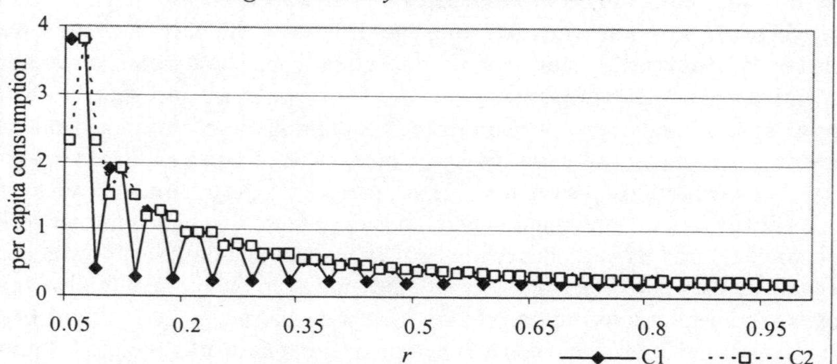

Figure 6.8: per cap. consumption vs. r, networks B

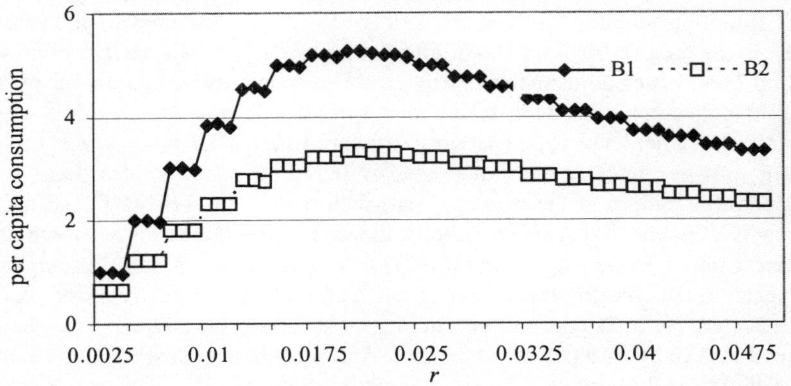

Figure 6.9: Per cap. consumption vs. r, n = (100,100,50)

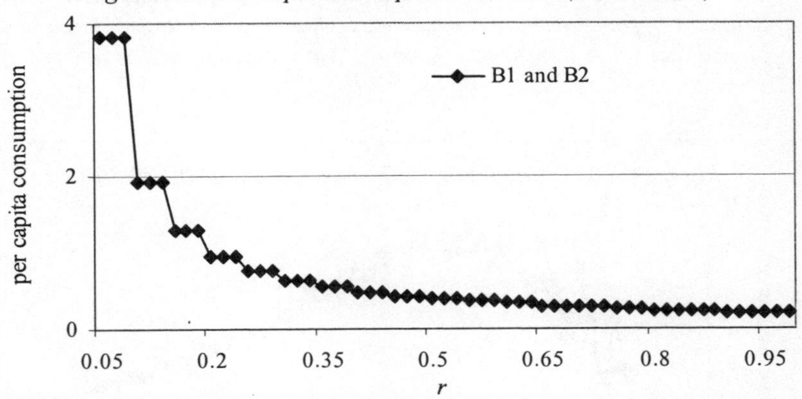

2: the node to which 3 is paired pulls it up to the level of consumption that the better performing node would achieve if it were disconnected from 3 (see network D in Figure 6.7a for what 1 and 2 achieve when wholly disconnected from 3). This is not, of course, what occurs in the previous chapter. There, if we compare Figures 5.6b and 5.2b, we see that although 3 exerts less influence and performs better (as measured by per capita consumption) when it is smaller, it nevertheless continues to exert some influence – the graphs of networks B1 and B2 are not identical in Figure 5.7b, whereas they are in Figure 6.9.

However, Figure 6.8 cautions us that the current model may be sensitive to parameters for a different range of values than what we found in the preceding chapter and should caution us against drawing definitive conclusions until we consider return on investments below 0.5. Figure 6.10, then, graphs per capita consumption against r for three network structures (B1, B2 and C2) and values of r in the range 0.0025 to 0.0475 after letting node 3 be half the size of nodes 1and 2 and letting adjustments in ES p made on the basis of total rather than per capita consumption so as to take account of size differences. The important thing to note here now is that network structure and relative group size is irrelevant only for interest rates in excess of 0.025. For lesser rates there is a discernible pattern to the results. First, with node 2 in network C2 showing what is attainable if there is no direct or indirect connection to the node with the lowest initial propensity to invest (# 3), it is evident that 3 has a discernible impact on the consumption of nodes to which it is either directly or indirectly connected. Node 2s consumption in these instances generally exceeds what 1 and 3 realize by as much as 50 percent. And second, except when 2 is wholly isolated from 3 (network C2), all three nodes do best in B1. This result, then, mirrors what we learned earlier; namely,

> *if all nodes of the network are connected, directly or indirectly (networks A, B1 and B2), social welfare is maximized when an 'advantaged' node (node 1) mediates the connection between the two remaining nodes.*

Figure 6.10: Per cap. consumption vs. r, n=(100,100,50)

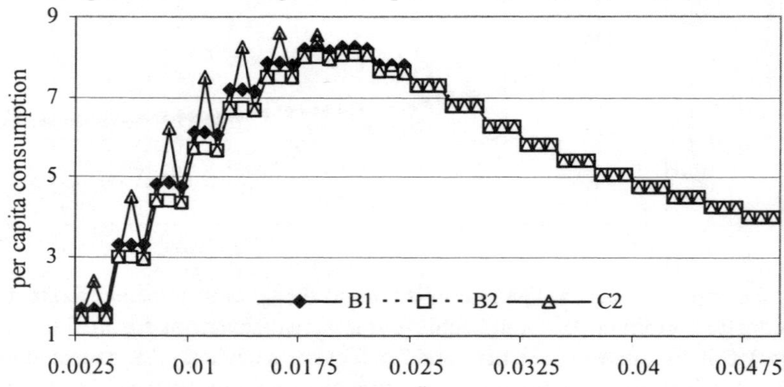

If we try now, as we did in the previous chapter when comparing Figures 5.2a–c and 5.11a–c and as summarized eventually in Figures 5.14 and 5.15, to gauge further the impact of group size by iterating on the rate of ES p change, our primary conclusion is that with n_3 equal to one-half or even three-quarters the size of nodes 1and 2 and with performance (and thus ES *p* adjustments) based on total consumption, node 3 has virtually no impact on 1and 2, regardless of network structure. As Figure 6.11 shows, it is only when n_3 exceeds .85 times the size of group 1 (or, equivalently, of 2) that 3 has a discernible impact on the long-term consumption of the nodes to which it is connected (and by implication on their eventual ES *p* values). Thus, Figure 6.11, whose construction parallels that of Figure 5.15, extends the range of sizes for 3 beyond what we consider in Chapter 5. This figure, moreover, provides an interesting contrast to 5.15. In both figures we see that among the networks in which node 1 is connected to 3 (B1, A and C2), 1 does best when it is connected to nodes 2 and 3 but 2 is not connected to 3 (network B1). However, note that if 3 is the same size as 1 in Figure 5.15, 1's per capita consumption is cut by a bit more than half over what it achieves when 3 is only half the size of 1. In Figure 6.11, in contrast, 1's per capita consumption drops from 3.81 to only 3.35 when 3's size is similarly doubled. Indeed, even if we increase 3's size so that it is three times as large as group 1, 1's per capita consumption in network B1 is reduced by less than a third to 2.78.

Figure 6.11: Per capita consumption of group 1 vs. the relative size of group 3 (rate of ES p change = 0.02)

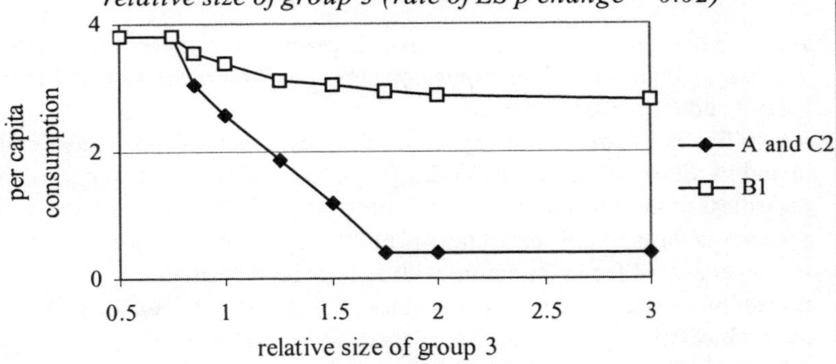

The relative insensitivity of consumption to the relative size of 3 in B1 does not mean, however, that consumption is similarly insensitive for all network structures. Figure 6.11 also portrays a situation in which consumption rates are far more sensitive to structure than in Figure 5.15. Node 1's per capita consumption differs across structures by a factor of little more than one in Figure 5.15 whereas in Figure 6.11 its consumption differs by a factor of up to seven. The explanation for this contrast lies primarily in the fact that since 1 no longer uses node 3's consumption to estimate the proportion of high and low *ds* within it, but instead operates on the basis of true proportions, as shown in Figure 6.5 it more readily tracks 2, which is not directly connected to 3. Node 1's per capita consumption, then, is not pulled down owing to the absence of investment to the same extent here as it is in the model of the previous chapter.

6.3 Conclusions and The Relevance of Network Structure

In at least one respect the algorithm we offer here to model the mechanism whereby network nodes adjust their internal proportions of high and low discount types provides little differentiation with the one we consider in Chapter 5. Specifically, as before, the initial starting point in terms of the initial proportion of high and low *d* types at each node seems of little consequence. Other parameters such as the return on investments, the pace of various adjustments, network structure and, of course, initial ES *p* values seem of far greater consequence. Thus, even if we take initial proportions as one dimension in the description of a population's 'culture', we cannot say that culture is irrelevant since it is reasonable to suppose that these other parameters also describe dimensions of that concept.

The relative insensitivity of outcomes to the initial proportion of types is not the sole pattern that fails to differentiate the two adjustment assumptions considered in this and the preceding chapter. Other patterns that hold across both models include:

- although there is an 'optimal' return on investments, *r*, there is a monotone decrease in equilibrium per capita consumption for large rates (see Figures 5.4a–b and 6.6a–b).
- for sufficient returns on investment, all networks perform equivalently, including those with isolated nodes (D, C1 and C2), which suggests that regardless of the adjustment assumption considered, the dominant impact of *r* when *r* is 'large' is the encouragement it gives to consumption that cannot be compensated for in the future with returns on investment.
- at least for the range of parameter values considered, between the two semi-hierarchical networks B1 and B2, all three nodes in B1 enjoy uniformly higher rates of consumption when mimicking is based on the per capita consumption and we consider rates of ES *p* adjustment above 0 (see Figures 5.2b and 6.1b).

It is unsurprising that nodes experience higher rates of consumption in network structure B1 than in B2. As we noted earlier, 2's relative isolation from 3 in B1 gives 1 the opportunity to mimic a node (2) that does not have its proportion of low *d* types increased as readily as it is when connected directly to 3. This logic applies to the analysis in this chapter as well as Chapter 5, and thus is another commonality of our two adjustment models. Perhaps the more interesting comparison, then, is between network structures A and B2, wherein we can ask whether 2's connection to 1 (and 1's to 2) helps offset the fact that both are connected to node 3. In Chapter 5 we concluded that if the return on investments and the rate of ES *p* adjustment are sufficiently great, the advantage afforded to nodes 1 and 2 by network B2 as opposed to A for lower values of these parameters disappears and ultimately becomes a liability. But no such pattern emerges in this chapter: the performance of all three nodes is at least as great in B2 as it is in A for all such values.

6.4 Appendix: A BASIC Program

The following is a simple BASIC program that is the basis of the analysis in this and the previous chapter. No attempt is made to provide the most efficient program as this one runs, with 5000 iterations, in under ten seconds. It nevertheless offers the choice of internal adjustment models, and the possibility of iterating on *r*, *ar*, *k,* rate of ES *p* change, the initial proportion of low *d*s and the initial ES *p*. However, when iterating on a parameter, that parameter assumes the same value for all three nodes of the network. When iterating on rate of ES *p* change, two ranges are allowed – [0, 0.045] and [0, 0.45]. Output files can include all periods or simply the last period, and in addition to per capita consumption, auxiliary files include consumption attributable to high and low *d*s within each node (cl.txt and ch.txt), ES *p* (p.txt), the node being mimicked (mimic.txt), the proportion of low *d*s 'seen' by each node (apl.txt), and the actual proportion of low *d*s at each node (pl.txt).

```
 CLS
 DIM apl(25, 3), con(25, 3), ch(25, 3), cl(25, 3), p(25, 3), c(5, 5), r(25), ttot(25)
 DIM pch(25, 3), pmax(25, 3), idi(25, 3), pl(25, 3), ih(25, 3), trl(25, 3), trh(25, 3)
 DIM incp(25), pt(25,3)
 LOCATE 3, 10
 PRINT "Iteration parameters:": PRINT
 PRINT " 1:  po(g) = initial proportion of low ds [0 to 1; step 0.05]"
 PRINT " 2:  k(g)  = supplemental income/period/person [0 to 1; step 0.05]"
 PRINT " 3:  r   = return on  investments [.0 to 1; step 0.05]"
 PRINT " 4:  in(g)  = size of incremental adjustment [0 to 1; step 0.05]"
 PRINT " 5:  p(.,g)  = ESS proportion of low ds [0 to 1 step 0.05]"
 PRINT " 6:  incp(.) = rate of adj. in ES p [0 to .95/K]":  PRINT
 PRINT "  Choose parameter for data set construction [enter 1, 2, 3, 4, 5 or 6] ";
 INPUT pmeter: PRINT
 INPUT "  Name of primary data file to be created:                  ", fle$
 INPUT "  Generate aux. Files? [y = yes]                            ", axx$
 OPEN  fle$ FOR OUTPUT AS #1
 IF axx$ = "y" THEN 101 ELSE 102
101  OPEN "pl.txt" FOR OUTPUT AS #2
 OPEN "apl.txt" FOR OUTPUT AS #3: OPEN "pl.txt" FOR OUTPUT AS #4
 OPEN "cl.txt" FOR OUTPUT AS #5: OPEN "ch.txt" FOR OUTPUT AS #6
```

```
102  INPUT "   Create a data file with all iterations [y or n]?                    ", iter$
  INPUT "   Maximum number of periods =                                       ", t
  INPUT "   Node proportions based on consumption [0] or p [1]?     ", intern
  INPUT "   Adjustments based on percap [1] or abs consume [0]?     ", percap
  PRINT
  INPUT " 1 connected to 2? [1 = yes, 0 = no]                 ", c(1, 2): c(2, 1) = c(1, 2)
  INPUT " 1 connected to 3? [1 = yes, 0 = no]                 ", c(1, 3): c(3, 1) = c(1, 3)
  INPUT " 2 connected to 3? [1 = yes, 0 = no]                 ", c(2, 3): c(3, 2) = c(2, 3)
  IF pmeter = 5 THEN GOTO 10
  INPUT " ES p values  p(1), p(2), p(3):                        ", pp(1), pp(2), pp(3)
  FOR j = 1 TO 3: ko(j) = 1: c(j, j) = 1:FOR xx = 1 TO 19
  p(xx, j) = pp(j)
  NEXT xx: NEXT j
10  INPUT " Size of groups, n(1), n(2), n(3)                    ", n(1), n(2), n(3)
  REM    input non-iterative parameter values
  IF pmeter = 1 GOTO 2
1   INPUT " initial proportion of low ds (0 < po(g) < 1)      ", po(1), po(2), po(3)
  IF pmeter = 2 GOTO 3
2   INPUT "supplemental income, (0 <= k(g) <= 1)              ", k(1), k(2), k(3)
  IF pmeter = 3 GOTO 4
3   INPUT " market rate of return, (0 < r <= 1)               ", rr
  IF pmeter = 4 GOTO 5
4   INPUT " evolutionary rate, (0 < ar(g) <= 1)               ", in(1), in(2), in(3)
5   IF pmeter = 6 THEN INPUT "  ES p  iteration range, 0 to .95/K              "; scle
  IF pmeter = 6 THEN GOTO 11
  INPUT " preset adj. rate for ESS ps  (0 <= incpp <=1)      ", incpp
11 REM    variable dictionary
  REM    ih = total investment of high ds, ill = total investment of low ds
  REM    cl = total consumption of low ds, ch = total consumption of high ds
  REM    pch=percap total consumption
  REM    apl = prop. of consumption attributable to low ds
  REM    trh = total resources of highds after investment
  REM    trl = total resources of lowds after investment (if any)
  REM    p = ESS proportion of low ds
  REM    pl = actual proportion of low ds
  REM    con = conversion proportion in accord with ES p
  REM    incp = adj rate for ES ps
  FOR j = 0 TO t: IF j > 0 THEN 12
  FOR g = 1 TO 3: FOR jj = 1 TO 19
  IF pmeter = 1 THEN pl(jj, g) = jj / 20 ELSE pl(jj, g) = po(g)
  NEXT jj: NEXT g
12  REM
  FOR px = 1 TO 19: para = px / 20
  FOR g1 = 1 TO 3: pmax(px, g1) = 0: idi(px, g1) = 0: NEXT g1
  FOR g = 1 TO 3
  IF pmeter = 2 THEN k(1) = para: k(2) = para: k(3) = para
  IF pmeter = 3 THEN r(px) = para ELSE r(px) = rr
  IF pmeter = 4 THEN in(1) = para: in(2) = para: in(3) = para
```

```
  IF pmeter = 5 THEN p(px, 1) = para: p(px, 2) = para: p(px, 3) = para
  IF pmeter = 6 THEN incp(px) = (para - 0.05) / scle ELSE incp(px) = incpp
  nlow(g) = n(g) * pl(px, g): nhigh(g) = n(g) - nlow(g)
  IF j > 0 THEN GOTO 13
  REM  assume only high ds invest
  ih(px, g) = (1 - r(px)) * ko(g) * nhigh(g)
  ch(px, g) = r(px) * ko(g) * nhigh(g)
  cl(px, g) = ko(g) * nlow(g)
13   pch(px, g) = 0
  IF n(g) * percap > 0 THEN pch(px, g) = (ch(px, g) + cl(px, g)) / n(g)
  IF percap = 0 THEN pch(px, g) = ch(px, g) + cl(px, g)
  NEXT g
  FOR g=1 to 3
  FOR g2 = 1 TO 3
  IF c(g, g2) = 0 THEN GOTO 14
  IF pch(px,g2)>pmax(px,g) THEN pmax(px,g)=pch(px,g2): idi(px,g) = g2
14 NEXT g2
  NEXT g
  IF intern = 1 then GOTO 9
  REM  compute ESS-relevant weight of low ds at each node
  x1 = cl(px, 1) + c(1, 2) * cl(px, 2) + c(1, 3) * cl(px, 3)
  x2 = cl(px, 2) + c(1, 2) * cl(px, 1) + c(2, 3) * cl(px, 3)
  x3 = cl(px, 3) + c(1, 3) * cl(px, 1) + c(2, 3) * cl(px, 2)
  tx1 = cl(px,1)+ch(px,1)+c(1,2) *(cl(px,2)+ch(px 2))+c(1,3)*(cl(px,3)+ch(px, 3))
  tx2 = cl(px,2)+ch(px,2)+ c(1,2)*(cl(px,1)+ch(px,1))+c(2,3)*(cl(px,3)+ch(px, 3))
  tx3 = cl(px,3)+ch(px,3)+ c(1,3)*(cl(px,1)+ch(px,1))+c(2,3)*(cl(px,2)+ch(px, 2))
  apl(px, 1) = x1 / tx1: apl(px, 2) = x2 / tx2: apl(px, 3) = x3 / tx3
9  REM  too many or too few low ds in group g?
  FOR g=1 to 3: pt(px,g)=p(px,g): NEXT g
  FOR g = 1 TO 3
  if intern = 1 then GOTO 8
 IF pch(px,g)<pmax(px,g)THEN p(px,g)=p(px,g)+incp(px)*(pt(px,idi(px g))-p(px,g))
  IF apl(px, g) <= p(px, g) THEN 15: REM  too few lows
  IF apl(px, g) > p(px, g) THEN 16: REM  too many lows
8  IF pch(px,g)<pmax(px,g) THEN
p(px,g)=p(px,g)+incp(px)*(p(px,idi(px,g))-p(px,g))
   IF pl(px, g) <= p(px, g) THEN 15: REM  too few lows
   IF pl(px, g) > p(px, g) THEN 16: REM  too many lows
15  REM     TOO FEW LOWS
  IF intern = 1 THEN 90
  con(px, g) = (p(px, g) * (ch(px, g) + cl(px, g)) - cl(px, g)) / ch(px, g): GOTO 91
90  con(px, g) = (p(px, g) -pl(px,g))/(1-pl(px,g))
91  IF con(px, g) < 0 THEN con(px, g) = 0
  IF con(px, g) > 1 THEN con(px, g) = 1
  REM    adjust transfer by incremental evolutionary rate
  con(px, g) = con(px, g) * in(g)
  pl(px, g) = pl(px, g) + con(px, g) * (1 - pl(px, g))
  trl(px, g) = con(px, g) * (1 + r(px)) * ih(px, g) + n(g) * k(g) * pl(px, g)
```

```
  trh(px, g) = (1-con(px, g))*(1+r(px))* ih(px,g) + n(g)*k(g)*(1-pl(px,g))
  REM   calculate investment and consumption for next period
  ih(px, g) = (1 - r(px)) * trh(px, g)
  ch(px, g) = r(px) * trh(px, g)
  cl(px, g) = trl(px, g)
  GOTO 17
16   REM    TOO MANY LOWS
  IF intern = 1 then 80
  con(px, g) = 1 - p(px, g) * (cl(px, g) + ch(px, g)) / cl(px, g): GOTO 81
80  con(px, g) = (pl(px,g)-p(px,g))/pl(px, g)
81  IF con(px, g) < 0 THEN con(px, g) = 0
  IF con(px, g) > 1 THEN con(px, g) = 1
  con(px, g) = con(px, g) * in(g): pl(px, g) = (1 - con(px, g)) * pl(px, g)
  trl(px, g) = n(g) * k(g) * pl(px, g)
  trh(px, g) = (1 + r(px)) * ih(px, g) + n(g) * k(g) * (1 - pl(px, g))
  REM   calculate investment and consumption for next period
  ih(px, g) = (1 - r(px)) * trh(px, g)
  ch(px, g) = r(px) * trh(px, g)
  cl(px, g) = trl(px, g)
17   REM
  NEXT g: NEXT px
  IF iter$ = "n" AND j < t THEN GOTO 18
  FOR w = 1 TO 19: FOR gg = 1 TO 3
  PRINT #1, USING "#######.##"; (cl(w, gg) + ch(w, gg)) / n(gg);
  IF axx$ = "y" THEN 103 ELSE 104
103   PRINT #2, USING "##.######"; pl(w, gg);
  PRINT #3, USING "##.######"; apl(w, gg);
  PRINT #4, USING "##.######"; p(w,gg);
  PRINT #5, USING "#######.##"; cl(w,gg)/(cl(w,gg)+ch(w,gg));
  PRINT #6, USING "#######.##"; ch(w,gg)/(cl(w,gg)+ch(w,gg));
104   NEXT gg: NEXT w
  IF axx$ = "y" THEN 105 ELSE 106
  PRINT #2, dum$: PRINT #3, dum$: PRINT #4, dum$: PRINT #5, dum
  PRINT #6, dum
106 PRINT #1, dum$
18    NEXT j
  CLOSE : END
```

7 Endogenous Networks: Part 1

7.1 Introduction

The model offered in Chapter 6 suffers from four limitations. First, it considers only three nodes. Even if a node is interpreted as a group, the group's members must all bear the same relationship to all others outside of it. Moreover, assuming that a group adjusts the proportion of high and low *d*s in the direction of its current ES *p*, we are in effect assuming that a node conceptualized as a group forms a complete network – each member is connected to every other member. Second, our analysis takes the networks structure as an exogenous input parameter. Thus, unlike the simulation models in Zeggelink (1995) or Hummon (2000), we cannot study the process by which networks evolve. Third, it assumes that nodes are either connected or not connected, and thus it fails to incorporate Granovetter's (1973) notion of 'strength of connection', that is, rendering the idea of a connection a continuous parameter. In this context it is important to keep in mind that a 'connection' in our analysis is also intended to model the likelihood or frequency of interaction or economic exchange between nodes (groups), in which case its representation by a continuous parameter is more appropriate. Finally, the preceding model assumes that all connections are symmetric – if node *i* is connected to *j*, then node *j* is connected similarly to *i*. This assumption may be viable for certain types of networks, such as when we study friendship and simple market exchanges, but it seems inappropriate for other relations such as those that describe the transmission of ideas, fads, norms, and information (see also Wasserman and Faust 1994, Volker and Flap 2001).

Thus, although the preceding analysis demonstrates the importance of network structure for the evolutionary development of time preferences, it precludes the simultaneous study of time preference and the dynamics of that structure. In particular, it does not allow us to consider the possibility that network structures are themselves a product of the parameters of individual preference as much as those parameters are a function of those structures. Moreover, because it does not consider strength of connection, it fails to admit the possibility that network structure is a continuous variable rather than a discrete one that can evolve incrementally and continuously over time. In this chapter, then, we extend our model to allow for endogenous network structures, and, at the same time, we allow for more than three nodes and connections that can vary in strength between 0 and 1 as well as be asymmetric.

7.2 Model Outline

Our extension of the 3-node model consists of the following additional elements:

1. Allowing for $n \geq 3$ nodes (the upper limit we in fact employ is 36, as dictated by the fact that we have written our computer code in BASIC and accept its constraints on array size and the use of memory. Conceptually, of course, there is no limit to the size of the network that can be considered aside from the practical matter that the length of time it takes to conduct a single run of the program increases exponentially with network size).
2. An $n \times n$ input matrix that corresponds to the strength of connection, c_{ij}, between pairs of nodes, where $0 \leq c_{ij} \leq 1$ and, where, allowing for asymmetric connections, we do not require that $c_{ij} = c_{ji}$.
3. Two connection transformation parameters, *t1* and *t2*, that operate thus:

$$c_{new} = \frac{t2 + c_{old}}{t1|p_j - p_i|} \tag{7.1}$$

   subject to the constraint that c_{new} not exceed 1 or be less than 0.
4. A parameter *S* that denotes the minimal strength of connection required for one node to 'consider' another in any adjustment it might make in the value of its ES *p*.

This extension, then, matches the model of the preceding chapters in the following ways. First, it allows the ES value of *p* for each node to change in accordance with the relative performance of the other groups to which it is connected with sufficient strength. However, since in our earlier analysis strength of connection was either 0 or 1, the parameter *S* was irrelevant there – a node 'pays attention' to all groups to which it was connected. Now a nodes ES *p* is influenced only by those to which it bears a strong enough connection. Second, a number of initial parameters of the preceding model are maintained here; namely, the return on investment (*r*), a small per period supplemental income (*k*), an initial ES *p* value for each node; an initial proportion of low versus high discounters for each node; an initial size for each node in the event that we choose to interpret a node as a collection of individuals, and a rate of evolutionary adjustment within each node (throughout the simulations reported here, unless otherwise noted, we let $r = 0.05$, $k = .2$, the initial proportion of low *d*s be 0.05, $S = .6$, and the within-node rate of convergence to the current ES *p* be 0.025. Since we consider only adjustments based on 'per capita' consumption, the size of groups becomes irrelevant here. On the other hand, the vector of initial ES *p*s is one of the critical parameters explored here).

7.3 Strength of Connection

Equation (7.1) requires some discussion. As noted earlier, the critical shortcoming of the model in Chapters 5 and 6 concerns the endogeneity and strength of network connections. The strengths we consider there are 0-1 constants so that the choice of a network from the six that can describe a 3-node system are assumed to be an input parameter of the analysis. As early as 1973, however, Granovetter offers a distinction between weak and strong ties, where weak ties serve the function of transferring information not otherwise held by all members of the network. Strong ties, in contrast, bond similar people together for whom this information is redundant (see also Krackhardt 1992: 216). The literature, moreover, offers a variety of ways to operationalize 'continuity' that suggest the formalization considered here. For example, Granovetter (1973: 1361) suggests that 'the strength of a tie is a (probably linear) combination of the amount of time, the emotional intensity, the intimacy (mutual confiding), and the *reciprocal services* which characterize the tie' (emphasis added). Similarly, Burt (1992: 74) distinguishes between strong and weak ties by the frequency of contacts. And as Krackhardt (1992: 217) notes 'tie strength has been measured many different ways. Some have measured strong ties as reciprocated nominations, weak ties as unreciprocated nominations and no ties as no nominations.' Other measures included recency of contact (Lin et al. 1978), while still others identify strong ties with the use of such words as 'friend', 'relative' or 'neighbor' (Ericksen and Yancey 1979, 1980; Lin et al. 1981; Unger and Wandersman 1985).

What each of these discussions suggests is that connection strength is not some discrete variable, but rather a continuous one that, presumably, can be set on some [0,1] scale. But in addition, they suggest that strength is endogenous to the network itself and the characteristics of a networks nodes. It is this endogeneity that expression (7.1) seeks to model, and in a way suggested by the literature. Specifically, as Watts et al. (2002: 1303) argue, 'the probability of acquaintance between individuals *i* and *j* decreases with decreasing similarity of the groups to which they respectively belong.' Similarly, Zeggelink (1995: 84), after arguing that 'people do not choose their friends randomly. Attraction is the very first basis of choice for a potential friend,' and after reviewing several theories that attempt to explain attraction (e.g. social exchange theory [Homans 1961, Blau 1964, Altman and Taylor 1973], reinforcement theory [Lott and Lott 1964], and social comparison theory [Festinger 1957, Schachter, 1959]), concludes that the critical variable is similarity: 'people are attracted to similar others, others who share the same beliefs, abilities, values and attitudes.' Equivalently, in her work on adolescent friendships, Kandel (1978: 427) stresses that 'individuals are consistently more likely to hold a particular attitude or engage in a particular behavior when their close associates also hold that attitude or engage in that behavior' (see also Duncan et al. 1972).

Equation (7.1) renders strength of connection endogenous by taking as our measure of similarity the inverse of the difference in the ES p of two nodes. Thus, network structure here evolves not simply, as in Zeggelink (1995), because nodes 'act' to minimize some measure of difference, but also because those differences themselves are subject to evolutionary change. This distinction is clearly made in Kandels (1978: 428) study of drug addiction: 'homophily between friends can result from a selection process leading to assortative pairing ... [or it] could result from a socialization process in which individuals who associate with each other, irrespective of their prior similarity, influence each other.' Thus, 'if there is a state of unbalance such that [a] friends attitude or behavior is incongruent with [an] adolescents, the adolescent will either break off the friendship and seek another friend or will keep the friend and modify his own drug behavior' (Kandel 1978: 435). But if both processes – selection and socialization – occur simultaneously, we can assume that each affects the other: who one selects (i.e., forms a tie with) is a function of attitudes (socialization) and the nature of socialization is a function of the networks to which one is connected. It is this simultaneity that our model seeks to capture.

Turning now to the specifics of expression (7.1) and referring to Figures 7.1a– f, if two groups are similar in their ES p, then the denominator of (7.1) is small, in which case, unless t_1 is very large, c_{new} will be larger than c_{old}. Note, however, that since we set $c_{new} = 1$ whenever the value of c_{new} calculated on the basis of (7.1) exceeds 1, this model is equivalent to our earlier one if $t2 = 0$ and $t1 << 1$.[1] In much of the analysis that follows, we set $t2 = 0$ or some small number such as 0.0001, and include it here to allow for the possibility that eveb if $c_{old} = 0$ then c_{new} can evolve to be some non-zero number. That is, setting $t2$ equal to some arbitrarily small number allows a connection to develop in the event that two nodes are similar even if an initial connection is nonexistent and in this way we can interpret $t2$ as a measure of how difficult it is in a social system to develop network connections where there are none to begin with.

The more critical parameter, however, is $t1$. Specifically, c_{new} will converge ultimately to 1 (i.e., $c_{new} > c_{old}$) if $t1 < |p_j - p_i|$ whereas c_{new} will converge to 0 if the opposite inequality holds. Of course, the inherent complexity of the model is that although $t1$ is treated as a fixed parameter, $|p_j - p_i|$ is not fixed and itself is a function of a dynamically changing matrix of connections (to the extent that the stronger a tie, the more likely a node is to adjust its ES p in the direction of a network partner).

We appreciate that in some sense the formal specification of the relationship between c_{new} and c_{old} is arbitrary: There are an infinity of functional forms that

1 Since the absolute value of $(p_j - p_i)$ is necessarily less than 1 owing to the fact that both p_j *and* p_i are between 0 and 1, the calculated value of c_{new} will necessarily exceed 1 if $t1 < 1$ and $t2 = 0$

Figure 7.1a: t1=2

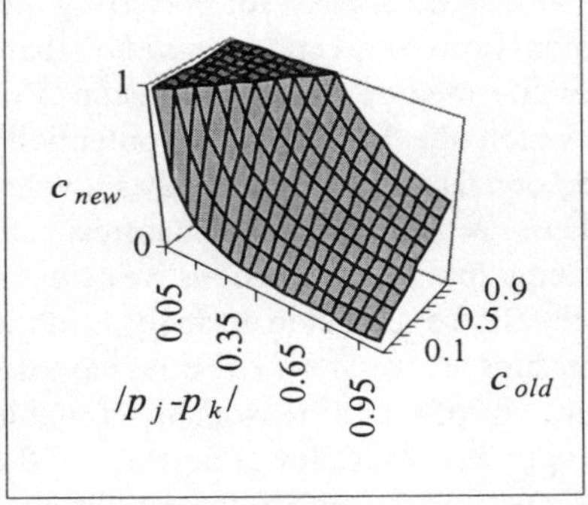

Figure 7.1b: t1=6

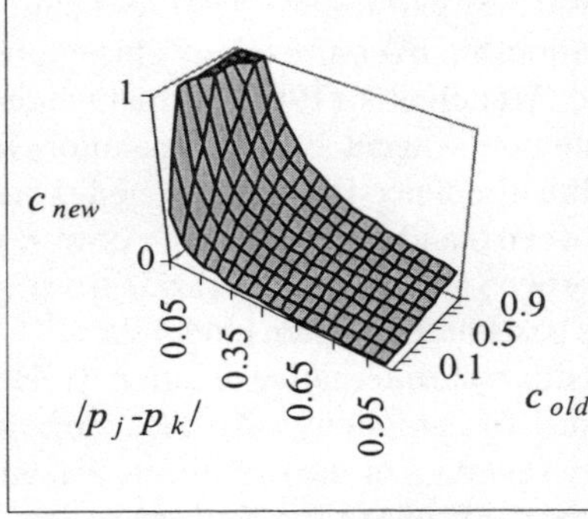

Figure 7.1c: t1=10

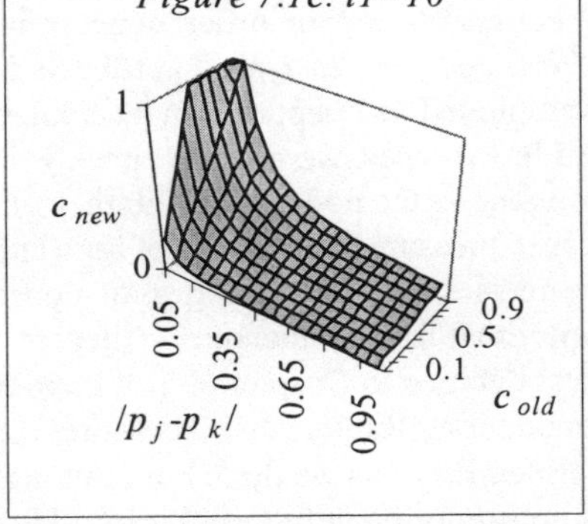

Figure 7.1d: t1=12

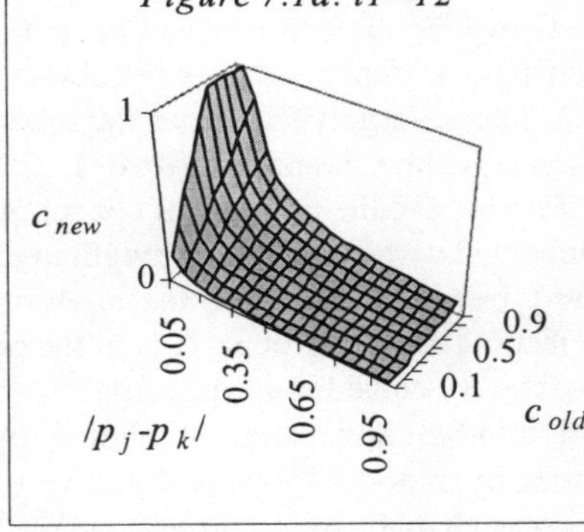

Figure 7.1e: t1=16

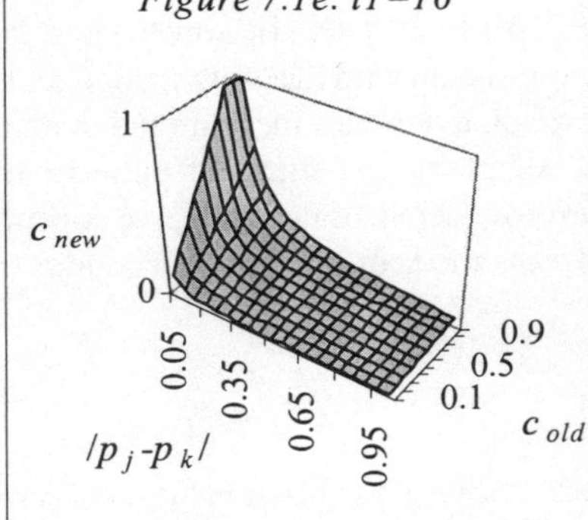

Figure 7.1f: t1=20

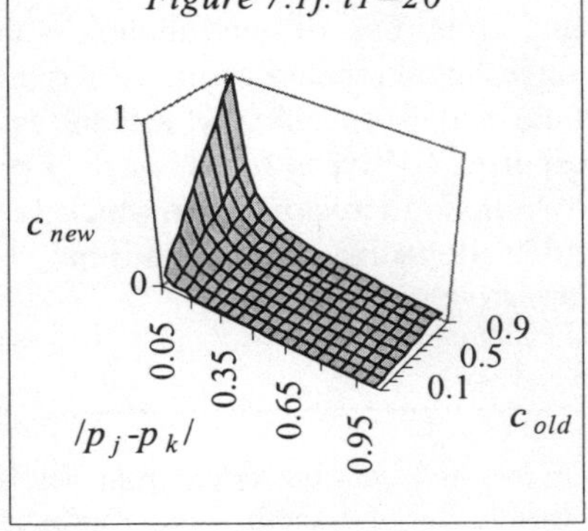

yield a relationship similar to or at least substantively equivalent to the one implied by expression (7.1). And there is no empirical basis for preferring one relationship over any other. This expression, however, is no less *ad hoc* than, say, Zeggelink's (1995) formalization of the idea of 'network tension',[2] or Hummon's agent based simulation in which the benefit of a (potentially indirect) connection between nodes i and j equals some exogenously set value between 0 and 1 raised to the t power, where t is the length of the shortest path between i and j. However, aside from agreeing that the conclusions we offer on this basis of the current model should eventually be subject to an analysis of the sensitivity to alternative functional relationships, expression (7.1) does have the virtue of simplicity. At the same time, it offers a reasonably flexible representation of the hypothesis that an important descriptor of networks – the strength of connection between nodes – is positively related to the similarity of those nodes.

However, before proceeding a few comments are in order concerning alternatives to the model we are about to consider. The analysis that follows in this chapter largely subsumes the model employed in Chapter 5. In particular, when adjusting the proportion of high and low d types, we assume once again that nodes use the patterns of consumption among the nodes to which they are connected with sufficient strength in order to measure proportions of high and low types. Thus, again, the analysis here is more in keeping with nodes conceptualized as groups. It is in the chapter that follows this one – Chapter 8 – where we consider the adjustment process detailed in Chapter 6. But there is still another dimension of choice in modeling. Rather than measure the similarity of nodes by their objectively defined ES ps as we do in both this and the next chapter, we could let patterns of consumption define similarity. That is, we could say that two nodes are similar to the extent that we attribute the same proportion of consumption within them to low ds. In some ways, of course, this alternative assumption is more consistent with the other components of the model – specifically, with the assumption that nodes measure the overall proportion of types to which they must adapt by the summed patterns of consumption across nodes to which they are connected. Later, then, we consider this alternative assumption if only to assess the sensitivity of our model to alternative specifications.

---

2 In that study tension derives from two 'issues': a divergence between the number of friends (ties of strength 1) and a nodes ideal number of friends; and, assuming that a node is either similar or dissimilar , the number of dissimilar friends. A nodes tension, then, is defined as *w1* times the first divergence plus *w2* times the number of dissimilar friends, where *w1* and *w2* are exogenously set parameters. Connections then develop or are broken in a dichotomous fashion as individual nodes seek to reduce their tension thus defined.

7.4 An Initially Complete Network

Twins: Aside from the addition of expression (7.1) to the analysis and an expansion in the number of nodes, the internal process of that simulation is otherwise identical to the one we describe in Chapter 5: on the basis of their current ES *ps*, nodes make their investment and consumption decisions in accordance with the model described in Chapter 3. Each node then 'looks at' the nodes to which it possesses a sufficiently strong connection (connections whose strengths exceed the exogenously set parameter *S*), and adjusts its ES *p* in the direction of the most productive partner (as measured by per capita consumption). Network connections are then adjusted in accordance with expression (7.1) and these new ES *ps*. At the same time, nodes (groups or memes within an individual) attempt to adjust the proportion of types they see in the direction of their ES *p*, at which point the process repeats for a fixed number of iterations (generally 4000, which is sufficient for convergence to occur for the parameter values considered).

Despite these similarities there are now two 'parameters' that need to be set as starting points : The initial strengths of connection and an initial vector of ES *ps*. Needless to say, we cannot consider all possibilities, but what we can do is explore extremes, beginning with an initial connections matrix of all 1s that corresponds to a complete network. And with respect to ES *ps*, we begin with a vector that spans the range [0.05, 0.95] but which at the same time pairs nodes in much the same way that nodes 1 and 2 are paired in network B1. Hence, our initial vector of ES *ps* is as follows:

P1 = (.05, .05, .20, .20, .30, .30, .40, .40, .50, .50, .70, .70, .80, .80, .95, .95)

The pairing of nodes here is dictated by our desire later to compare this situation to one in which nodes are initially unpaired so as to assess the impact of 'twins' – the impact of having at least one other node that is wholly similar. Of course, it might seem that with P1 and an initial connection matrix in which all elements equal 1, we are in effect merely exploring an 8-node network since nodes 1 and 2 are identical in all respects, 3 and 4 are identical and so on. However, this is not strictly the case. Although we can assume that twins will evolve identically given the symmetry of the situation, it is not the case that the analysis is equivalent to an 8 node model since here each node can 'refer' to its twin not only when it attempts to assess the proportion of types in the population, but it can also look to that twin when adjusting its ES *p* parameter.

Two parameters in particular, now, are the focus of our first set of simulations – *t1* and the rate at which individual nodes adjust their ES *p* values on the basis of relative rates of per capita consumption. Figures 7.2a–d, then, portray per capita consumption against the rate of adjustment of ES *p* for four values of *t1* (9, 7, 6, and 2) and reveal the following:

Figure 7.2a: all c s = 1, t1 = 2

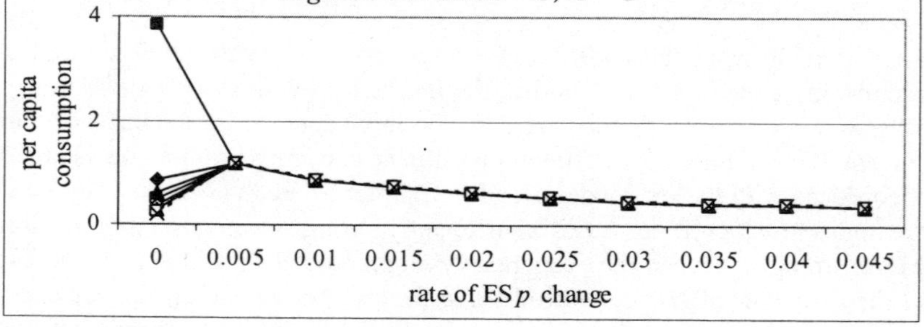

Figure 7.2b: All c s = 1, t1 = 6

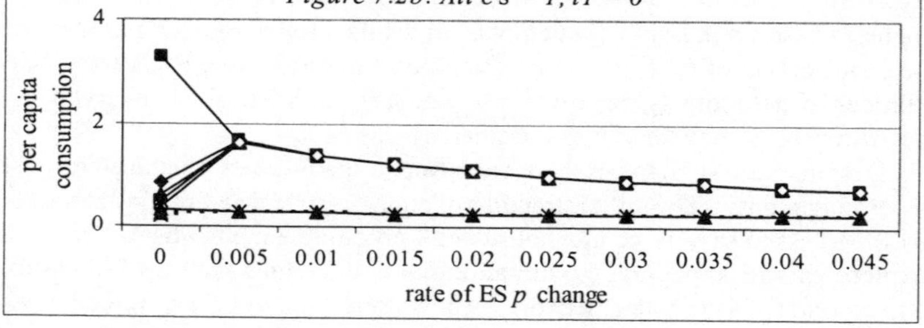

Figure 7.2c: all c s = 1, t1 = 7

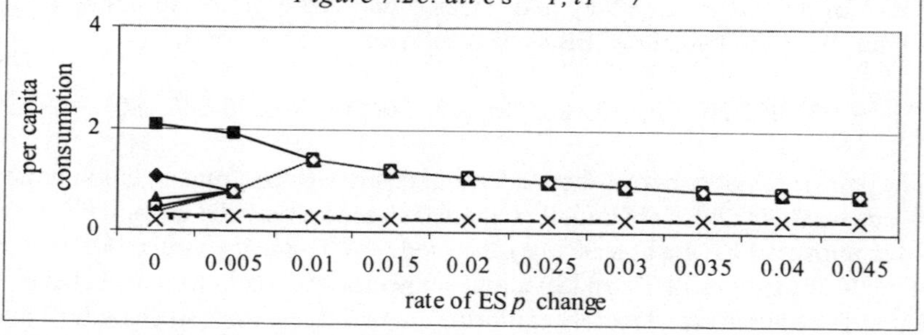

Figure 7.2d: all c s = 1, t1 = 9

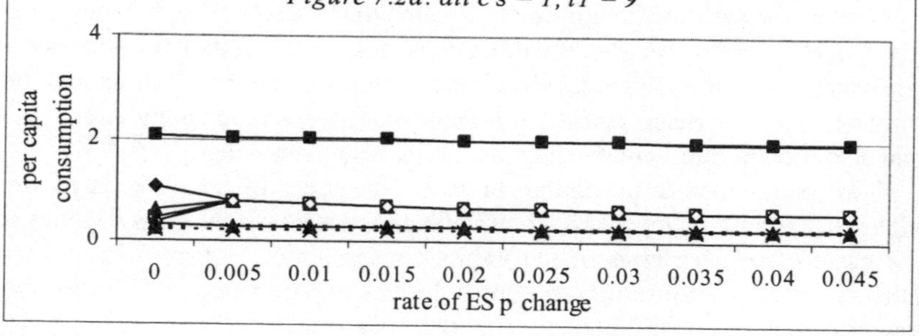

- If that adjustment rate is 0, then as expected, we have 8 distinct sub-networks that consist of the 8 nodes distinguished solely by their initial ES *p* values. The per capita consumption rates here correspond identically to what we observe in Chapter 3, where ES *p* is a fixed input parameter.
- What occurs once the adjustment rate of ES *p* exceeds 0 depends critically on *t1*– the parameter that controls the rate at which connections adjust upwards or downwards. If *t1* is small (e.g., 2) so that c_{new} is allowed to grow above c_{old} or to maintain a strong connection for even dissimilar nodes (as characterized by *p*), then any non-zero rate of adjustment in ES *p* yields a single wholly connected network. Increasing *t1* to, say, 6, results in two sub-networks. Increasing t1 to 7 yields either 3 or 2 sub-networks, depending on the rate of ES *p* change, and increasing this parameter to 9 yields four distinct sub-networks for any non-zero adjustment rate.

This second result is not wholly surprising since it confirms simply that the more readily we allow strengths of connection to increase or be sustained, the fewer will be the number of sub-networks. The change across Figures 7.2a through 7.2d, moreover, follows a logical progression: the easier a connection can be maintained, the more readily relatively poor performing nodes can influence and pull down otherwise better performing ones. But if *t1* is increased so that connections are more readily dissolved when ES *p*s differ, those well performing nodes are no longer compelled to mimic those with higher ES *p* and lower rates of investment.

Figure 7.2c is especially interesting, for here we see not only the interaction of parameters – *t1* and rate of ES *p* adjustment – in determining sub-network structures, but also a specific form of path dependence. If we look across Figures 7.2a-c, we see that the number of subnetworks depends self-evidently on t1. But Figure 7.2c also shows that of *t1*= 7, that number also depends on the rate of ES *p* adjustment: Three subnetworks form if that rate is 0.005 whereas two form for higher rates. Thus,

> *the number of subnetworks that form depends both on the ease with which connections are formed or maintained and the rate at which ES p changes.*

Clearly, alternative rates of change here can impact the number of periods a node mimics one as opposed to another and, therefore, its ultimate level of consumption and connections. But aside from this, we can also offer the following two observations:

- Figures 7.2a–d offer a counter-example to any presumption that there is necessarily a monotonic relationship between overall social welfare and the number of sub-networks; and

- for various rates of change in ES *p*, diversity in sub-networks yields a relatively impoverished sub-population whereas homogeneity yields a society that does not reach down to the level of the poorest sub-group in the heterogenous case.

Figure 7.3 graphs total average consumption for each value of *t1* against alternative values of rates of change and here we see that if we let the rate of ES *p* change equal 0.01, then the highest total of per capita consumption is achieved when *t1* = 6 or 7 (which corresponds to the intermediate number of sub-networks) whereas if we consider the highest rate of ES *p* change, the highest total occurs when *t1* = 9 and four subnetworks prevail.

Figure 7.3: Total per capita consumption vs. t1

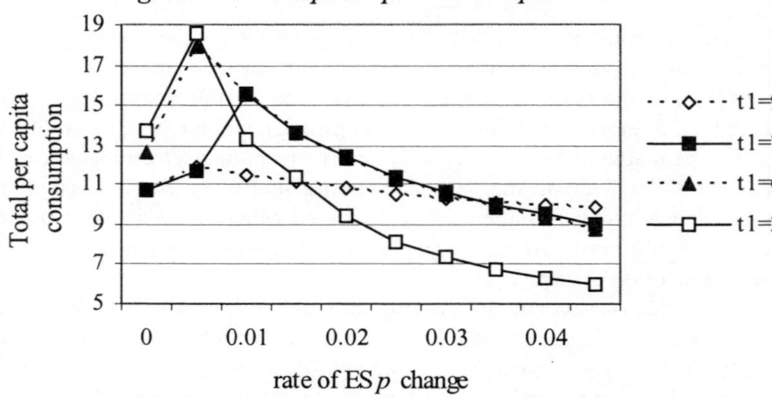

It is, though, the second of the preceding two observations combined with the first that yields a more substantively interesting implication. Consider Table 7.1, which, for various rates of ES *p* change and *t1*, reports three numbers – the mean level of consumption and two measures of diversity: the variance of consumption and the difference between the highest and lowest consuming nodes. Note first that regardless of which measure of diversity we consider, the relationship between diversity and overall social welfare (as measured by average per capita consumption) follows no pattern for intermediate rates. On the other hand, for the very lowest rate of ES *p* change (0) and the highest one considered (0.045) there is an inverse relationship between the average level or per capita consumption achieved and the variance of consumption across nodes. Regardless of our diversity measure, then, the missing pattern here is a negative relationship between diversity and social welfare. That is, if there is a monotonic relationship here it is where diversity correlates positively with welfare.

Table 7.1: Social welfare and diversity

	rate of ES *p* change			
tl	0	.020	.030	.045
2	0.86 1.14, 3.62	0.51 0, 0	0.46 0, 0	0.37 0, 0
6	0.79 0.96, 3.06	0.77 0.40, 0.82	0.66 0.31, 0.65	0.55 0.25, 0.49
7	0.67 0.59, 1.84	0.78 0.40, 0.83	0.66 0.32, 0.66	0.56 0.26, 0.51
9	0.67 0.59, 1.84	0.68 0.54, 1.78	0.64 0.53, 1.76	0.67 0.59, 1.72

It is, at this point, tempting to draw social implications from the absence of any positive relationship between diversity and social welfare and our finding that social welfare can actually correlate positively with diversity, which we can interpret as a measure of income inequality. We might, for example, cite the general impoverishment but homogeneity of the former Soviet Union as compared to the United States with its greater per capita wealth but also significant inequalities in income. The suggestion, then, is that income equality can be purchased only at the expense of overall general welfare.

Drawing conclusions from such comparisons, though, is dangerous given the varied political–economic systems. But we can offer a more compelling analysis by considering the data on national income inequality collected by the World Bank in conjunction with the 'Human Development' (HD) rank of countries offered by the United Nations.[3] If we take all 124 countries given a rank and for which the World Bank offers a Gini Index measure of income inequality, we find, contrary to the implication of our analysis, that the correlation between HD and income inequality equals *negative* 0.46. That is, income inequality declines with this measure of 'social productivity'. However, this correlation entails a comparison of 'apples and oranges' in terms of underlying political–economic–

---

[3] All data are from http://www.undp.org/hdr2003/indicator. For the list of indicators used to construct their 'human development' index and the methodology of the index see http://www.undp.org/hdr2003/indicator/index_indicators.html. Briefly, that index considers such variables as life expectancy, infant mortality, percent of population in poverty, various measures of the quality of sanitation and health care, literacy rates, expenditures on education, public and private debt, energy consumption, human rights conditions, unemployment rates and HIV cases as a percent of the population.

social structures. So suppose instead that we divide countries into three categories: the 31 developed Western states – primarily states of the EU, including new members from Central Europe, plus Japan, Canada, the United States, New Zealand and Australia;[4] the 15 countries of the former Soviet Union; and all others. If we calculate the correlation between HD and income inequality for the 'developed' group, we get a positive but weak 0.14. Inspection of this data reveals a pattern in the outliers – Sweden, Finland, Denmark and Norway. If we exclude them, that correlation increases to 0.30 as shown in Figure 7.4a. Thus, as Human Development increases, income inequality increases as well. A similar pattern is revealed if we look at the states of the former USSR. The data portrayed in Figure 7.4b yields a correlation coefficient of 0.13 when considering only the overall ordinal ranking of states on HD and increases to 0.17 if we consider a states ranking relative to others in the former USSR. Only for the category of 'others', where the data are notoriously unreliable and includes countries with widely varying social systems, does this correlation go to zero (0.01 actually).[5]

Figure 7.4a: Human Development vs Gini Index, West

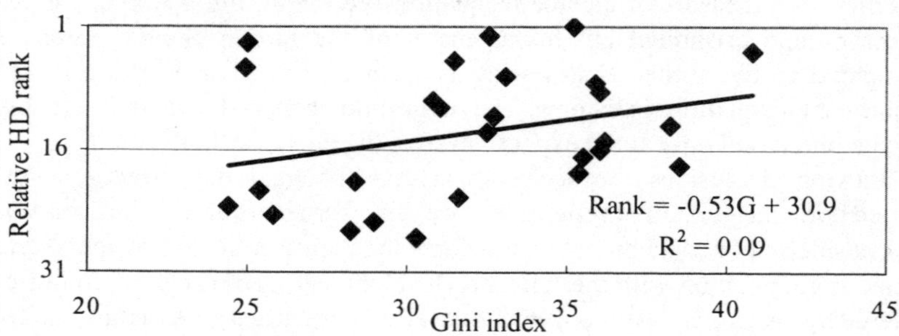

[4] The countries in category 1 are: Norway, Denmark, Finland, Sweden, Australia, Netherlands, Belgium, United States, Canada, Japan, Switzerland, Ireland, United Kingdom, Luxembourg, Austria, France, Germany, Spain, New Zealand, Italy, Israel, Portugal, Greece, Slovenia, Czech Republic, Poland, Hungary, Slovakia, Croatia, Macedonia, Romania. The states for the former Soviet Union are: Estonia, Lithuania, Latvia, Belarus, Russian Federation, Ukraine, Kazakhstan, Turkmenistan, Georgia, Azerbaijan, Armenia, Uzbekistan, Kyrgyzstan, Moldova, Tajikistan.

[5] That a positive correlation becomes negative when we divide the sample of countries into three parts suggests considerable 'aggregation error' and there is little reason to believe that this error does not persist in the larger sample of poor countries we include in the category of 'other'.

Figure 7.4b: Human Devel. vs Gini Index, former USSR

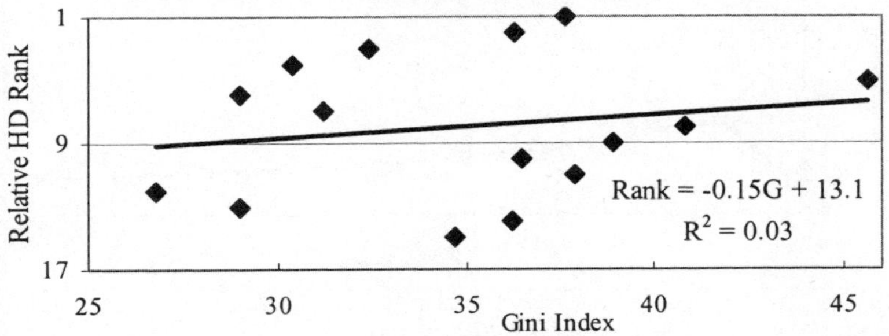

We find in this data, then, some empirical support for at least one implication of our analysis; namely, that by discouraging connections, we induce a greater range in the relative long-term welfare of the various subnetworks and an overall increase in average welfare. Of course, the data here are merely suggestive. Any supposition that our model explains the pattern we see in the relationship between human development and income inequality must be tempered by the understanding that, barring evidence to the contrary, our results should be assumed to be sensitive to parameter values (as well, of course, to the models structural specifications). And in that spirit it is worthwhile looking more closely at the process whereby subnetworks develop. To begin, consider Figures 7.5a–c, which parallel Figures 5.2a–d in their presentation, but which consider four values of *t1* in a small range, *t1* = 5.2, 5.3, 5.5 and 5.6. With *t1* = 5.2 (Figure 7.5a) we see the network of the whole begin to disassemble when the rate of change in ES *p* is low, at least in terms of a uniformity of rates of consumption. For *t1* = 5.3 (Figure 7.5b), a new sub-network of nodes {11, 12, ... 16} forms when the rate of ES *p* change is less than 0.015. For *t1* = 5.5 (Figure 7.5c), this subnetwork develops across an even greater range of rates of ES *p* change, and for *t1* = 5.6 (Figure 7.5d) this sub-network exists for the full range of rates considered. Thus, once again we see the interaction of parameters: the impact of *t1* depends on the rate of change and the impact of this rate depends on *t1*. But Figures 7.5a–c also reveal another interesting phenomena; namely,

> *nodes within a subnetwork need not achieve the same level of per capita consumption.*

This might seem a curious property of our model until we look at the pattern of connections that emerge within subnetworks. Specifically,

> *the connections within subnetworks need not be complete.*

Figure 7.5a: all c s = 1, t1 = 5.2

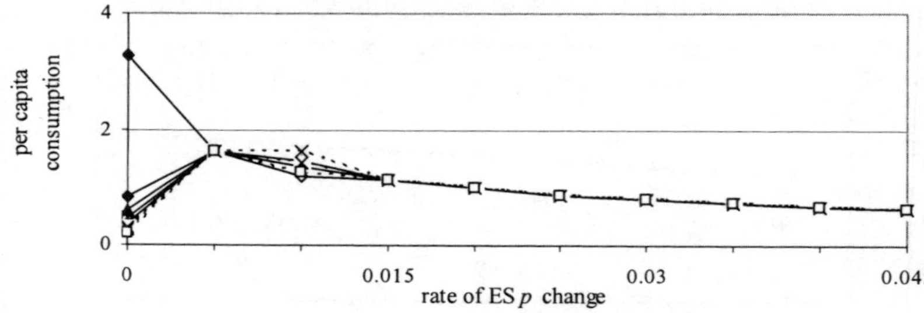

Figure 7.5b: all c s = 1, t1 = 5.3

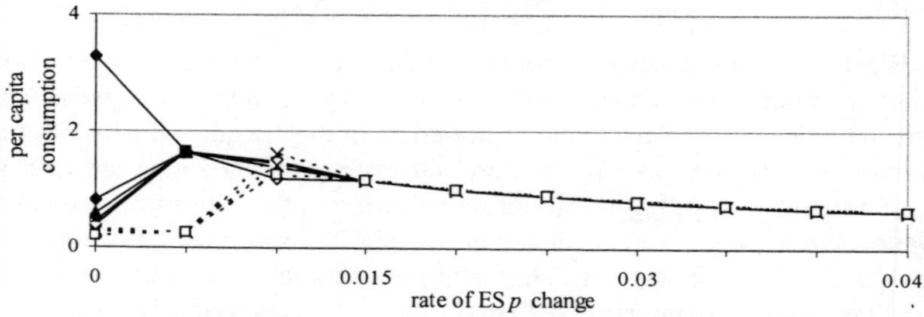

Figure 7.5c: all c s = 1, t1 = 5.5

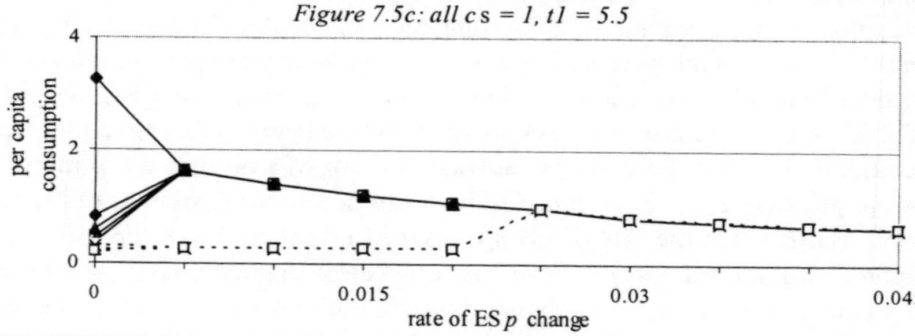

Figure 7.5d: all c s = 1, t1 = 5.6

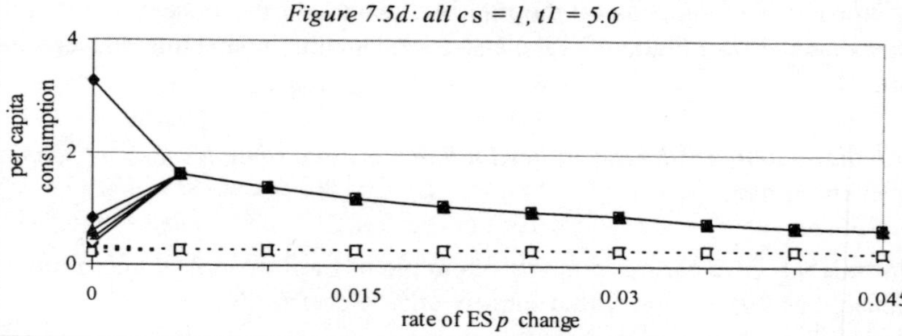

To see this feature of our model in bold relief, consider Figure 7.6, which, keeping in mind that with 16 nodes, the maximum number of connections is $16^2 = 256$ and occurs when all nodes form a single sub-network, graphs the number of connections against *t1* for the rates of ES *p* change considered in Figures 7.2a–d and 7.5a–d.

Figure 7.6: Number of connections vs t1 for various rates of ES p change

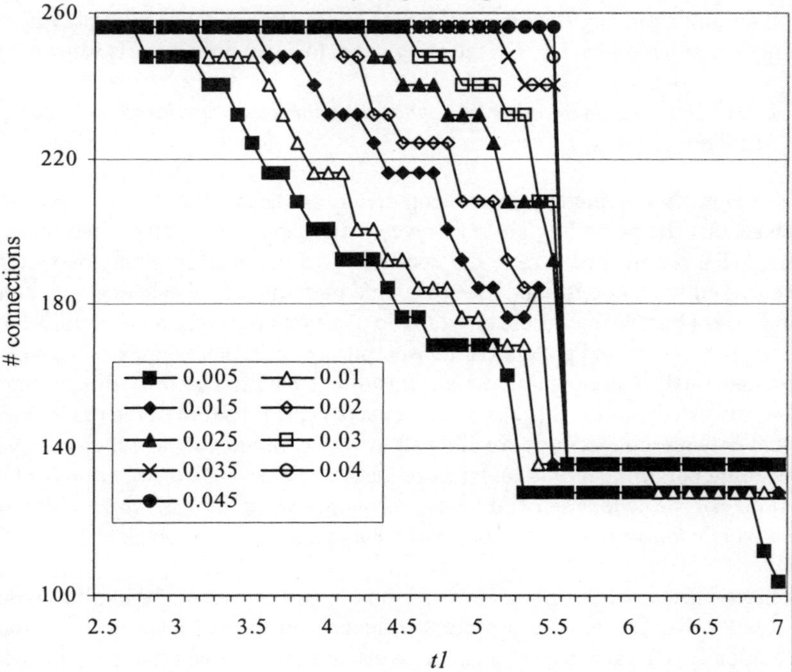

Note first the general decline in the number of connections at all but the highest rates of ES *p* change for all *t1* in the range [2.5, 5.2]. However, at the lowest rate (0.005) and for *t1* = 5.3, there is a precipitous drop in the number of connections (from 160 to 128) and it is at this point, in fact, that nodes 11 through 16 become a separate subnetwork. This separation continues for still greater values of *t1* as the rate of ES *p* change increases until, at *t1* = 5.6, nodes 11-16 are a separate subnetwork for all rates considered. This process then repeats as *t1* continues to increase so that at *t1* = 6.8 we begin to see the number of connections decrease within both subnetworks for the lowest rate of ES *p* change. And, of course, were we to consider still greater values of *t1,* we would find for example that for $t1 = 9$, there are four distinct subnetworks ({1, 2}, {3, 4,..., 10}, {11, 12, 13, 14}, {15, 16} as in Figure 7.2d) with between 88 and 80 connections total across all 16 nodes.

It might seem now that the number of connections within subnetworks decreases as *t1* increases. However, Figure 7.6 offers a single counter example to this generalization: with the rate of change in ES *p* equal to 0.015, the number of connections actually increases as *t1* increases from 5.3 to 5.4. That this counter example exists immediately before a new subnetwork forms opens the door to speculation as to when violations of monotonicity arise, but we refrain at this point from any such generalization given the small range of parameter values considered thus far. On the other hand, one generalization is warranted:

> ceteris paribus, *the number of connections increases as the rate of change in ES p increases*.

We explore the completeness and incompleteness of subnetworks later, and note instead that the preceding analysis reveals a complex interactive relationship among ES *p* adjustment rate, the sensitivity of connection strengths to differences among nodes, and the resultant patterns of investment and consumption. Thus, we have here an example of a system in which the equilibrium network structure and patterns of consumption within them depend on several simultaneously operating interdependent adjustment rates so as to reveal a complex network dynamic. It is also evident that a critical feature of the model here is that network connections are allowed to vary continuously in which case the preceding patterns serve to underscore the fact that *models of social networks that do not allow for the endogenous development of network structures will necessarily fail to reveal the potential complexity of those networks*.

Discontinuities: At this point we need to note two features of the initial input vector P1 that can have a profound influence on the subnetwork evolution. Specifically, P1 pairs nodes to create twins and at the same time it established an initial 'gap' between various nodes. Although the difference between ES *ps* is 0.05 in most cases, the gap between 1and 2 versus 3 and 4 is 0.15, between 9 and 10 versus 11 and 12, it is 0.20, and between 13 and 14 versus 15 and 16 it is 0.15 again. Note moreover that the subnetworks that form in all of the preceding figures form around these gaps. Clearly and quite naturally, then,

> *if there are 'discontinuities' in the initial distribution of ES p in the form of gaps and if subnetworks form, they will tend to form with these gaps acting as boundaries between subnetworks.*

To explore the consequences of twins and gaps, we first consider the following alternative to P1, which eliminates twins but maintains the same gaps:

P2 = (.01, .05, .20, .24, .28, .32, .36, .40, .44, .48, .68, .72, .76, .80, .95, .99)

Figures 7.7a–c rerun our simulation for *t1* = 6 using three alternative within-group adjustment rates, *ar*: 0.005, 0.025 (the rate we employ in all previous runs) and 0.05.The results are not altogether unexpected insofar as network structure is concerned, but somewhat unexpected in terms of patterns of consumption. First, after eliminating twins with P2 and for rates of ES *p* change greater than 0.01, two subnetworks continue to form around the gap in initial ES *p* values between nodes 10 and 11. Second, Figures 7.7a and 7.7b reveal a pattern of differential rates of consumption when the rate of change of ES *p* equals 0.01 within the (incompletely connected) subnetwork {1,2,...,10} that is even more varied than what we observe in Figures 7.5a and 7.5b. Thus, the possibility of differential consumption rates is not restricted to the parameter values considered earlier. Finally, and perhaps more substantively interesting, is that an otherwise coherent pattern of consumption exists across all values of *ar* considered here. This contrasts sharply with what we observe in Chapter 5 when considering only a 3-node network. Of course there we consider a considerably greater range of values for *ar* (0 to 0.9) and thus Figures 7.8a-c portray what occurs when we let *ar* = 0.25, 0.5 and 0.9. Once again, however, we see a coherent pattern in which overall equilibrium consumption declines as *ar* increases. The suggestion here, then, is twofold:

> *patterns of consumption are a more coherent function of internal adjustment rates to the extent that network connections are endogenous and the number of nodes is large.*

We appreciate that we have again considered only a small subset of parameter values. Nevertheless, the oscillations and consequent appearance of 'instability' that was apparent in Chapters 4 and 5 seems more difficult to replicate here with a greater number of nodes. This finding is logical. With nodes that are more uniformly distributed in terms of initial ES *p*s and all nodes initially connected, the proportion of types based on rates of consumption will necessarily correspond to some more moderate average, in which case the smaller will be the magnitude of adjustments in the proportion of types that individual nodes attempt. We find here, then, a form of stability within populations that, initially at least, are not represented solely by extreme types (in terms of initial ES *p*s).

It is tempting to draw a multitude of inferences from this pattern about social processes generally. If, for example, we consider the evolution of street gangs in otherwise closed urban neighborhoods, we might speculate that social change will proceed more smoothly to the extent that the neighborhood is populated by a variety of types. This is not to say that some elements of the population will not gravitate to gang membership while others pursue more acceptable 'careers',

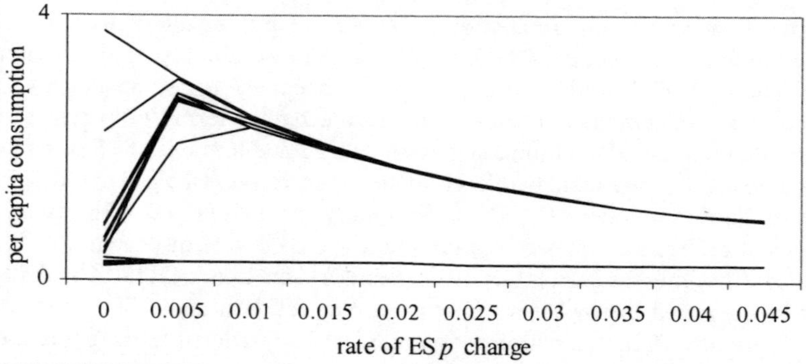

Figure 7.7a: All c s = 1, P2, ar = 0.005

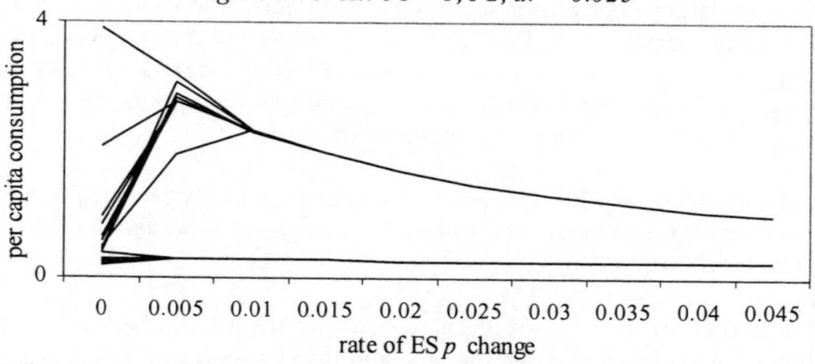

Figure 7.7b: All c s = 1, P2, ar = 0.025

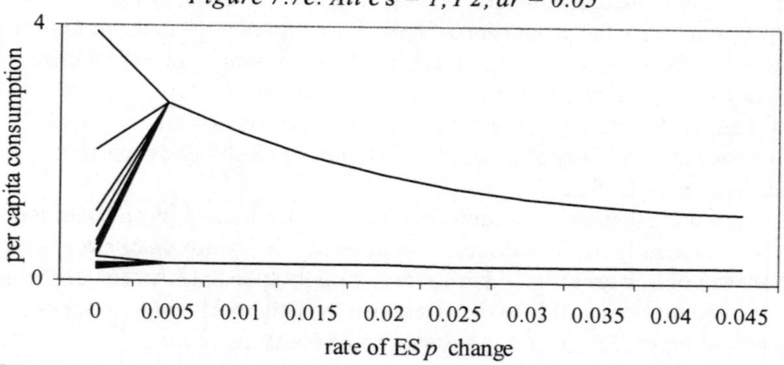

Figure 7.7c: All c s = 1, P2, ar = 0.05

Figure 7.8a: All c s = 1, P2, ar = 0.25

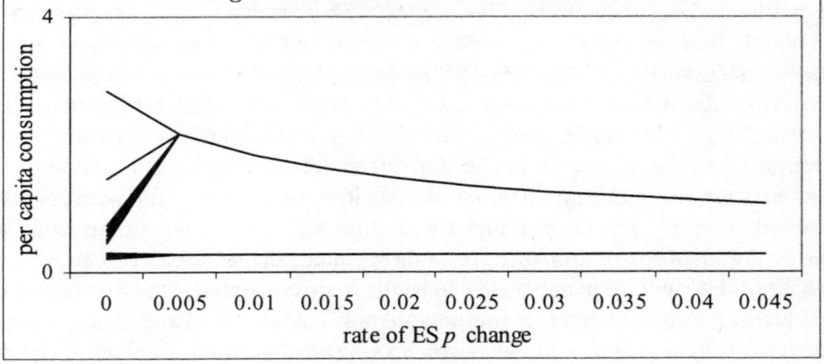

Figure 7.8b: All c s = 1, P2, ar = 0.50

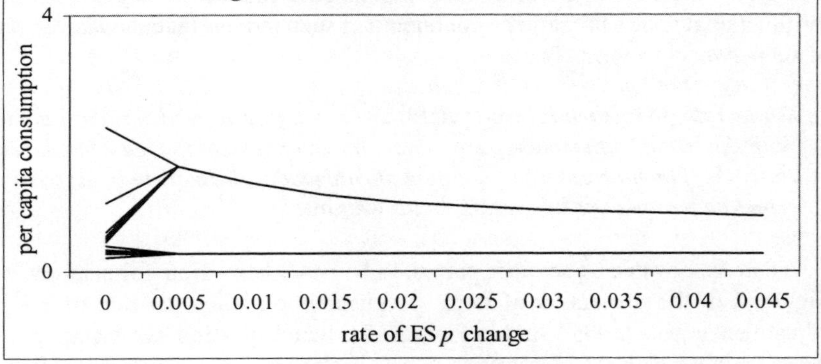

Figure 7.8c: All c s = 1, P2, ar = 0.90

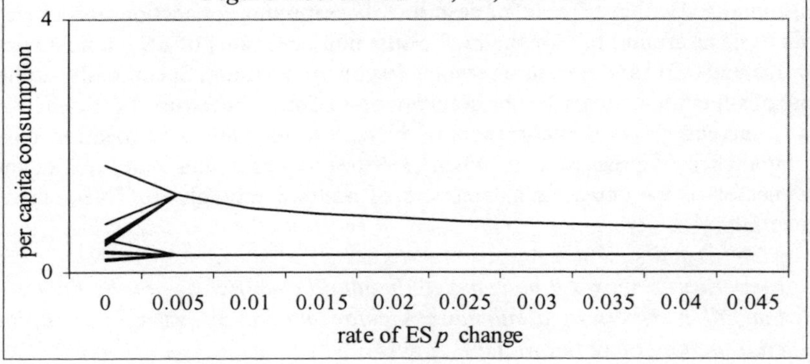

but only that we are less likely to see wild swings in the character of specific individuals. The substantive policy suggestion here is that the choices presented to youth in inner city neighborhoods ought not to be simply two: gang membership versus college. Instead, an attempt should be made to provide a variety of alternative and intermediate peer models, including for instance, the pursuit of a viable trade, sports, and clerical– professional occupations that require less than a college degree. Or, for another example, if we contrast a society with, say, some initial unimodal distribution of time preferences against one with a nearly discrete bimodal distribution, and if the object of our study is the evolution of corruption, then we might speculate that although we cannot say which society will ultimately prove to be the more corrupt, it is the first that will evolve in a more coherent or predictable way. And it is also the first that is most likely to avoid the extreme in which a single subnetwork evolves in which all members of society are equally acceptant of corrupt behavior. This idea, then, adds a quantitative dimension to the notion of role models. Specifically, it is not the mere existence of a role model that can move a society away from dysfunctional action but rather a continuum of such models that allow a variety of subnetworks to form. Thus,

while a single role model might yield the most highly desired social outcome with the 'right' adjustment parameters, it can also yield the least preferred outcome, in which case a continuum of role models is likely to be less risky from the perspective of overall social welfare.

There is, however, one other pattern to be noted here. Note for example in Figure 7.7b the dispersion of rates of consumption when the rate of ES *p* adjustment equals 0.005. And in Figure 7.7a, that dispersion exists also when the same rate equals 0.01. However, in both instances nodes 1 through 10 form a subnetwork. The dispersion, though, is explained by the fact that the subnetwork is not complete – even in equilibrium, various nodes remain disconnected within it. Table 7.2 describes these missing connections for several values of *tl* around 6.0 for the two lowest non-zero rates of ES *p* adjustment (0.005 and 0.01) and reveals a reasonably coherent pattern. Specifically, as the rate of adjustment increases, the completeness of the subnetwork increases. But as *tl* increases – as it becomes more difficult to maintain a connection – its completeness decreases until, when *tl* = 6.9, nodes 1 and 2 are no longer connected to the complete subnetwork of nodes 3 through 10. Thus, stated generally:

subnetworks need not be complete in equilibrium nor need it be the case that all members of a subnetwork enjoy the same level of per capita consumption or social welfare.

Table 7.2: Missing connections in {1, 2, ..., 10} for Figure 7.7a

ES *p* adj =.005	*t1*=5.0	*t1*=5.5	*t1*=5.7	*t1*=6.3	*t1*=6.7	*t1*=6.9
node #1	–	10	9, 10	8–10	7-10	3–10
2	–	–	10	9, 10	9, 10	3–10
3	–	–	–	–	–	1, 2
4	–	–	–	–	–	1, 2
5	–	–	–	–	–	1, 2
6	–	–	–	–	–	1, 2
7	–	–	–	–	1	1, 2
8	–	–	–	1	1	1, 2
9	–	–	1	1, 2	1, 2	1, 2
10	–	1	1, 2	1, 2	1, 2	1, 2
total missing	0	2	6	10	12	32
ES *p* adj = 0.01						
node #1	–	–	–	10	9, 10	3–10
2	–	–	–	–	9	3–10
3	–	–	–	–	–	1, 2
4	–	–	–	–	–	1, 2
5	–	–	–	–	–	1, 2
6	–	–	–	–	–	1, 2
7	–	–	–	–	–	1, 2
8	–	–	–	–	–	1, 2
9	–	–	–	–	1	1, 2
10	–	–	–	1	1, 2	1, 2
total missing	0	0	0	2	6	32

To explore further the internal structure of subnetworks and the idea that a continuum of role models has a salutary effect on overall welfare, we note that if there is a pattern that holds throughout the preceding simulations it is that subnetworks tend to be defined by the gaps in the initial vector of ES *p*. Although we have not yet explored fully the extent to which other initial parameters can dictate how many subnetworks form, it is nevertheless useful to consider yet another modification in the vector of initial ES *ps*; namely one that eliminates these gaps and employs a uniform difference across all numerically adjacent pairs of nodes. Thus, consider the following vector, which varies over the same range of values as P1:

P3 = (.05, .11, .17, .23, .29, .35, .41, .47, .53, .59, .65, .71, .77, .83, .89, .95)

Figures 7.9a–d portray the results of our simulations for a range of values of *tl* (2 to 17) and here we note a number of newly emergent patterns. First, note that Figures 7.9a through 7.9d all look much like Figure 7.2a, which is based on an initial ES *p* vector that contains both twins and gaps and employs a value of *tl* = 2. Thus, even when we make it relatively difficult to sustain connections (e.g., *tl* = 14), the absence of gaps yields a single subnetwork for all but the lowest rates of change in ES *p*. This is not to say, however, that *tl* is irrelevant even for the range of values considered here. When *tl* = 2, the network is complete – all 256 connections exist at strength 1. When *tl* = 7, completeness declines to 130; at *tl* = 9 it becomes 104 and at *tl* = 14, it declines to 74. Indeed, at *tl* = 14, no node is connected to more than four other nodes, and the 'boundary' nodes of 1and 16 are connected to only two other nodes. Nevertheless, despite the incompleteness of connections, this much is clear:

> *subnetworks will be more inclusive,* ceteris paribus*, to the extent that there are no gaps in initial conditions ... in this instance, in the initial propensity of groups or individuals to save versus invest.*

On the other hand, if we make it even more difficult to sustain connections – say by setting *tl* equal to 17 – the system, unsurprisingly, becomes nearly wholly atomistic: nodes 1 through 13 are not connected to anyone, whereas nodes 14 through 16 form a unique subnetwork (and for still greater values of *tl* even this subnetwork disappears).[6] This contrasts sharply to what occurs with gaps: if, instead, we use P2 as our input vector of initial ES *ps*, it isn't until *tl* approaches 30 that the system becomes atomistic. Thus,

---

6 It remains, however, that as the rate of ES *p* change increases, this single multi-node subnetwork increases in size. At 0.0150, nodes 13–16 form a completely connected subnetwork, whereas at 0.030, nodes 12-16 form such a subnetwork.

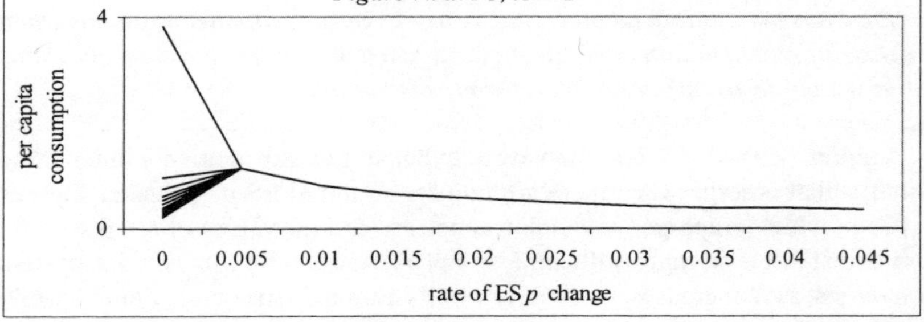

Figure 7.9a: P3, t1 = 2

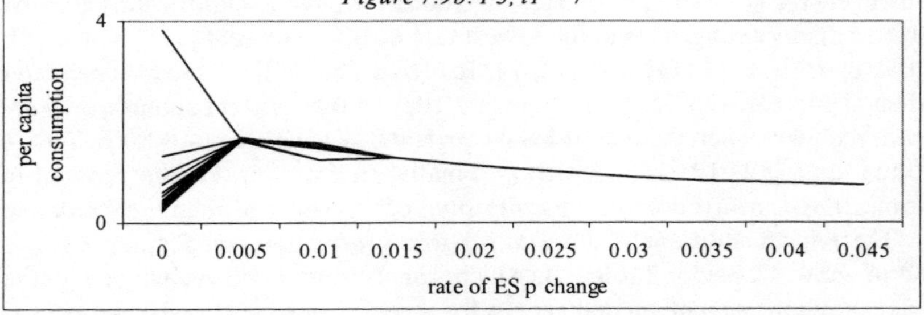

Figure 7.9b: P3, t1 = 7

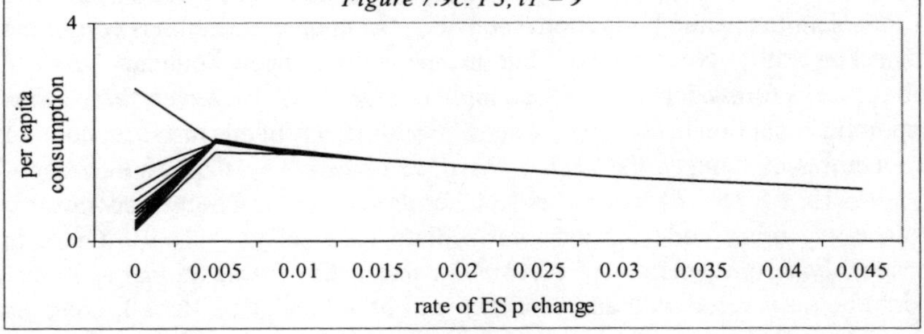

Figure 7.9c: P3, t1 = 9

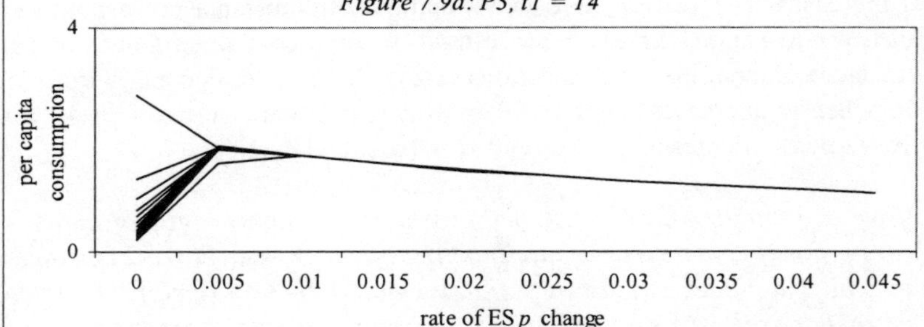

Figure 7.9d: P3, t1 = 14

while gaps in the initial configuration of ES ps encourage clustering when it is easy to maintain connections, gaps discourage atomistic systems when it is difficult to maintain connections, at least with respect to configurations in which ES ps are initially uniformly distributed.

Figures 7.9a–d do not, however, exhaust the substantively interesting patterns that emerge when there are no gaps in initial ES *p*. Consider Figures 7.10a-b, which graph per capita consumption for four values of *t1* all slightly less than 17. These figures illustrate the process whereby our 16 node system converges, as *t1* increases, to a nearly wholly atomistic structure. For *t1* = 16.7, Figure 7.10a looks much like what we observe when *t1* =14 (see Figure 7.9d). However, for *t1* = 16.8, a number of groups 'fall away' significantly to form several distinct subgroups at the lowest rate of ES *p* change ({1, 2, 3, 4}, {5}, {6, 7, 8, 9, 10, 11}, {12} and {13, 14, 15, 16}). This 'falling away' accelerates when *t1* increases to 16.9 (see Figure 7.10c) so that distinct subnetworks are observed even when the rate of ES *p* equals 0.01 ({1, 2, 3}, {4, 5}, {6, 7, 8, 9, 10, 11}and (12, 13, 14, 15, 16}). Finally, in for *t1* = 17, the process of atomization is nearly complete for all values of ES *p* rates of change considered.

There is an additional feature of subgroup formation that Figures 7.10a–d fail to show. Consider Table 7.3, which, for the rate of ES *p* change equal to 0.01, shows the subnetworks that exist for several values of *t1* in the range 16.8-17.0 (subnetwork boundaries are given by the dark solid lines while the numbers in cells denote the number of nodes to which the node in question is connected other than itself). Note that with but one exception, once a boundary between subnetworks forms, it persists for all higher values of *t1*. However, there is one important exception to this rule – when *t1* equals19.91. In this instance, node 11 is a member of subnetwork {6, 7, 8, 9, 10, 11} when *t1* = 16.9, is a member of {11, 12, 13, 14, 15, 16} when *t1* is 16.92 or greater, but is wholly disconnected from every other node for the intermediate value of *t1* = 16.91. One can reasonably assume, moreover, that similar instances of 'transitional isolation' would be discovered with an even finer grid of values of *t1*. Recall, now, the one instance in Figure 7.6 whereby the number of connections fails to decrease as *t1* increases. The analysis here shows that this circumstance corresponds to the case where a node crosses over to join a different preexisting subnetwork (in this instance, the number of connections across all 16 nodes decreases from 56 to 46 when *t1* increases from 16.90 to 16.91, but then increases to 48 for *t1*=16.92 before decreasing to 44 with *t1* = 16.93). Thus,

if we can imagine a circumstance in which the maintenance of network connections becomes incrementally more difficult, we can observe instances in which a nodes stands in isolation until either it remains isolated thereafter as the difficulty of maintaining connections increases further or stands in isolation only until it joins some other subnetwork.

Figure 7.10a: t1 = 16.7, P3

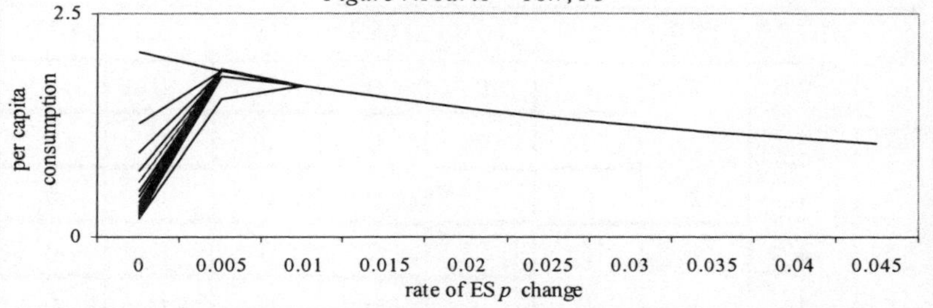

Figure 7.10b: t1 = 16.8, P3

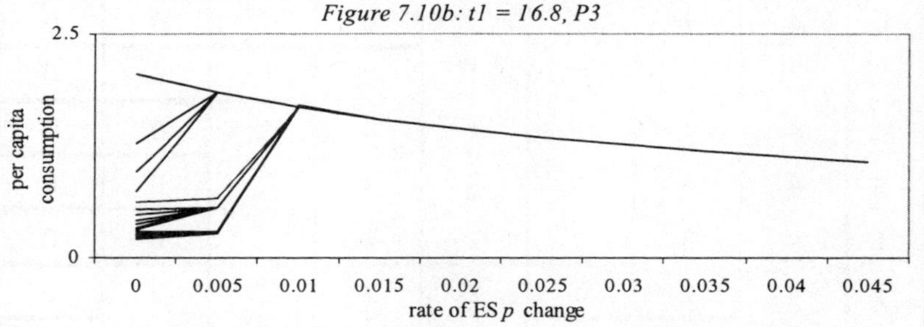

Figure 7.10c: t1 = 16.9, P3

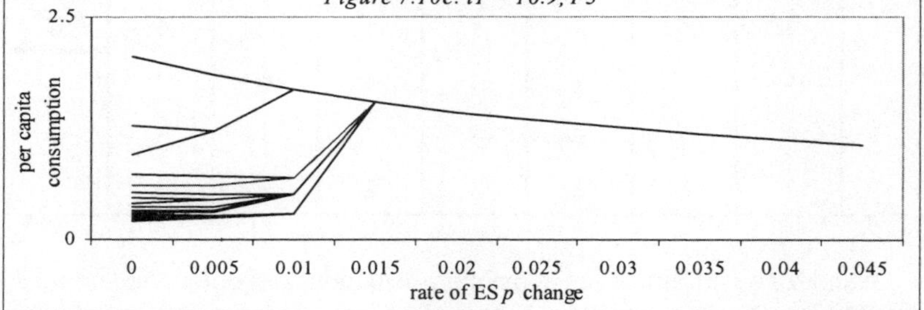

Figure 7.10d: t1 = 17, P3

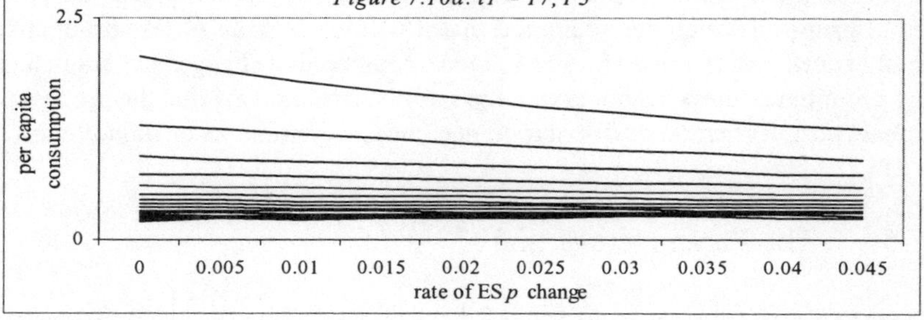

Table 7.3: Subnetworks and connections for values of t1

	16.8	16.9	16.9	16.9	16.9	16.9	16.9	17.0
1	2	2	0	0	0	0	0	0
2	3	2	0	0	0	0	0	0
3	4	2	0	0	0	0	0	0
4	4	1	1	0	0	0	0	0
5	4	1	1	0	0	0	0	0
6	4	2	2	2	0	0	0	0
7	4	3	3	3	2	0	0	0
8	4	4	3	4	3	2	0	0
9	4	4	3	3	3	2	0	0
1	4	3	2	2	2	2	0	0
1	4	2	0	2	2	2	2	0
1	4	2	2	3	3	3	3	0
1	4	3	3	4	4	4	4	0
1	4	4	4	4	4	4	4	2
1	4	3	3	3	3	3	3	2
1	2	2	2	2	2	2	2	2

Admittedly, instances of temporary isolation are not common in our simulations. Their occurrence, nevertheless, suggests a social phenomenon wherein the isolation of a group from society is an otherwise unstable equilibrium in which the likelihood that it will join a more or less productive subnetwork is determined by small, even imperceptible changes in parameters. At a minimum, these instances of temporary isolation reveal that the process of subnetwork formation or dissolution need not be continuous or monotonic.

7.5 The Parameters *t1* and *t2*

The preceding analysis includes one other critical parameter – the initial matrix of connections. For example, rather than use an initial connections input matrix that looks as follows:

$$C = \begin{bmatrix} 11111.... \\ 11111.... \\ 11111.... \\ .... \end{bmatrix}$$

we might employ one that looks thus:

$$C' = \begin{bmatrix} 101010... \\ 010101... \\ 101010... \\ .... \end{bmatrix}$$

However, by way of imparting some understanding of the internal operation of our model and the impact of pairing nodes via the choice of an initial vector of ES *p*s, we note that such an initial and symmetric matrix should occasion no change in our results thus far whenever nodes are otherwise initially paired in terms of their initial ES *p*s as they are with P1. One node impacts another in two ways. First, it can influence another's ES *p* by exhibiting a greater level of per capita consumption. In this instance a node conducts a series of pair-wise comparisons between its rate and the rate of each other node to which it is connected. Suppose, when all *c*s = 1that node *j*'s consumption is at least as great as that of all other nodes. Then we would expect node *i* to move its ES *p* in the direction of *j*s ES *p* value. Now suppose we disconnect twins by using the connection matrix *C'*. Although node *i* and *j* are no longer connected, *i* is still connected to $j - 1$ and $j + 1$, and one of these two nodes will be equivalent in all respects to *j*. Hence, *i*'s incentive to change its ES *p* is unchanged. The second impact a node can have on another is in the calculation of the proportion of total consumption that it attributes to high and low *d*s. But splitting the population of nodes in half via the substitution of *C'* for *C* can have no effect in that per capita consumption cannot be effected by whether each node is paired with an identical twin.

This discussion, though, presumes, as in all previous simulations, that *t2* = 0, in which case, with *C'*, no odd-numbered node can impact any even-numbered one or vice versa – connections between them cannot develop. But surely the more general and substantively reasonable case is when connections can develop even if they are initially nonexistent. In effect, letting *t2* = 0 is equivalent to assuming that there are technological as opposed to purely social barriers to the development of network connections, which largely defeats the purpose of this study. Hence, to explore the issues that arise when *t2* > 0, we proceed with P3 as the input vector of initial ES *p* values, and setting *t1* not altogether arbitrarily equal to 16.7, Figures 7.11a–c graph per capita spending for four values of *t2* – 0.5, 0.01, 0.00125 and 0.001 (in fact, setting *t1* = 16.7 is not altogether arbitrary as it corresponds to a value such that the characteristics of the system are sensitive to variations in *t2* over a reasonably narrow range). These figures appear to reveal the obvious; namely, that as *t2* decreases – as it becomes more difficult to establish a connection where there is none initially – the atomization of the system increases as measured by the number of subnetworks. At *t2* = 0.5, only one subnetwork prevails for essentially any positive rate of ES *p* adjustment; when *t2* = 0.01, the rate of ES *p* adjustment needs to equal 0.01 or greater before the subnetwork {15,16} joins the remaining nodes; when *t2* = 0.00125, the system quickly converges to three wholly connected subnetworks ({1, 2, 3, 4, 5}, {6, 7, 8, 9}, {10, 11, 12, 13, 14, 15, 16}), and when *t2* is reduced still further to 0.001, the system becomes wholly atomistic with no node connected to any other node.

It might seem that this is the end of the story and that we have merely reproduced what we found in the preceding section: the easier it is to establish a connection (or, as in the preceding section, to maintain one), the more likely we are to see that connection in equilibrium. This monotonicity, though, need not be a property of our simulation and network structures can in fact change as a function of *t1* and *t2* in more complex ways. To see what we mean, note first that although a single subnetwork prevails when *t2* equals either 0.5 or 0.01, the character of these subnetworks differ somewhat. Whereas with *t2* = 0.5 per capita consumption continues to decline as the rate of ES *p* change increases, consumption rates are a constant when *t2* = 0.01 (or greater). Thus,

> *if connections are relatively easy to establish, the influence of nodes that discount the future greatly and save little on other nodes increases as the rate of change in ES p increases; however, if connections are more difficult to establish but not so difficult so as to preclude them altogether, nodes that value the future more highly and save more succeed in bringing all other nodes up to a higher level of per capita consumption that is invariant with the rate of ES p adjustment.*

To some extent this result might seem counter-intuitive. Although we might otherwise suppose that social welfare is maximized when network connections are relatively easy to form, here we see the opposite. The reason is straight-forward. If high investment nodes are isolated for a time so that they are allowed to invest, they will of necessity achieve a relatively high level of per capita consumption. But if at the same time the difficulty of forming connections is not so great as to exclude them altogether, then ultimately those nodes that achieve a high level of per capita consumption through investment will impact those that initially saved little. Overall social welfare and homogeneity are maximized for intermediate values of *t2*. That is,

Figure 7.11a: C', P3, t1 = 16.7, t2 = 0.50

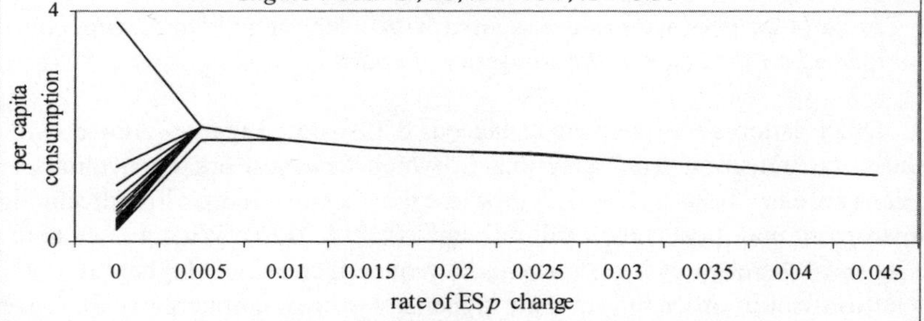

Figure 7.11b: C', P3, t1 = 16.7, t2 = 0.01

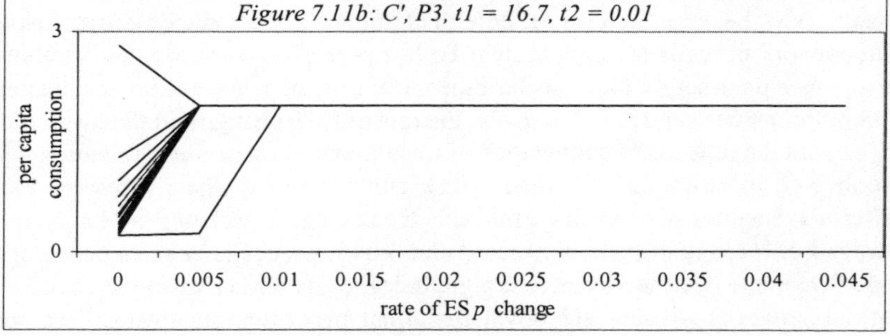

Figure 7.11c: C', P3, t1 = 16.7, t2 = 0.00125

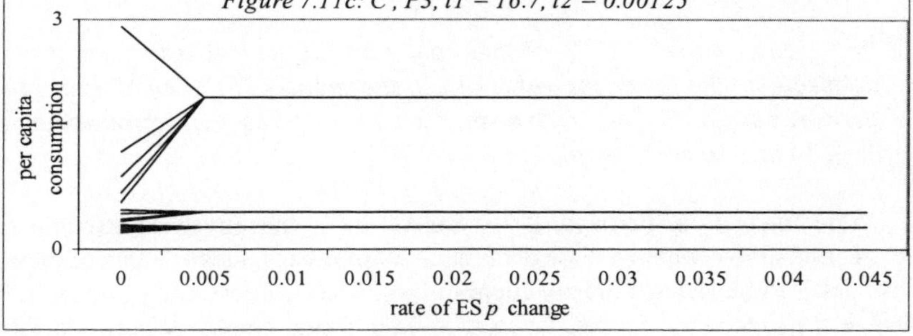

Figure 7.12d: C', P3, t1 = 16.7, t2 = 0.001

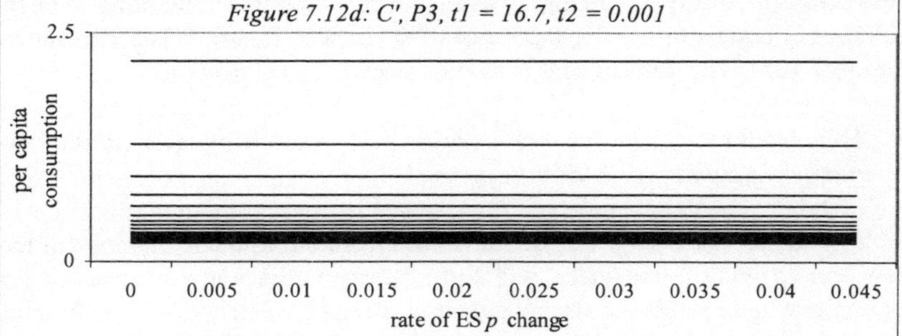

at least for the range of parameter values considered, it is as if there is a socially optimal value for the ease with which new connections can be formed on the basis of the similarity of nodes.

Other patterns emerge once *t2* exceeds 0. Consider Figures 7.12a–d, which show the transition from a system in which new network connections are relatively easy to form (*t2* = 0.1) to where they become more difficult and high investment nodes exert their full influence (*t2* = 0.0125). What we see here is a process whereby low investment nodes require increasingly higher rates of ES *p* adjustment in order to join their higher investment counterparts. However, once some critical value of *t2* is passed (0.0125 in this instance), high investment nodes become, in a sense, dominant and a fully connected all-inclusive subnetwork prevails for lower rates. To this point, however, we have ignored two other parameters that can be especially critical when initial connection strengths are zero and *t2* = 0; namely, the rate of ES *p* change and the threshold, S, at which a note takes cognizance of the actions of other notes to which it is connected at strengths less than 1.0. Figures 7.13a–d, then, consider four alternative values of *t2* when we multiply the rate of ES *p* change by 10 (hence, instead of looking at rates between 0 and 0.045, we consider rates between 0 and 0.45), and what we see here is a great deal more regularity in the character of subnetworks. Absent are any transitional processes and instead we see immediate convergence to subnetworks that follow a predictable pattern:

For greater rates of ES p change, the more difficult it is to form a new connection (the lower the value of t2), the greater is the number of subnetworks in equilibrium (with respect to Figures 7.13a–d, the progression is 1, 4, 14 and 16 subnetworks).

More interesting, perhaps, is the impact of *S*, the threshold strength of connection above which a node takes account of other nodes. To this point we have set S somewhat arbitrarily equal to 0.6. What is interesting, however, is that in the preceding figures, when *t2* = 0.0012, the boundary between subnetworks is precisely 0.6. Figures 7.14a–d, then; compare simulations when the value of S equals 0.6, 0.59, 0.55 and 0.50 (hence, Figure 7.14a reproduces Figure 7.13c). The pattern here is again somewhat as expected:

As S declines – as nodes more readily take account of other nodes – the number of subnetworks decline as well.

Of course, when we lower the threshold from 0.6 to 0.59, it is logical that new and stronger connections will form between adjacent subnetworks. For example, when *S* = 0.6, the strength of connection between nodes 1 and 2 equals precisely 0.6, and when the threshold *S* is lowered to 0.59, that strength

Figure 7.12a: C', P3, t1 = 16.7, t2 = 0.1

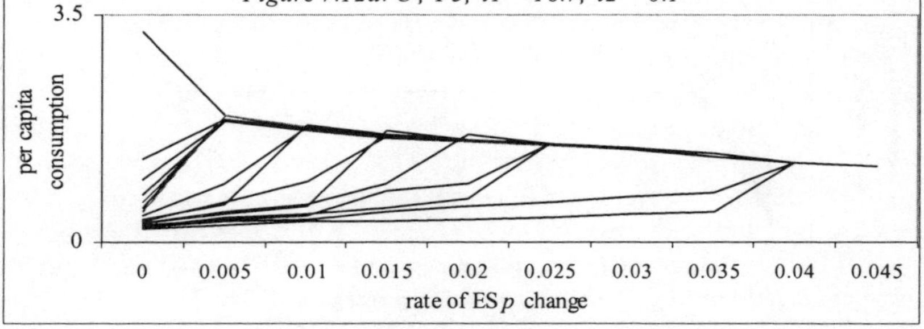

Figure 7.12b: C', P3, t1 = 16.7, t2 = 0.05

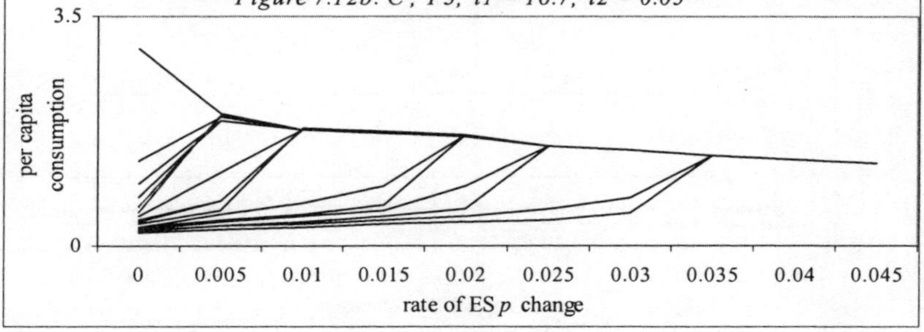

Figure 7.12c: C', P3, t1 = 16.7, t2 = 0.025

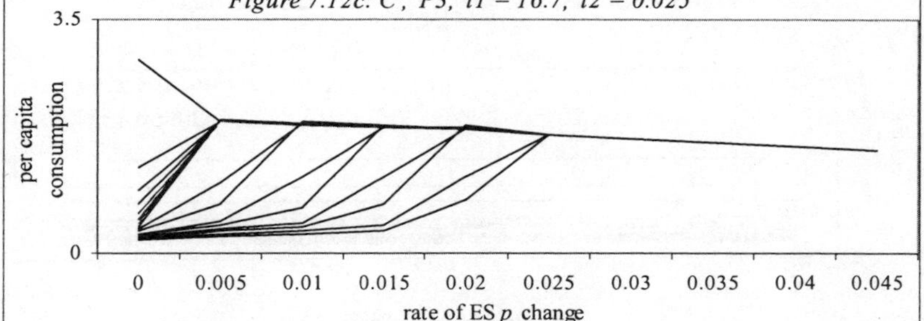

Figure 7.12d: C', P3, t1 = 16.7, t2 = 0.0125

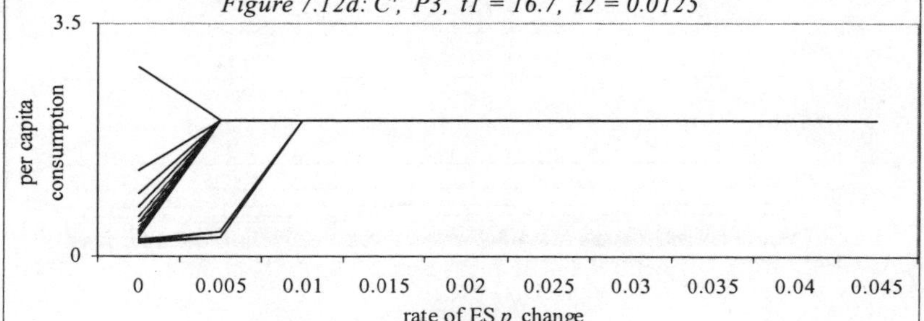

Figure 7.13a: S = 0.6, P3, C', t1 = 16.7, t2 = 0.002

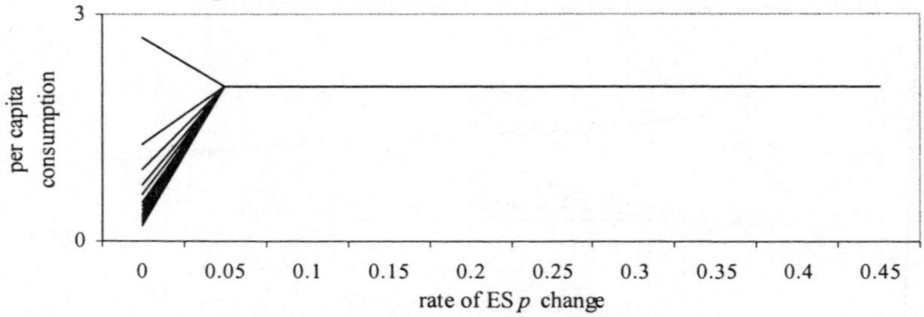

Figure 7.13b: S = 0.6, P3, C', t1 = 16.7, t2 = 0.00121

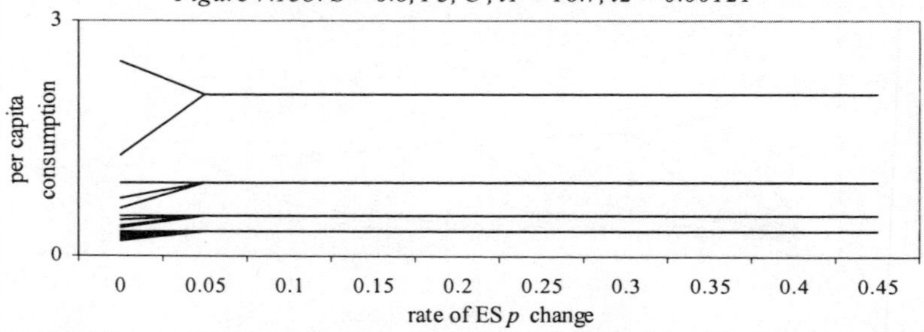

Figure 7.13c: S = 0.6, P3, C', t1 = 16.7, t2 = 0.0012

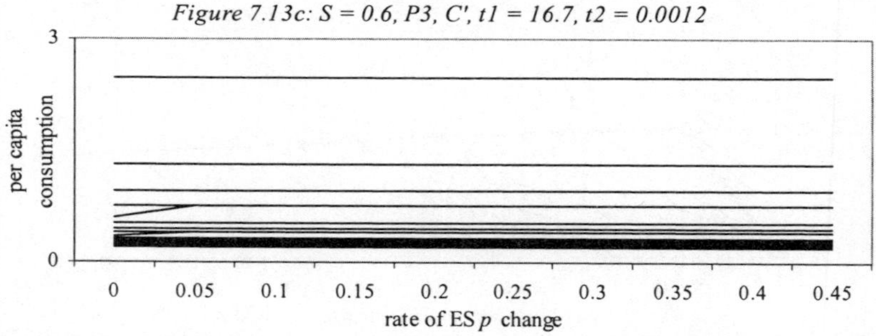

Figure 7.13d: S = 0.6, P3, C', t1 = 16.7, t2 = 0.001

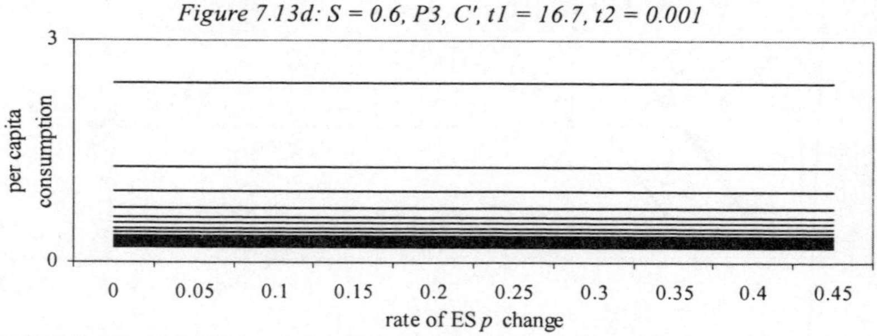

becomes 1.0 in equilibrium. But 2's initial strength of connection to node 3 is also 0.6. However, rather than join 3, 2 joins 1, whereas node 3 forms a three-node subnetwork with nodes 4 and 5. At this point, the strengths of connection between subnetworks does not exceed 0.001. Nevertheless, if we lower S further, to 0.55, subnetworks continue to consolidate (to two), and consolidate further still when S is lowered to 0.50. Thus,

> *connection strengths between subnetworks need not be at or near threshold values for subnetworks to consolidate when thresholds are decreased or to dissolve when thresholds are increased*

Surely this result seems counter-intuitive. Nevertheless it has a logical explanation. Specifically, we need to bear in mind that this discussion here concerns outcomes in equilibrium. However, suppose two nodes have some connection at or near the beginning of an equilibration process – say 0.56. With $S = 0.60$ those two nodes do not take into account each others rates of consumption in adjusting their respective ES *p*s and proportions of types, in which case, *ceteris paribus*, they are likely to 'drift apart' and form distinct subgroups. On the other hand, if $S = 0.55$, they each pay full heed to the other and are thereby more likely to converge join in the same subnetwork and arrive at equivalent levels of consumption in equilibrium.

Necessarily, the influence of *S* is determined by the functional forms within our model. Nevertheless, the preceding pattern holds social implications. Suppose we take *S* as a measure of the mutual visibility of nodes. The suggestion, then, is that even if a system appears wholly disjoint in terms of connections across subnetworks, small changes in visibility can have profound consequences for both social welfare and the subnetworks that evolve. We appreciate that the discrete 'triggering' character of our model (i.e., when S = 0.60, a node pays no attention to another in which its strength of connection is 0.59, but pays full heed to one if that strength equals 0.61) is unlikely to match reality. The model, though, does alert us to the possibility that the 'technologies of visibility' (e.g., exposure through the mass media) can have profound effects and we know that even models with a more continuous structure can yield discrete processes. A node, for example, that bears the strength of connection X to adjacent groups might move in the direction of one set of nodes to form a subnetwork whereas if that initial strength is $X + \epsilon$, it might move in a wholly different direction so as to yield a distinctly different subnetwork structure.

7.6 An Atomistic Initial State

The analysis in Section 7.4 begins at one extreme – where all nodes are initially connected at maximum strength and in the section that follows we consider an

Figure 7.14a: C', P3, t1=16.7, t2=.0012, S = 0.60

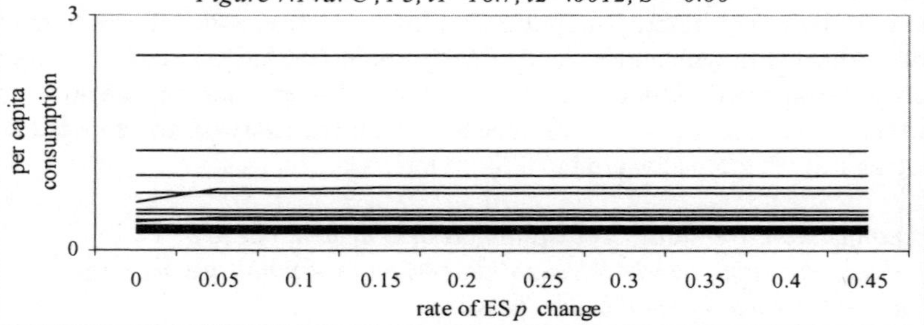

Figure 7.14b: C', P3, t1 = 16.7, t2 = 0.0012, S = 0.59

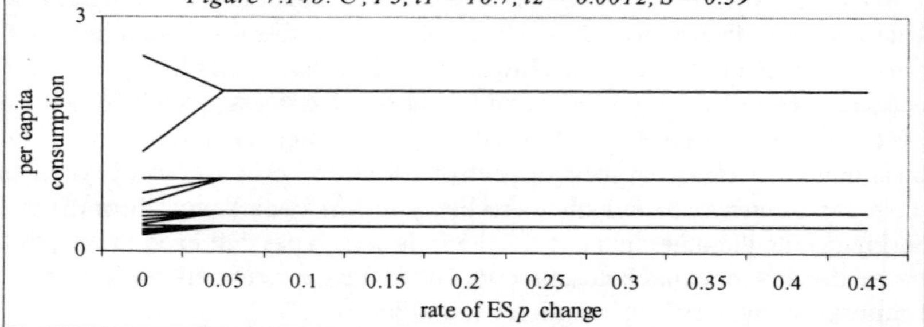

Figure 7.14c: C', P3, t1 = 16.7, t2 = 0.0012, S = 0.55

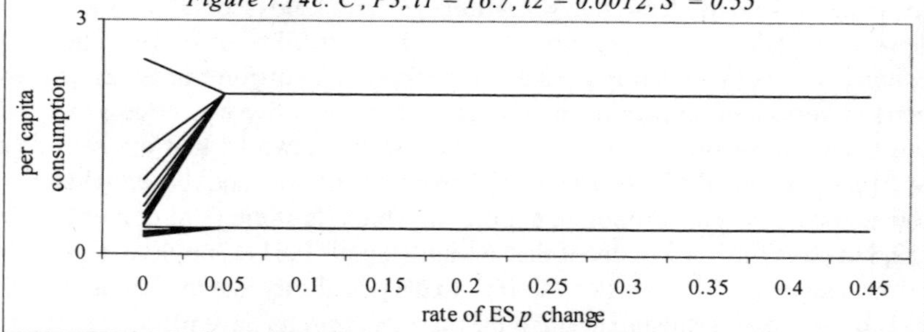

Figure 7.14d: C', P3, t1 = 16.7, t2 = 0.0012, S = 0.50

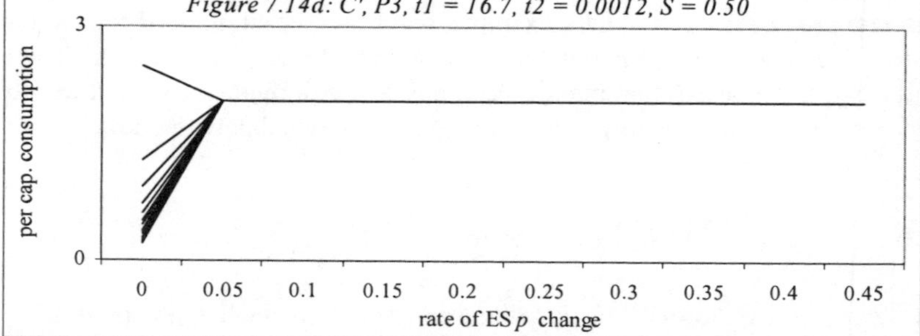

intermediate case. Here we want to consider the opposite extreme of where no node is connected initially to any other node (i.e., where the 16 × 16 input matrix of connections is simply the identity matrix **I**). And although we know that *t2* must be some number greater than zero for any alternative network structure to develop, we can begin by considering the range of values for *t1* that we consider in section 7.4 along with the vector P3 of initial ES *ps*. However, before proceeding, a bit of algebra is in order. Note that if $c_{old} = 0$ as our initial condition, we can rewrite expression (7.1) as

$$c_{new} = t2/t1\delta p$$

where $\delta p = |p_i - p_j|$. Moreover, if $\delta p = 0.06$ initially and if $S = 0.6$, then the initial condition for two nodes to develop a connection is ,

$$t2/t1\delta p > 0.6 \Rightarrow t2/t1 > 0.036.$$

Surely this is an easy condition to satisfy, provided only that *t2* is large enough and *t1* small enough. We cannot, however, assume that a connection will not develop if this inequality is reversed. Specifically, note that expression (7.1) defines a recursive equation. Letting $c(j)$ denote the strength of connection in iteration or period *j, if* this strength is 0 initially ($j = 0$), then in period 1 it equals

$$c(1) = t2/t1\delta p.$$

Letting $K = t1\delta p$, then in period 2, if δp is unchanged because of the disconnect between nodes, we have

$$c(2) = (t2 + t2/K)/K = (Kt2 + t2)/K^2$$

in period 3,

$$c(3) = [t2 + (Kt2 + t2)/K^2]/K = (K^2t2 + Kt2 + t2)/K^3$$

and in period *n*,

$$c(n) = t2(K^{n-1} + K^{n-2} + ... + K + 1)/K^n$$

which can be rewritten as

$$c(n) = t2/K + t2/K^2 + t2/K^3 + ... + t2/K^n$$

Note, now that if $K < 1$, the most relevant term (in terms of magnitude) is the last, whereas if $K > 1$, the most relevant is the first. Hence, if $K > 1$, the value of $t2$ is critical for determining whether a connection will ultimately develop that exceeds the threshold S (0.6 throughout this analysis thus far with the exception of Figures 7.14a-d). But if $K < 1$, we need only consider a sufficiently large enough number of iterations before $c(n)$ achieves this threshold. And in fact if, as is the case with P3, $\delta p = 0.06$, then $K = 1$ if $t1 = 16.667$. Thus, it is no accident that in describing the sensitivity of per capita consumption and subnetwork structures to variations in $t2$, we focus in Figures 7.12 through 7.14 on the case when $t1 = 16.7$. It is here that we are likely to find the eventual equilibrium most sensitive to other parameter values, especially $t2$. We emphasize, however, that none of this implies that disjoint subnetworks cannot develop for values of $K < 1$. Even if $t1\delta p$ is initially less than 1, it is possible that ES *p*s evolve so that δp grows to reverse this inequality (which establishes that if $t1 < 1$, a single wholly connected subnetwork prevails ultimately in our model since in this case K can never equal or exceed 1). But clearly, such an occurrence is more difficult than when $K > 1$.

Of course, this discussion is largely irrelevant when all initial connections are set equal to 1.0 as in Section 7.4.[7] And it is only partially relevant to the preceding section, which focuses on the case of two disjoint and complete subnetworks. However, in this section, with $C = \mathbf{I}$, it is critical to understanding or interpreting the sensitivity of our evolutionary model to parameter values.

With these considerations in mind, let us turn now to our simulations when we once again use P3 = (0.05, .11, .17, .23, .29, .35, .41, .47, .53, .59, .65, .71, .77, .83, .89, .95) as our initial vector of ES p values. First, for purposes of comparison, let $t1 = 16.7$ and consider Figures 7.15a-c, which graph per capita consumption for three values of $t2$, 0.1, 0.05 and 0.0125. The important thing to note here is that these three figures are in fact identical to Figures 7.12a, b and d. Thus, the initial disconnect between odd and even number nodes in the previous section is of no consequence, at least when compared to the case where all nodes are initially disconnected. A word of caution is in order here, however, before we infer that it matters little whether we begin with C' or $C = \mathbf{I}$. There are, in fact, subtle differences in per capita consumption rates. It is simply that these differences are not large enough to impact subnetwork structures for the parameter values considered. There is, though, no formal reason for concluding that C' and $\mathbf{I}$ are equivalent starting points. Figure 7.16, for example, illustrates

---

[7] If $t2 = 0$ and if we set $c(0) = 1$, then $c(n) = 1/K^n$, provided that K is constant. It might seem, then, that we have modeled a wholly predetermined erosion of connection strength or its maintenance at 1, depending on whether K is less than or greater than 1. However, the assumption of constant K is untenable in this case since if a connection initially exists between nodes, each will impact the others ES p and K will change. It is the dynamic of the resulting complex process of change and its interaction with other parameters and nodes that the simulations in Section 7.4 explore.

Figure 7.15a: C′ = **I**, *P3, t1 = 16.7, t2 = 0.1*

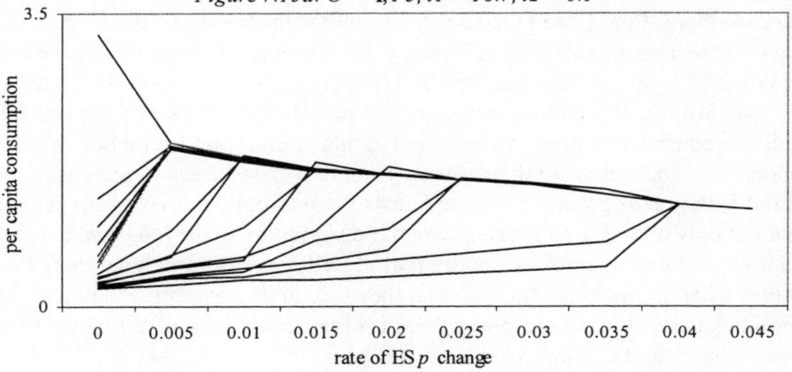

Figure 7.15b: C′ = **I**, *P3, t1 = 16.7, t2 = 0.05*

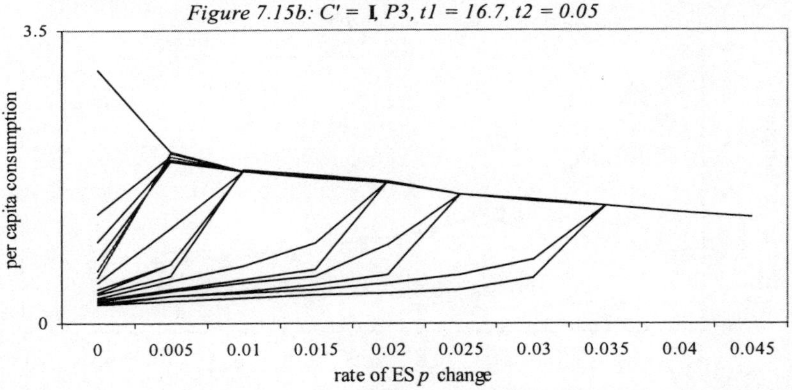

Figure 7.15c: C′= **I**, *P3, t1 = 16.7, t2 = 0.0125*

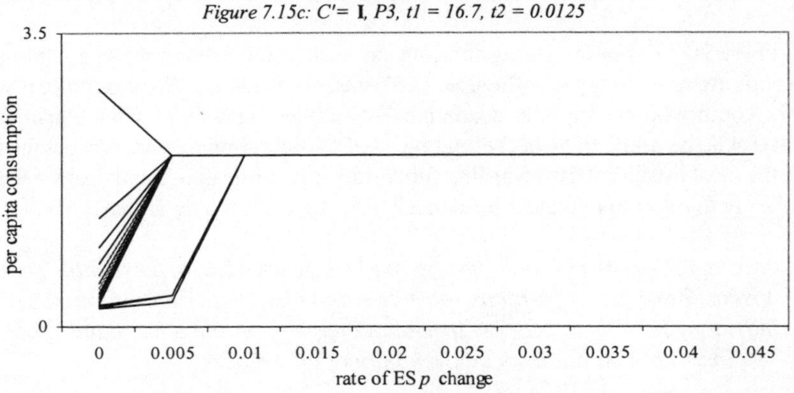

the difference in per capita consumption that arises with *C′* versus **I**. It is also the case that *C′* and **I** can give rise to different subnetworks (e.g., if we choose the same parameter values as in Figure 7.16, then when the rate of ES *p* change equals 0.020, we get the subnetworks {1, ..., 12}, {13, ..., 16} with *C′* and {1, ..., 13}, {14, ..., 16} with **I**). Admittedly, these differences are slight, but they follow a consistent pattern; namely, per capita consumption is higher with **I** as against C'. The reason for this difference, moreover, seems straightforward: the initial isolation of groups that save allows them to build a level of investment that not only allows for a greater level of consumption in the long run, but their relative performance induces nodes that initially save little to raise their ES *p* values when mimicking others so that they too, in the long run, enjoy a higher equilibrium level of consumption. We observed this feature of our model when comparing *C* and *C'*, and we see it again here.

Figure 7.16: C = I versus C', P3, t1=11, t2=0.10

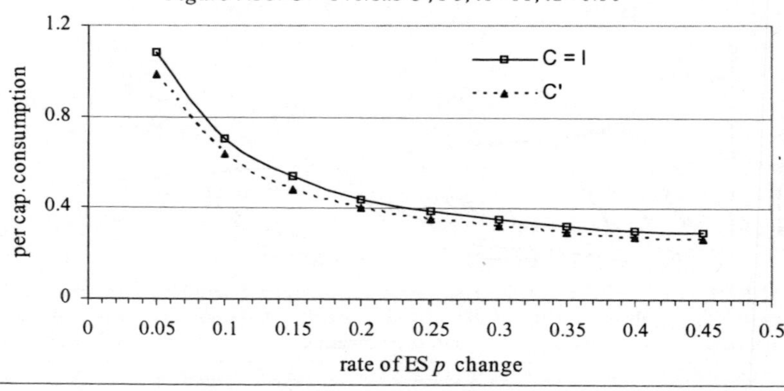

There is, of course, a longstanding issue in sociology as to when 'good' groups are most likely to influence 'bad' ones or vice versa. For example, if we mix youth who are initially academically inclined with those whose primary interest is the immediacy of 'belonging' – of participating in whatever seems to be the most immediately rewarding subculture, including gangs and drugs – who is likely to exert the greatest influence? The suggestion here is that

> *mixing of 'good' and 'bad' groups ought to be avoided, at least until 'good' groups have the opportunity to realize and display the long-term relative advantages of their actions. Immediate mixing, on the other hand, lowers the long-term equilibrium consumption of all groups.*

Earlier, in Section 7.4, we argued that overall social welfare is maximized when groups are wholly isolated in equilibrium. Advocating such an outcome as a goal of policy, on the other hand, may be untenable. In the event that it is – in the event that eventual homogeneity is the objective, then the best policy is to allow the most potentially profitable or efficient sectors of society the 'independence' or 'distance' to establish themselves first.

Before concluding this assessment of the performance of our model when $C = \boldsymbol{I}$, we want to consider one additional set of simulations – those that pertain to letting *t1* take a relatively extreme value, 20, so that K, at least initially, exceeds 1 and *t2* becomes especially critical. Briefly, with *t1* = 20, if *t2* $\geq$ 0.45, the system converges to a single wholly connected subnetwork regardless of the rate of change of ES *p* (and again, with S = .6 and ar = 0.025). On the other hand, if we let *t2* $\leq$ 0.11, we sustain the initial atomistic structure of **I**. But now, consider the four graphs in Figures 7.17a–d. Clearly, the impression here is that of incoherence – per capita consumption fails to abide by any monotonic relationship with rates of ES *p* adjustment and nodes seem to shift almost randomly among subgroups. However, there is more of a pattern here than initially meets the eye. First,

> *a good share of the apparent incoherence in Figures 7.17a-d is a product of the fact that even in equilibrium, nodes that are members of the same subgroup need not enjoy the same level of per capita consumption.*

For example, although it might seem in Figure 7.17c that there are at least three and possibly four subnetworks once the rate of ES *p* change equals or exceeds 0.02, there are in fact only two subnetworks ({1, ..., 13},{14, ..., 16} and then {1, 2}, {3, ..., 16}).

There are, though, additional sources of the apparent incoherence here: subnetwork structures are not a monotonic function of either the rate of ES *p* change or of *t2*. First, Table 7.4 describes the subnetworks that form for each rate of ES *p* change between 0.005 and 0.045 versus nine values of *t2* (with *t1* = 20). For example, then, with *t2*=0.18 and the rate of ES *p* change equal to 0.010, we have four subnetworks: {1, ..., 9}, {10, 11}, {12, 13}, {14, ..., 16}. The first thing to note now is that although there is a tendency for the number of subnetworks to increase as the rate of ES *p* change increases, that tendency is not a universal rule: It is violated in several instance when *t2* = 0.30, 0.25, 0.20 and 0.125. The same is true with respect to the relationship between the number of subnetworks and *t2*: although there is a general tendency for that number to increase as *t2* increases, there are at least two instances where monotonicity does not hold (when *t2* = 0.19 and the rate of change is 0.010 and when *t2* = .18 and that rate equals 0.015). Finally, the shaded cells in Table 7.4 denote points of 'subnetwork structure' discontinuity. Recall in Table 7.3 the single instance of what we labeled a discontinuity – of where a boundary between subnetworks disappeared and a new one appeared as *t1* increased. The discontinuities to which we refer in Table 7.4 are of a similar sort, except that

Figure 7.17a: $C = \mathbf{I}$, $t1 = 20$, *P3*, $t2 = 0.25$

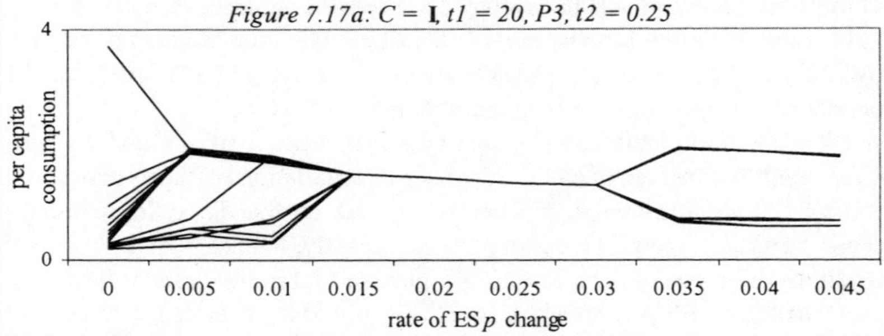

Figure 7.17b: $C = \mathbf{I}$, $t1 = 20$, *P3*, $t2 = 0.20$

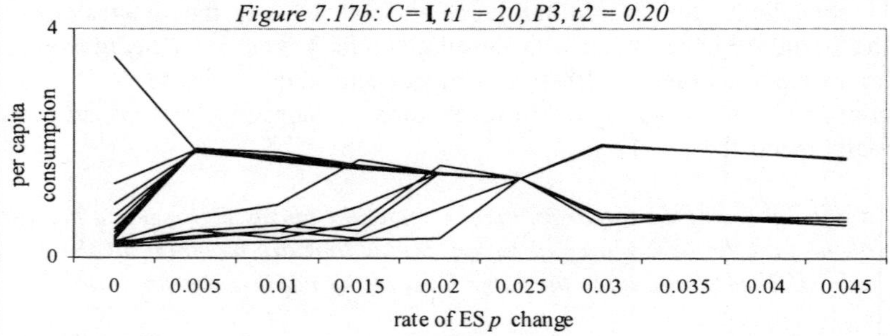

Figure 7.17c: $C = \mathbf{I}$, $t1 = 20$, *P3*, $t2 = 0.175$

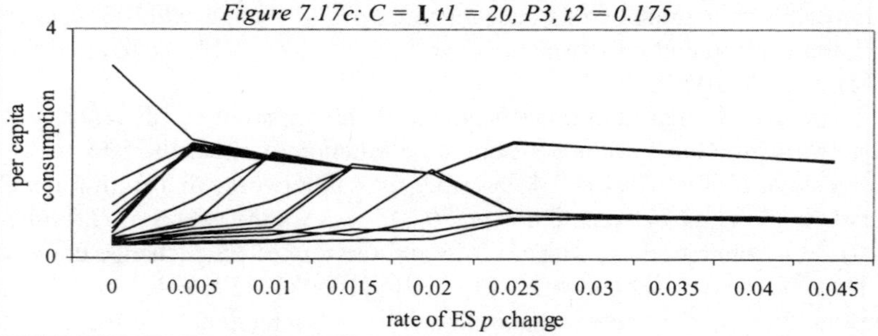

Figure 7.17d: $C = \mathbf{I}$, $t1 = 20$, *P3*, $t2 = 0.125$

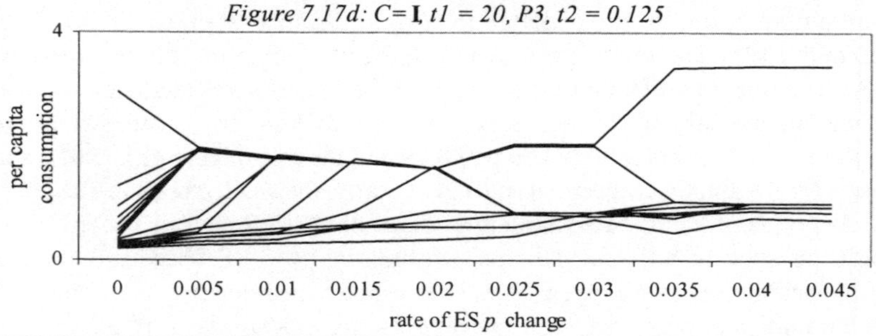

now we are considering changes in *t2*. For example, when the rate of ES *p* adjustment equals 0.010 and we decrease *t2* from 0.20 to 0.19, the boundaries between {1, ..., 10} and {11, 12}, between {11, 12} and {13, 14} and between {13, 14} and {15,16} disappear since node 11 splits off from {11, 12} to join {1, ..., 10}, 12 joins 13, and 14 splits off from {13, 14} to join {15, 16}.

Table 7.4: Subnetworks vs t2 and rates of ES p change, P3, t1 = 20

	.30	.25	.20	.19	.18	.175	.15	.125
.005	1-11 12-16	1-10 11-12 13-16	1-10 11-12 13-16	1-9 10-11 12-13 14-16	1-7 8-9 10-11 12-16	1-7 8-9 10-11 12-13 14-16	1-7 8-9 10-11 12-13 14-16	1-8 9-10 11-12 13-14 15-16
.010	1-16	1-11 12-13 14-16	1-10 11-12 13-14 15-16	1-11 12-13 14-16	1-9 10-11 12-13 14-16	1-9 10-11 12-13 14-16	1-9 10-11 12-13 14-16	1-10 11-12 13-14 15-16
.015	1-16	1-16	1-11 12-13 14-16	1-11 12-13 14-16	1-13 14-16	1-12 13-14 15-16	1-12 13-14 15-16	1-11 12-13 14-16
.020	1-16	1-16	1-14 15-16	1-13 14-16	1-13 14-16	1-13 14-16	1-12 13-14 15-16	1-11 12-13 14-16
.025	1-16	1-16	1-16	1-14 15-16	1-2 3-16	1-2 3-16	1-2 3-14 15-16	1-2 3-14 15-16
.030	1-16	1-16	1-2 3-16	1-2 3-16	1-2 3-16	1-2 3-16	1-2 3-16	1-2 3-16
.035	1-16	1-2 3-16	1-2 3-16	1-2 3-16	1-2 3-16	1-2 3-16	1-2 3-16	1 2-14 15-16
.040	1-2 3-16	1-2 3-16	1-2 3-16	1-2 3-16	1-2 3-16	1-2 3-16	1-2 3-16	1 2-16
.045	1-2 3-16	1-2 3-16	1-2 3-16	1-2 3-16	1-2 3-16	1-2 3-16	1-2 3-16	1 2-16

What warrants emphasis here is that these seemingly peculiar 'patterns' arise largely because we are considering small rates of ES *p* adjustment – rates that allow for the complex interplay of the variety of parameters that are endogenously determined (i.e., strengths of connection, the nodes that each node considers when making any adjustment in its ES *p*, the ES *p*s themselves, consumption and investment rates, and the proportion of high and low *d* types). If, on the other hand, we consider higher rates of change, then as Figures 7.18a–d show, a more coherent and wholly anticipated pattern prevails. Indeed, as even these figures show, whatever 'peculiarities' appear in the relationship between subnetwork structure at rate of ES *p* adjustment occurs only at the lower rates.

7.7 Asymmetric Connections

There are, of course, a nearly endless list of combinations and permutations of initial connections that might be explored using our simulation mode. But one additional possibility in particular warrants attention - that in which initial connections are asymmetric. Asymmetry in the current context means that if node 1 bears a strong connection to 2 but 2 bears none to 1, then node 1 can take account of node 2's consumption when adjusting its ES *p*, whereas 2 (initially) pays no heed to 1. And of the many asymmetric configurations we might consider, one in particular – the hierarchy – seems especially important. Here, then, we consider a 15-node network and the general structure portrayed in Figure 7.19, where the notation 'x —> y' means that, initially, node y is connected to x (i.e., x can influence y) , but y is not connected to x (i.e., $c_{xy} = 0$ but $c_{yx} = 1$).

Figure 7.19: Hierarchical Structure of the Network

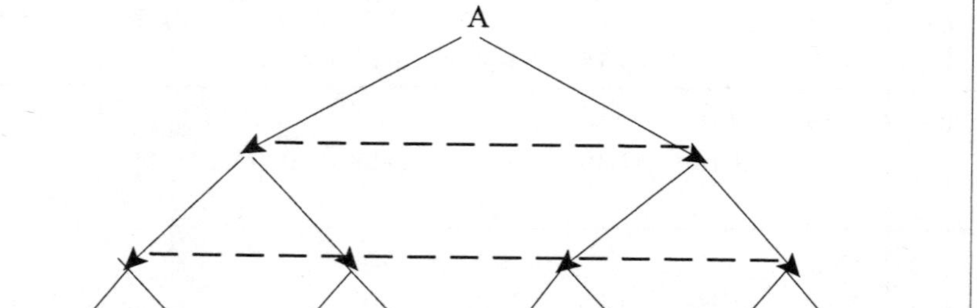

Figure 7.18a: C = **I***, P3, t1 = 20, t2 = 0.1*

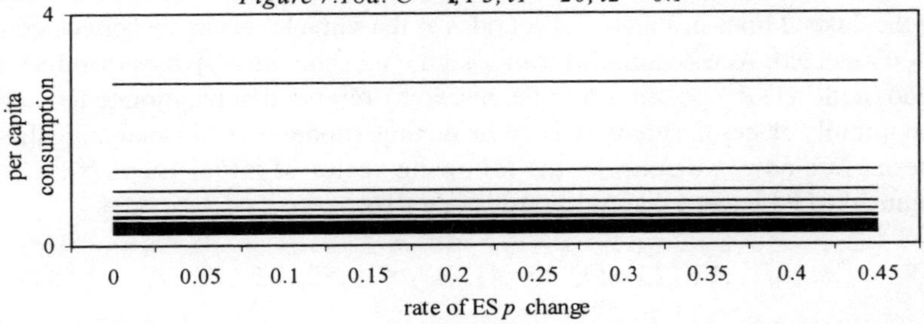

Figure 7.18b: C = **I***, P3, t1 = 20, t2 = 0.35*

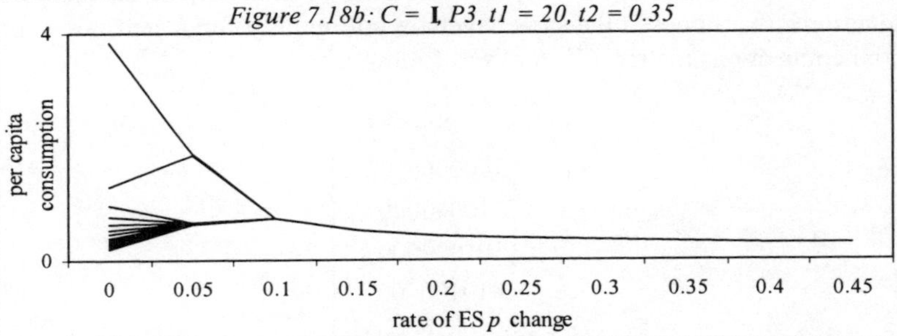

Figure 7.18c: C = **I***, P3, t1 = 20, t2 = 0.40*

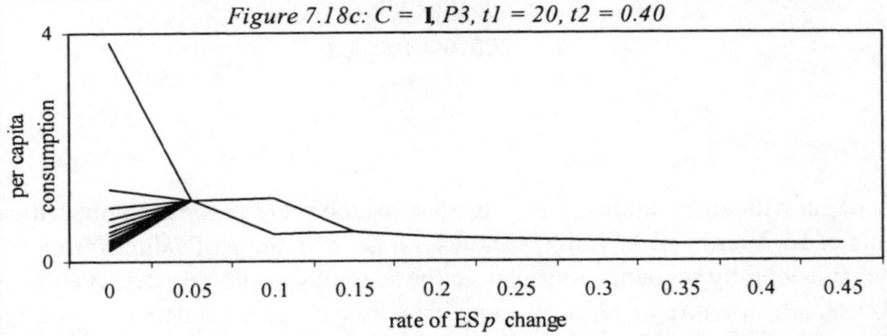

Figure 7.18d: C = **I***, P3, t1 = 20, t2 = 0.50*

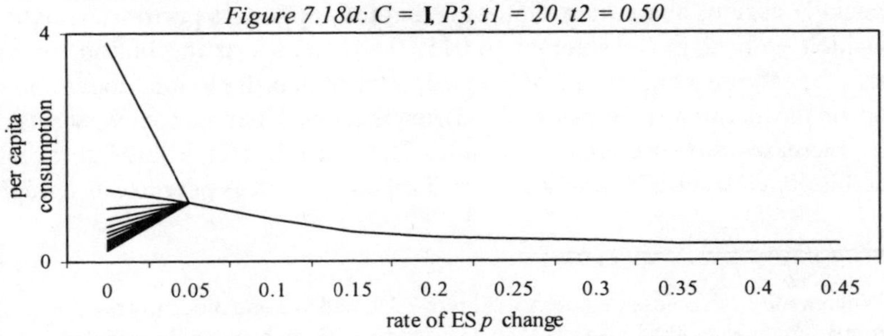

Even a hierarchical structure allows for several variations. First, as illustrated by the dashed lines in Figure 7.19, nodes at the same level can be connected or not connected. And second, we can assign either the initially most productive node (node #1) the position A in the hierarchy or give that position to the node that initially places the greatest discount on time (node #15).[8] In analyzing these alternatives now, we consider the following vector of initial ES *ps*, which is identical to P3 except that it discards node #16:

$$P4 = (.05, .11, .17, .23, .29, .35, .41, .47, .53, .59, .65, .71, .77, .83, .89)$$

Our first series of simulations, now, set node #1 at the top of the hierarchy and assume that nodes at the same level are not initially connected. Thus, our initial connections matrix, *C1*, looks as follows:

$$C1 = \begin{bmatrix} 100000000..... \\ 110000000..... \\ 101000000..... \\ 010100000..... \\ 010010000..... \\ 001001000..... \\ 001000100..... \\ 000100010..... \\ .... \end{bmatrix}$$

We begin with eight figures, 7.20a–h, all of which set *t1* at the high but critical value of 16.7, and which, with *C1* and P4, consider a range of values for *t2*. We offer this lengthy sequence to illustrate the seemingly complex dependence of subnetwork structure on *t2* as we move from two distinct subnetworks (Figure 7.20a) to a single all-encompassing sub-network (Figure 7.20h). What is especially curious about these figures is that they appear to portray a situation in which with *t2* in the interval [0.015, 0.04], nodes at the bottom of the hierarchy require a higher rate of ES *p* adjustment in order to join nodes above them on the hierarchy (Figures 7.20a–d), but that this dependence reverses itself as *t2* increases further thereafter (Figures 7.20e–h). In fact, Table 7.5 reveals that this dependence of network substructure on *t2* is not as perverse as it might

---

[8] When node 1 occupies position A in Figure 7.19, nodes 2 and 3 occupy the positions immediately beneath it, nodes 4, 5, 6 and 7 the positions beneath them, and nodes 8 through 15 the positions at the lowest level of the hierarchy.

Figure 7.20a: C1, P4, t1 = 16.7, t2 = 0.015

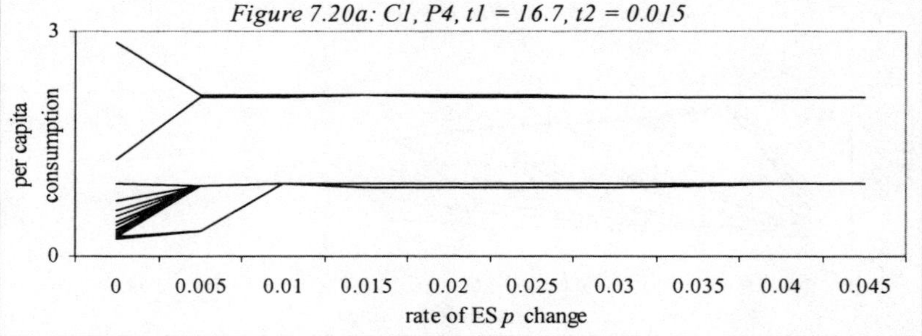

Figure 7.20b: C1, P4, t1 = 16.7, t2 = 0.02

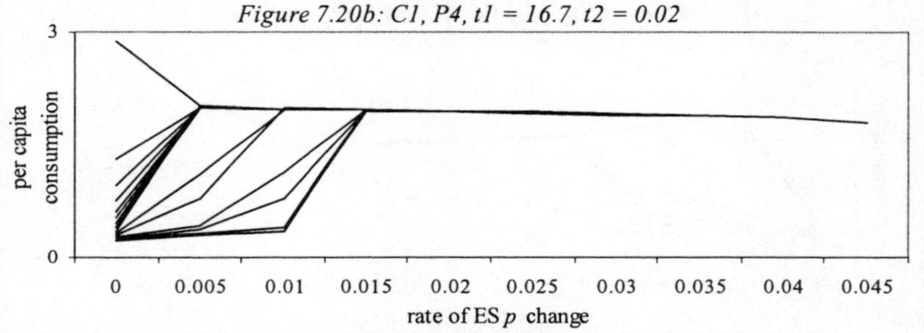

Figure 7.20c: C1, P4, t1 = 16.7, t2 = 0.03

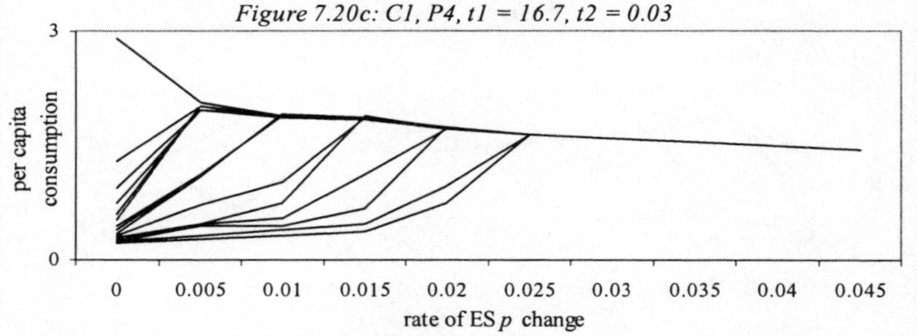

Figure 7.20d: C1, P4, t1 = 16.7, t2 = 0.04

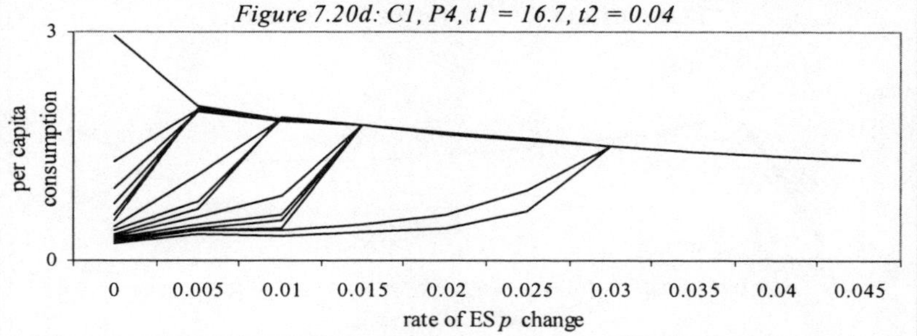

Figure 7.20e: C1, P4, t1 = 16.7, t2 = 0.05

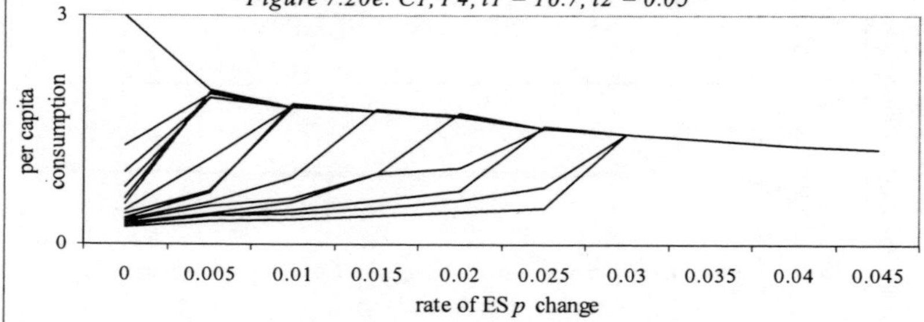

Figure 7.20f: C1, P4, t1 = 16.7, t2 = 0.10

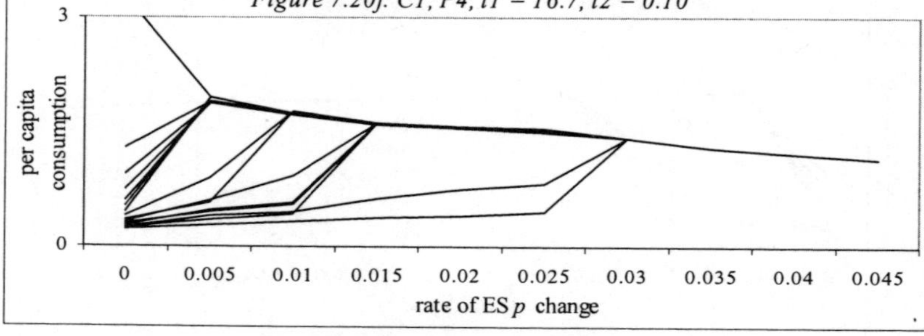

Figure 7.20g: C1, P4, t1 = 16.7, t2 = 0.14

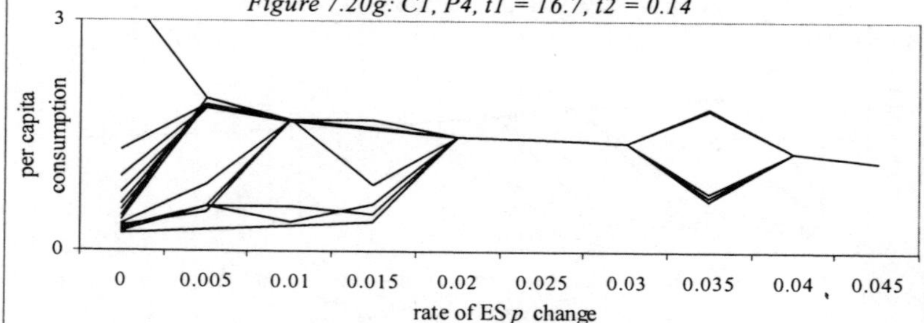

Figure 7.20h: C1, P4, t1 = 16.7, t2 = 0.20

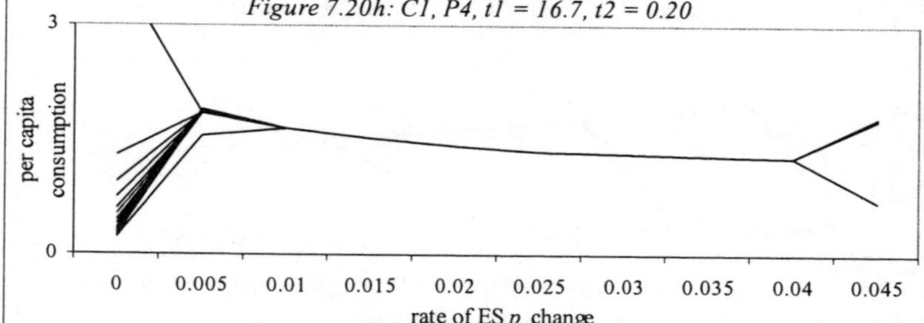

appear in Figures 7.20a-h. Specifically, note that for values of *t2* in the interval [0.12,0.02], it is the nodes at the bottom of the hierarchy (11 through 16) that form distinct subnetworks and are ultimately isolated from those at the top who form a single subnetwork, whereas once *t2* drops below 0.02, it is the two top nodes that are separate from the rest (the solid line corresponds to rate of ES *p* change = 0.005, the dashed line to 0.010 and the dark heavy line to 0.015). Moreover, the subgroup fracturing for the lowest nodes in the hierarchy are, in a sense, merely a precursor to the development, at higher values of *t2*, of a single wholly connected subnetwork. Any apparent incoherence in Figures 7.20a -h is largely a function of the fact that at low rates of ES *p* adjustment, the mix of simultaneously determined and endogenously changing parameters can interact to yield seemingly discontinuous adjustments at the boundaries of subnetworks.

Table 7.5: Subnetwork boundaries versus t2 with P4, t1=16.7, C1

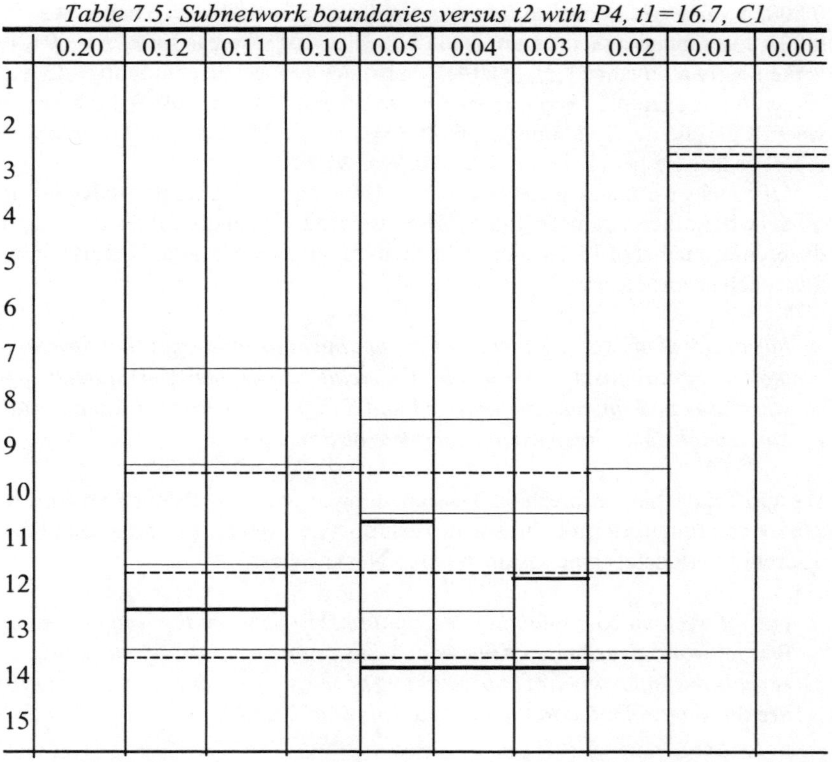

Table 7.5, however, does alert us to an interesting and important question. We know, of course, that if *t2* = 0, new connections cannot form between nodes *i* and *j* unless ES p_i = ES p_j. In other words, the system should become wholly

atomistic at some point as *t2* drops below 0.01. Thus, the question is what does this progression look like since the opposite seems to happen for the range of values of *t2* considered in Table 7.5 – there in fact seems to be a consolidation of subnetworks. The answer is both interesting and somewhat unanticipated. First, we note that at $t2 = 0.0001$ the system does in fact become nearly wholly atomistic – only nodes 1 and 2 form a subnetwork. All other nodes are isolated from each other. However, adjacent nodes do have non-zero strengths of connection. Specifically, nodes 3 through 15 have a strength of connection to their numerically adjacent node of 0.05 (with the exception of node 3, as its connection to 2 equals 0.0001). Thus, these nodes fail to pay heed to each other simply because we have set the threshold for doing such at 0.60. But now, as we increase t2 from 0.0001 to 0.001, these auxiliary strengths increase from 0.05 to 0.50 at $t2 = 0.0009999$. Thus, for *S* – the strength of connection above with a node 'pays attention to another' – greater or equal to approximately 0.500, a value of *t2* equal to 0.001 is a point of discontinuity. That is, if $t2 \geq 0.001$, the subnetwork structure is (1, 2), (3, ..., 15) whereas for $t2 < 0.001$, it is the nearly atomistic (1,2), (3), (4),..., (15). Moreover, discontinuities do not end with this example. Suppose we set S = .50 and $t2 = 0.0009999$. In this case we still get the nearly atomistic (1, 2), (3), (4),..., (15) system. But with $S = 0.499$ or less we get (1, 2), (3, 4, 5), (6, 7, 8, 9), (10, ..., 15).

Of course, we cannot guarantee that such discontinuities are pervasive in our model when other parameter values are considered, especially values of *t1* much different from that of 16.7 which is the focus of our analysis here. Nevertheless, this much seems clear:

> *hierarchical networks, at least for the endogenous processes that form our model, reveal discontinuities in the relationship between subnetwork structures and the parameters of t2 and S that we do not find when initial connections are either wholly complete or atomistic.*

We appreciate that the threshold S is something of an artificial construct. Surely a more continuous or probabilistic process operates in reality as opposed to the discrete 'thermostat' type assumed here. Nevertheless,

> *even if we somehow smoothed the relationship between the weight a node in a network gives to other nodes in adjusting its ES p value, our analysis reveals the importance of this weighting process when connection strengths are no longer limited to the discrete values of 0 and 1.*

Thus, to reiterate a point made earlier, models of networks that assume discrete connection strengths are necessarily incomplete.

However, before we attempt any additional general conclusions, we should first look at a hierarchy in which the initially least productive node (#15) stands at the top. In this instance our input matrix of initial connections is as follows:

$$C2 = \begin{bmatrix} 100000001000000 \\ 010000001000000 \\ 001000000100000 \\ 000100000100000 \\ 000010000010000 \\ 000001000010000 \\ 000000100001000 \\ 000000010001000 \\ 000000001000100 \\ 000000000100100 \\ 000000000010010 \\ 000000000001010 \\ 000000000000101 \\ 000000000000011 \\ 000000000000001 \end{bmatrix}$$

Our first set of simulations, shown in Figures 7.21a–d, replicate four of the parameter configurations portrayed in Figures 7.20a–h when node 1 stood at the top of the hierarchy. And at first glance it might seem that substituting the connections matrix *C2* for *C1* makes little difference – indeed, Figures 7.20h and 7.21d are identical. Thus,

> *if new connections are relatively easily made, it matters little who is at the top and who is at the bottom of a network hierarchy.*

Nevertheless, there are important and wholly expected differences in the remaining figures once we lower *t2* so that making new connections becomes more difficult. Consider, for instance, Figures 7.20a and 7.21a. In both cases the eventual equilibrium consists of two subnetworks except for the smallest rate of ES *p* change. But those two subnetwork structures differ drastically and the result is not entirely as expected. When node 1 stands at the top of the hierarchy (Figure 7.20a), the subnetwork structure is {1, 2},{3, ..., 15} whereas when node 15 stands at the top, that substructure becomes {1, ..., 13},{14, 15}. Moreover, since each node in the subnetwork {1, 2} realizes the same level of per capita consumption as does the subnetwork {1, ..., 13}, overall social welfare is increased when the least productive node stands highest in the hierarchy. This implication for overall social welfare, however, only holds if *t2* is sufficiently small. Note that if we compare Figures 7.21b and c to Figures

Figure 7.21a: C2, P4, t1 = 16.7, t2 = 0.015

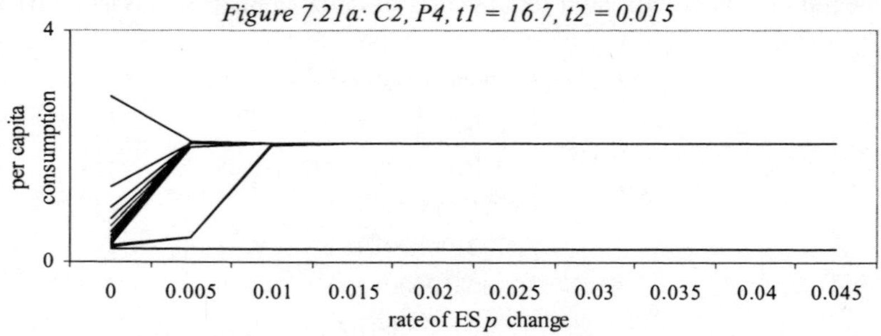

Figure 7.21b: C2, P4, t1 = 16.7, t2 = 0.05

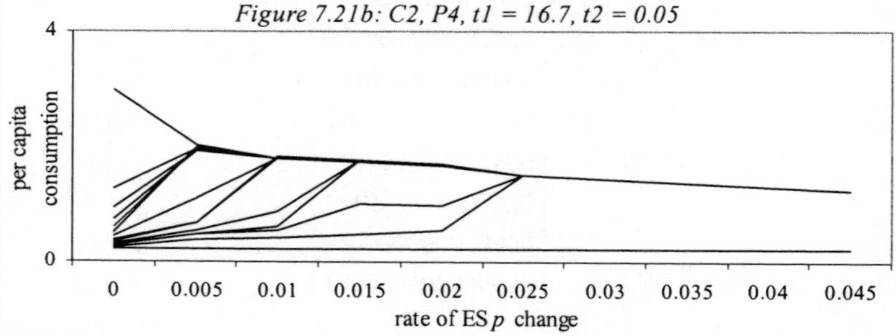

Figure 7.21c: C2, P4, t1 = 16.7, t2 = 0.10

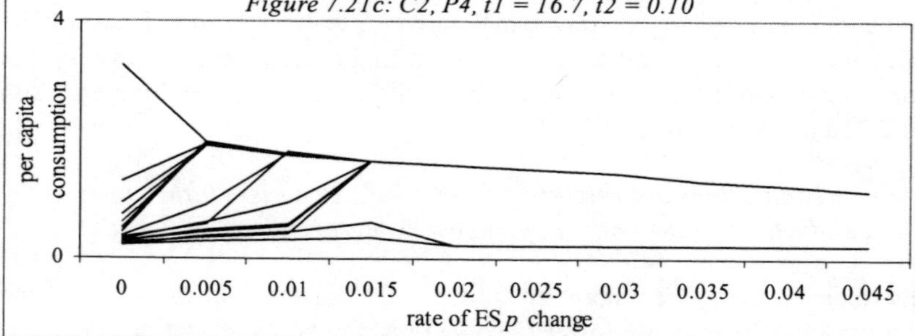

Figure 7.21d: C2, P4, t1 = 16.7, t2 = 0.20

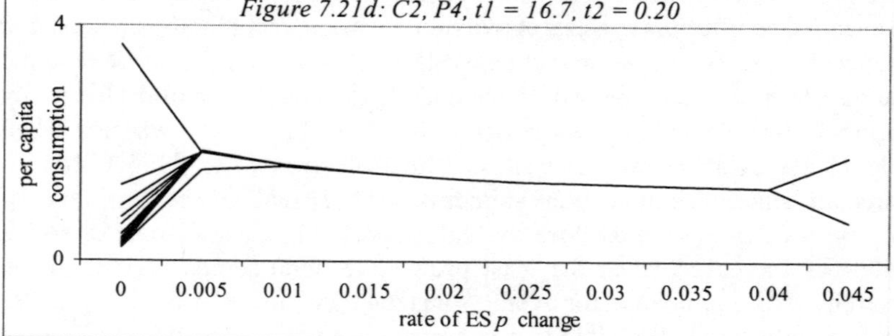

7.20e and f, placing node15 at the top of the hierarchy (*C2*) leaves nodes 15 and 14 isolated from the rest when the rate of ES *p* change is sufficiently great whereas when node 1 is at the top, a single subnetwork forms at this rate. And since the per capita consumption of this single network is the same as when nodes 14 and 15 are disconnected from the rest (Figures 7.21b and c), social welfare is now greatest when node 1 is at the top of the hierarchy. In general, then,

> *the social welfare implications of hierarchical networks in terms of who stands at the top of the hierarchy depends on the ease with which new connections can be made.*

It needs to be appreciated, of course, that any implications here for overall social welfare depend on whether the initially least productive nodes are isolated in equilibrium from the rest, whether they are compelled by parameter values to join the rest in a single subnetwork, or whether they can compel others to join them.

Table 7.6 offers a further comparison on our two hierarchical network types in terms of subnetwork structure (and here we consider rates of ES *p* adjustment in the range 0.05 to 0.45, in which case the subnetwork structure that prevails is generally the same for all rates considered)[9]. For 6 values of *t2* between 6 and 20 and for three values of *t2* between 0.01 and 0.05, the first subnetwork structure in each cell corresponds to what prevails when *C2* is the input connection matrix whereas the second entry describes what occurs when *C1* is the initial matrix. Discerning a pattern here is difficult, but largely because (in conformity with our formal analysis of expression (7.1) in Section 7.6) what occurs for values of *t1* less than 17 differs from what we observe when *t1* exceeds 16. If *t1* is 16 or less, then setting node 1 at the top of the hierarchy yields a single all-encompassing and wholly connected subnetwork; but if *t1* is greater than 17, then the system atomizes into fifteen wholly disconnected nodes. If, on the other hand, we set node 15 at the top of the hierarchy, we experience up to three subnetworks when $t1 \leq 16$, whereas for $t1 \geq 17$ we see the atomization of the system but not to the same extent as when C1 is the input connection matrix: even if nodes 1 through 13 are wholly disconnected, nodes 14 and 15 form a subnetwork.

Perhaps more informative is Table 7.7, which reports the per capita consumption summed across all 15 nodes and which adds the case of $t2 = 0$ so that connections can only be broken in the hierarchy but not formed. In addition,

---

9 All data for this table as well as Tables 7.7 and 7.8 that follow are taken from the case in which the rate of ES *p* adjustment is set at the maximum value we consider in this text, 0.45. It is generally the case, however, that what occurs at lower rates in the range [0.05, 0.45] is equivalent.

Table 7.6: Subnetworks for C2 and C1, P4

t1	*t2* = 0.01	0.025	0.05
2	(1,...,15) (1,...,15)	(1,...,15) (1,...,15)	(1,...,15) (1,...,15)
6	(1,...,7), (8,9), (10,...,15) (1,...,15)	(1,...,7), (8,...,15) (1,...,15)	(1,...,7), (8,...,15) (1,...,15)
9	(1,...,9),(10,11),(12,...,15) (1,...,15)	(1,...,9),(10,11),(12,...,15) (1,...,15)	(1,...,9),(10,11),(12,...,15) (1,...,15)
14	(1,...,11), (12,13),(14,15) (1,...,15)	(1,...,11), (12), (13,...,15) (1,...,15)	(1,...,11), (12), (13,...,15) (1,...,15)
16	(1,...,13), (14,15) (1,...,15)	(1,...,13), (14,15) (1,...,15)	(1,...,13), (14,15) (1,...,15)
17	{*i*}, (14,15) (1,2), {i}	(1,2), (3,...,13), (14,15) (1,2), (3,...,15)	(1,...,13), (14,15) (1,...,15)
20	{*i*}, (14,15) {*i*}	{*i*}, (14,15) {*i*}	{*i*}, (14,15) {*i*}

Table 7.7: Cumulative per capita consumption for C2 and C1, P4

t1	*t2* = 0	0.01	0.025	0.05
2	3.38 / 30.75	3.45 / 3.45	3.45 / 3.45	3.45 / 3.45
6	8.39 / 12.18	5.68 / 3.90	5.49 / 3.75	5.05 / 3.75
9	8.82 / 9.77	5.29 / 4.65	5.20 / 4.65	5.08 / 4.50
14	8.99 / 9.77	7.43 / 5.85	5.65 / 5.40	5.29 / 4.95
16	8.99 / 9.77	9.02 / 9.90	6.42 / 6.15	5.51 / 5.40
17	8.99 / 9.01	9.41 / 9.90	13.14 / 15.48	6.16 / 5.85
20	8.99 / 9.01	9.06 / 9.08	9.17 / 9.19	9.35 / 9.38

Table 7.7 shades those cells in which social welfare is greatest when node 1 stands at the top of the hierarchy as opposed to node 15. The pattern here is evident when we look across these two tables. Specifically,

> *the easier it is to form new connections, the better is the performance of the system when node 15 is at the top of the hierarchy relative to when node 1 is at the top. And when it is impossible to form new connections, setting node 1 at the top uniformly yields the higher level of overall social welfare.*

Of course, whatever advantage *C1* might hold over *C2* when *t1* increases can be expected to disappear when *t1* is so large that no node can influence any other node. Moreover, any advantage *C2* might hold over *C1* as *t2* increases can also be expected to disappear when connections between all nodes can be readily made and maintained (see, for instance the upper right-most cell in Table 7.7). Thus, in the shaded cells of Table 7.7, note that *C1*s advantage shrinks as *t1* increases, whereas in the other cells, *C2*s advantage decreases as *t2* increases.

The other feature of Table 7.7 to be noted is that when *t2* = 0, summed welfare decreases for *C1* but increases for *C2* at *t1* increases. On the other hand, when *C2* gives the greater overall level of welfare, summed welfare increases for *both C1* and *C2* when *t1* increases. The explanation for this and the general circumstances under which setting node 1at the top of the hierarchy (*C1*) yields

Table 7.8: Surviving connections with t2 = 0, P4

t1 =	2	6	9	14	16	17	20
#connections surviving *C2* / *C1*	14/14	7 / 3	3 / 1	1 / 1	1 / 1	1 / 0	1 / 0
C2 surviving	all	8,...,15	12,...,15	14,15	14,15	14,15	14,15
C1 surviving	all	1,...,4	1,2	1,2	1,2	none	none

the greater level of welfare can be seen more readily if we look at Table 7.8. Briefly, that table gives the number of surviving unidirectional network connections (the maximum is 14, the minimum 0) and the connections that survive when we set *t2* = 0 (e.g., if we indicate '1, 2, 3, 4', then that means that the unidirectional connection between 1 and 2, 1 and 3 and 2 and 4 survived). The source of the better performance of *C1*, then, is that with *t2* = 0 node 1 (and perhaps others that are relatively high on the hierarchy) cannot be influenced by the initially high rate of consumption of node 15 – node 15 cannot induce node 1 (and those in between) to increase the proportion of low *d* types within them (and thereby ultimately lower their long-term per capita consumption). Moreover, if *t1* is sufficiently small so that connections are readily maintained, if node 1 stands at the top of the hierarchy it ultimately induces all other nodes to fully mimic its investment/consumption decisions so as to yield the greatest possible level of social welfare. But if node 15 stands at the top, it induces all other nodes to mimic it, in which case social welfare is minimized. In this way, then, a hierarchical network yields exactly what we might otherwise expect it to yield. Tables 7.6 and 7.7 simply show, though, that expectation is merely an extreme point in the range of a rather complex set of possibilities.

There remains now one final assessment of a hierarchical network to be explored here: the situation where nodes at the same level are bidirectionally connected, where nodes at the same level influence numerically adjacent nodes so that *C1* and *C2* become:

$$C10 = \begin{bmatrix} 100000000000000 \\ 111000000000000 \\ 111000000000000 \\ 010111100000000 \\ 010111100000000 \\ 001111100000000 \\ 001111100000000 \\ 000100011111111 \\ 000100011111111 \\ 000010011111111 \\ \text{.......}etc. \end{bmatrix}, C20 = \begin{bmatrix} 111111111000000 \\ 111111111000000 \\ 111111110100000 \\ 111111110100000 \\ 111111110010000 \\ 111111110010000 \\ 111111110001000 \\ 111111110001000 \\ 000000001111100 \\ 000000001111100 \\ 000000001111010 \\ 000000001111010 \\ 000000000000111 \\ 000000000000111 \\ 000000000000001 \end{bmatrix}$$

Table 7.9, now, summarizes the summed per capita consumption of all fifteen nodes as well as the number of connections that remain in equilibrium (keeping in mind that the maximum is 225 and the minimum 15) when the rate of ES *p* change equals (as in the previous tables) 0.45. The first thing we see, of course, is that overall social welfare is uniformly maximized by setting node 1 at the top of the hierarchy. Recall that in Table 7.7, setting node 1 at the top of the hierarchy dominates (in terms of summed social welfare) setting node 15 there only if *t1* is sufficiently great or *t2* sufficiently small. The reason why setting node 1 at the top is universally dominant can now be seen if we contrast a specific cell in Tables 7.7 and 7.9 – namely, when *t1* = 14 and *t2* = 0.01. First,

Table 7.9: Total per cap consumption and surviving connections, C20 / C10

t1	t2 = 0	0.01	0.05
2	3.38 / 30.75 99 / 99	3.45 / 3.60 225 / 225	3.45 / 3.60 225 / 225
6	3.46 / 9.75 99 / 87	3.45 / 6.52 225 / 89	3.60 / 6.75 225 / 89
9	4.86 / 7.63 91 / 85	4.89 / 7.69 113 / 85	4.82 / 7.88 113 / 89
14	5.30 / 7.63 87 / 85	5.14 / 7.69 89 / 85	5.14 / 7.94 89 / 85
16	5.30 / 7.63 87 / 85	5.30 / 7.69 89 / 85	5.22 / 7.88 89 / 85
17	5.30 / 7.63 87 / 85	5.30 / 7.69 89 / 85	5.30 / 7.86 89 / 85
20	7.26 / 8.16 31 / 29	7.30 / 8.20 33 / 29	6.46 / 8.03 45 / 41

as Table 7.7 shows, if *C2* is the initial connections matrix, the subnetworks (1, ..., 11), (12) and (13, 14, 15) form. Hence, the negative influence (on long term investment) of the least productive nodes are isolated somewhat from those that can be most productive. But if *C1* is the initial input, *t2* is not so small as to preclude the formation of a single fully connected subnetwork, in which case the least productive nodes exert their influence on those that might, initially at least, choose to save for future consumption. Of course, the question arises here as to why the least productive nodes seem able to exert a greater influence with *C1* is the input connections matrix as opposed to when *C2* is that matrix. The reasons, unfortunately, are complex and depend, as Table 7.7 suggests, on parameter values – on the ease with which connections can be made or broken.

But now, consider the comparable cell in Table 7.9. In this instance, when *C20* is the initial connections matrix the resulting subnetwork structure is (1, ...,

8), (9, ..., 12), (13, 14, 15) whereas if C10 in the initial matrix, that subnetwork structure becomes (1), (2, 3), (4, ..., 7), (8, ..., 15). In other words, with the exception of nodes 13 and 14 connecting to 15 with *C20*, both subnetwork structures mimic the levels of the respective hierarchies. And when that occurs, the nodes that are most productive in the long run (nodes 1, 2, 3 etc.) isolate themselves from the rest, invest, and ultimately consume at a higher rate than when they are connected to less productive nodes (as occurs in the subnetwork (1, ..., 8)). It is true, of course, that a number of nodes consume, in equilibrium, less when node 1 is at the top of the hierarchy, but the gains in long-term consumption enjoyed by nodes 1, 2 and sometimes 3 are of such magnitude as to outweigh any losses incurred by other nodes. Such is the consequence, of course, of inequality discussed in Section 7.4. At least the following generalization seems warranted, then, by the preceding simulations:

> *setting the potentially most productive nodes at the top of a hierarchy improves overall social welfare (at the expense of equality), but only to the extent that the connections within the hierarchy are 'extensive'. Otherwise, the relative advantage of setting the most as opposed to the least productive nodes near the top of a hierarchy depend on parameter values – specifically the difficulty of forming new connections.*

At the very least, the preceding analysis illustrates the general complex relationship between the equilibrium natures of networks and the parameters that influence their evolution.

8 Endogenous Networks: Part 2

8.1 An Alternative Connection Strength Model

The model of the preceding chapter entails one potential inconsistency. Although we assume that nodes (or groups, if that is the interpretation we choose to apply) estimate the proportion of high and low *d* types in the population on the basis of rates of consumption as opposed to the actual proportions, we also assume that connections between nodes are based on the similarity of actual current ES *p* values. The alternative, then, is to assume that connections are based on the similarity of consumption patterns. Specifically, the relevant part of the BASIC program in the appendix of this chapter that dictates the readjustment of connection strengths in Chapter 7 is

```
    FOR k1 = 1 TO ngroups: FOR k2 = 1 TO ngroups
    FOR k3 = 1 TO 10: IF c!(k3, k1, k2) + t2 = 0 THEN GOTO 110
    kv1 = p!(k3, k1)
    kv2 = p!(k3, k2)
    IF kv1 = kv2 THEN c!(k3, k1, k2) = 1: GOTO 110
    c!(k3, k1, k2) = (c!(k3, k1, k2) + t2) / (t1 * ABS(kv1–kv2))
109 IF c!(k3, k1, k2) < 0 THEN c!(k3, k1, k2) = 0: GOTO 110
    IF c!(k3, k1, k2) > 1 THEN c!(k3, k1, k2) = 1
110 NEXT k3: NEXT k2: NEXT k1
```

where p(x, y) denotes node y's current ES *p* value at the xth value of whatever parameter the program iterates on (usually ES *p* rate of change) and c(x, y, z) denotes the strength of connection of node y to z at that value. The alternative, then, is to substitute the following for the definition of kv1 and kv2 (labeled alternative A1):

```
kv1 = cl!(k3, k1)/(cl!(k3, k1)+ch!(k3, k1))
kv2 = cl!(k3, k2)/(cl!(k3, k2)+ch!(k3, k2))
```

To some extent, of course, the choice is arbitrary as we have no empirical basis for preferring one assumption over another. Using the objective value of ES *p* can be justified by the argument that even if individuals misjudge the true state of affairs by relying on out-of-equilibrium observable behavior, there nevertheless exist underlying evolutionary (though here, unmodeled) processes that compel a system to abide by objective measures. The counter argument is to point out that we cannot say what those processes might be or whether they in fact are unbiased. In this chapter, then, we explore the alternative to see if and when it matters which assumption is the most appropriate.

There is, in fact, a second alternative. Rather than assume that connections are adjusted on the basis of actual or inferred proportions of high and low *d* types, we can assume alternatively that they are based simply on the per capita consumption of a node. In this case we would substitute the following two expressions for the calculation of kv1 and kv2 in our BASIC program (labeled alternative A2):

```
kv1 = (cl!(k3, k1)+ch!(k3, k1))/n%(k1)
kv2 = (cl!(k3, k2)+ch!(k3, k2))/n%(k2)
```

where n%(x) denotes the size of node x.

Before proceeding, a word of warning is in order. Thus far we have found that the ultimate equilibrium is relatively insensitive to the initial proportion of high and low *d* types. Thus, throughout Chapter 7 we simply set that initial proportion at 0.05 for all nodes. But in the case of this second alternative especially, kv1 and kv2 are immediately impacted by initial proportions. In addition, strengths of connection adjust incrementally only insofar as kv1 and kv2 adjust incrementally. Hence, for our core model (Chapter 7), connection strengths adjust incrementally because ES *ps* do so. On the other hand, consumption is allowed to adjust fully to the presumed optimum, given a nodes proportion of types. And even though that proportion adjusts incrementally according to the rate *ar*, we have seen in Chapter 4 how total consumption can oscillate widely as an equilibrium is approached (see, for instance, Figure 4.7). Thus, although alternatives A1 and A2 will both be impacted somewhat by initial proportions, alternative A2 should prove to be more sensitive to this parameter. One 'solution', of course, is to directly require that connection strengths change incrementally. But this merely adds another parameter to an analysis already replete with parameters. Instead, we will focus here on alternative assumption A1, with an occasional glance at A2 and with the understanding that A2 in particular doubtlessly overstates the sensitivity of real processes to parameters.

8.2 The Initially Complete Network Revisited

Much of the difference between the alternative A1 and our previous analysis is told by Figure 8.1 when compared to Figure 7.6. The essential thing to note here is the marked decrease in the sensitivity of subnetwork structures to variations in *t1*. Whereas the total number of connections drops to less than 140 once *t1* exceeds 5.5 in Figure 7.6, that number is not universally achieved until *t1* exceeds 14 here (it should be kept in mind, moreover, that the horizontal scale of Figure 8.1 is more than double that of Figure 7.6).

Figure 8.1: Number of connections vs. t1 for various rates of change of ES p, P1

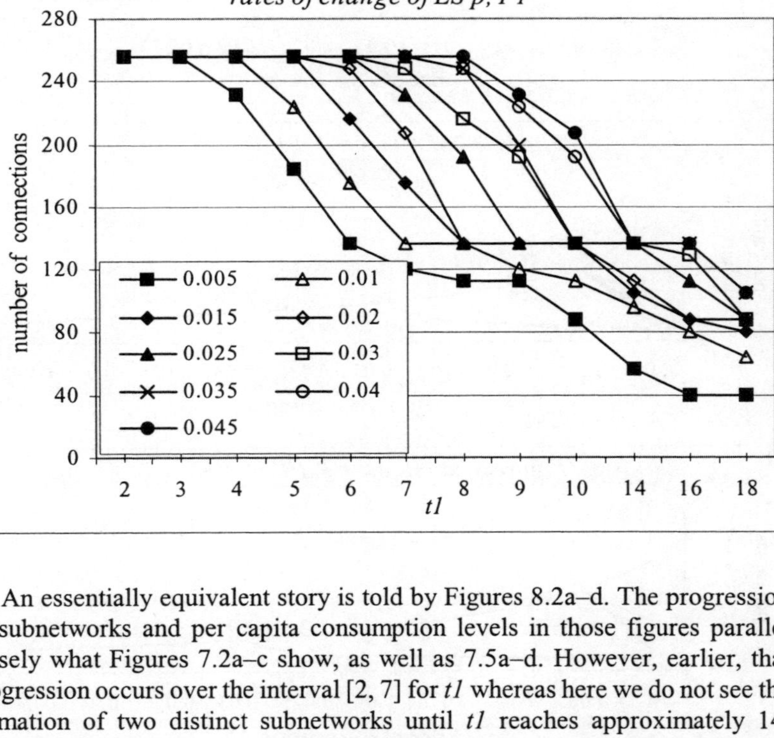

An essentially equivalent story is told by Figures 8.2a–d. The progression of subnetworks and per capita consumption levels in those figures parallel closely what Figures 7.2a–c show, as well as 7.5a–d. However, earlier, that progression occurs over the interval [2, 7] for *t1* whereas here we do not see the formation of two distinct subnetworks until *t1* reaches approximately 14. Nevertheless, three primary qualitative conclusions from Chapter 7 remain valid here. Specifically, with all connection strengths initially at 1,

- *the number of subnetworks depends not only on the ease with which connections are maintained but also on the rate at which ES p changes;*
- ceteris paribus, *the number of connections increases as the rate of change in ES p increases;*
- *connections within a subnetwork need not be complete.*

Insofar as this last conclusion is concerned, we note, for example, that if *t1* = 7, and the rate of ES *p* change equals 0.015, nodes 1 trough 14 form a subnetwork, but nodes 13 and 14 are not connected to nodes 1 through 6. A

*Figure 8.2a: All c*s *= 1, P1, t1 = 4*

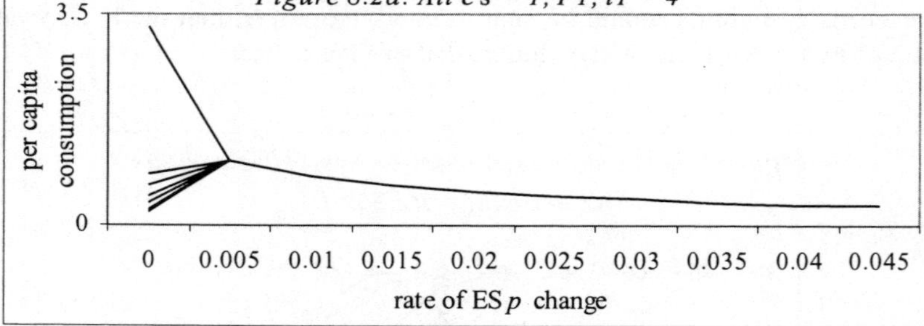

*Figure 8.2b: All c*s *= 1, P1, t1 = 7*

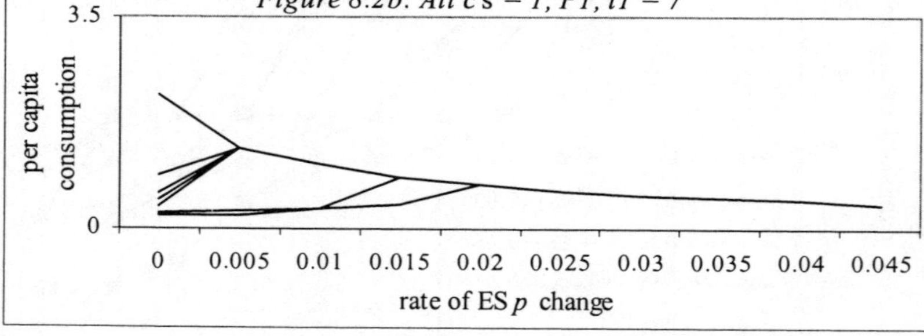

*Figure 8.2c: All c*s *= 1, P1, t1 = 9*

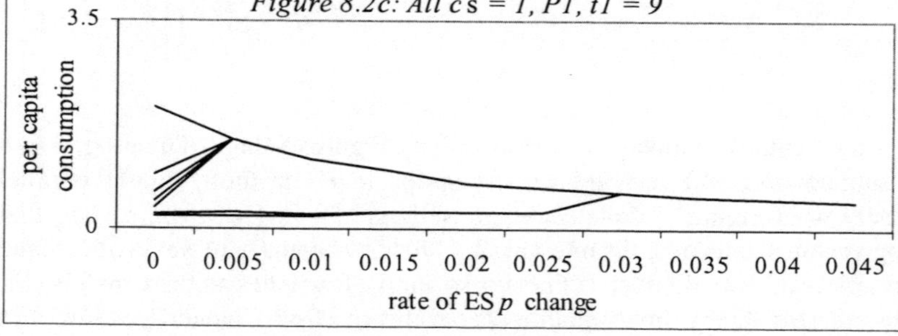

*Figure 8.2d: All c*s *= 1, P1, t1 = 14*

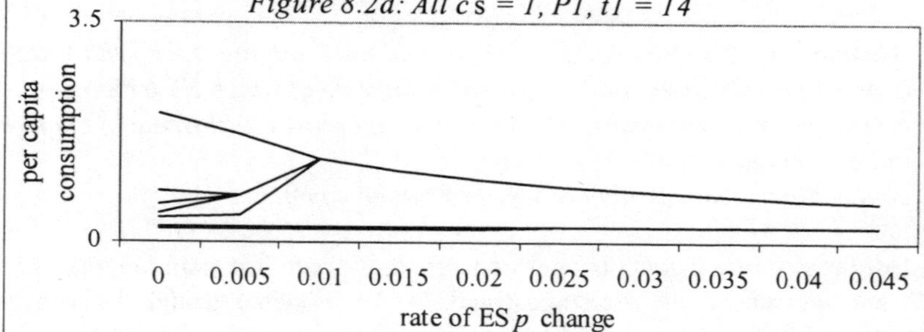

single subnetwork then forms once this rate equals 0.020, but the total number of connections increases from 208 to 232 to 248 and finally, once the rate equals or exceeds 0.40, the maximum of 256. However, one conclusion from Chapter 7 no longer seems to hold; namely, in equilibrium all nodes now enjoy the same level of per capita consumption. It remains to be seen whether this is an essential feature of this version of our model or a function of the initial input parameters – notably the initial connections matrix.

At this point, however, we should consider the influence of the initial proportion of low *d*s at each node since we know formally that the effect here on network connections will be greater than in Chapter 7. Recall, then, that the vector of initial proportions, p_{init}, employed thus far is simply

$$p = (0.05, 0.05, ..., 0.05).$$

However, as an alternative, consider the following, which matches p_{init} with each node's initial ES *p* value:

$$p' = (.05, .05, .20, .20, .30, .30, .40, .40, .50, .50, .70, .70, .80, .80, .95, .95)$$

Thus, nodes whose initial stable proportion of low *d*s is high, begin with a high proportion of low *d*s and those that begin with a low ES *p*, begin as well with a correspondingly low proportion of low *d*s. Figures 8.3a–d, now, rerun our simulations for the same parameter values as in Figures 8.2a–d, except that here we substitute p′ for p. The difference is evident and as anticipated:

> *with modeling alternative A1, unless the rate of ES p adjustment is slow and unless t1 is low so that the initial connection matrix is readily maintained, the system with* p′ *consists of far more subnetworks than it does with* p.

The system is even more fractured (disconnected) than Figures 8.3a–d suggest – a fact that can be seen by considering once again the total number of connections in our 16-node system. Briefly, plotting the total number of connections against *t1* for the different range of values of the ES *p* adjustment rate, Figures 8.4a and b essentially reproduce Figure 8.1 (indeed, the data for Figure 8.4a are taken directly from Figure 8.1) except that the data from Figure 8.4b are taken from Figures 8.3a–d and thereby sets $p_{init} = p'$. What we see here, then, is that with p′, although connections still depend critically on the ease with which connections can be maintained (*t1)*, subnetwork connections are now relatively insensitive to the rate of change in ES *p*. Of course, this is as expected. In Chapter 7 connection strength depended only indirectly on consumption patterns and directly on an incrementally changing ES *p* value. But now, since connections depend directly on consumption patterns and only indirectly on ES *p*, the impact of rate of ES *p* change is markedly less.

*Figure 8.3a: All c*s =1, P1, t1=4, A1, $p_{init} = p'$

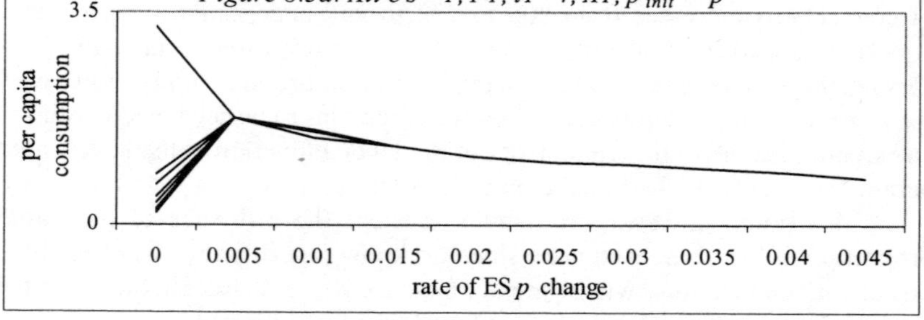

*Figure 8.3b: All c*s =1, P1, t1=7, A1, $p_{init} = p'$

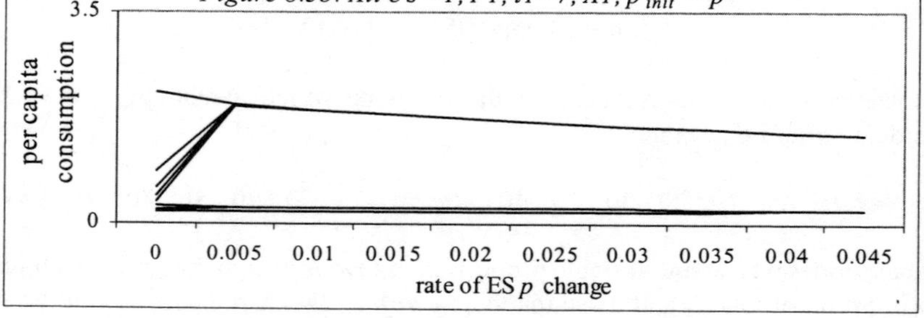

*Figure 8.3c: All c*s =1, P1, t1=9, A1, $p_{init} = p'$

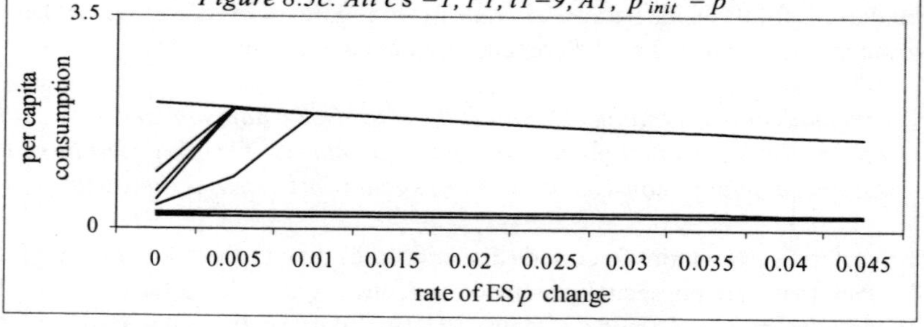

*Figure 8.3d: All c*s =1, P1, t1=14, A1, $p_{init} = p'$

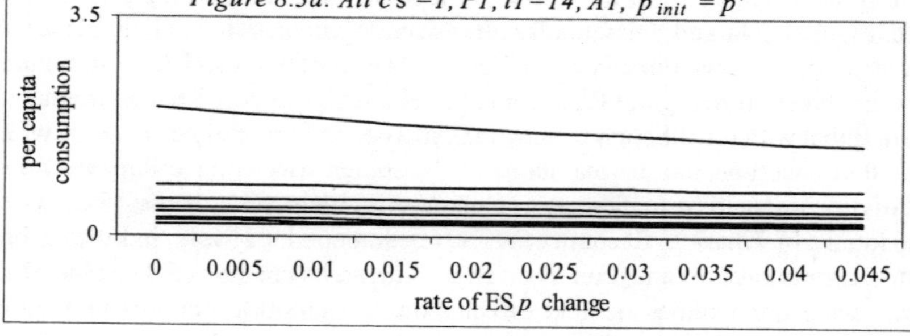

Figure 8.4a: Number connections with A1, P1, $p_{init} = p$

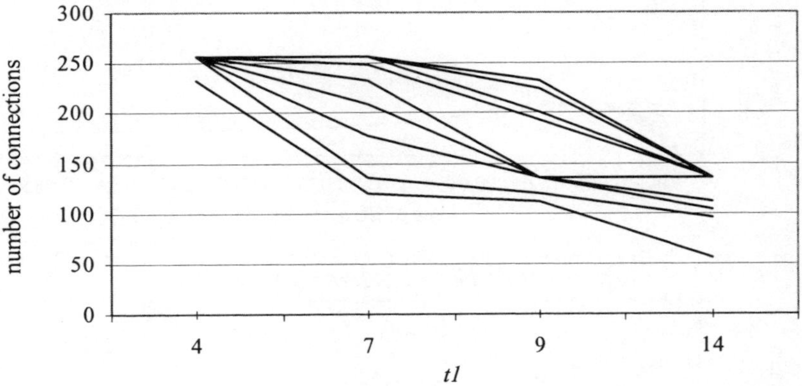

Figure 8.4b: Number connections with A1, P1, $p_{init} = p'$

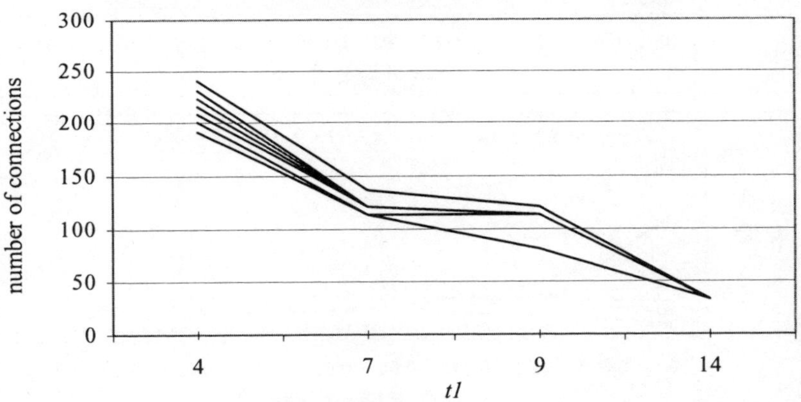

This story is largely repeated for alternative A2 except it occurs in perhaps a more interesting way. First, Figures 8.5a–d reproduce Figures 8.3a–d except we substitute alternative A2 for A1 as a means of adjusting connection strength. And although Figures 8.5a and 8.3a are comparable, the similarity seemingly ends there, with *t1* at 4 or less. Under A1 as *t1* increases, the less productive nodes drop away to form their own subnetworks much as they did in Chapter 7 (see Figures 7.5a–d). In the case of A2, on the other hand, the least productive nodes continue to impact the rest even when *t1* = 14 and the process of 'dropping away' is far less evident, if only because the more productive nodes enjoy a considerably lower level of per capita consumption than they do under

Figure 8.5a: All cs = 1, A2, t1=4, P1, $p_{init} = p'$

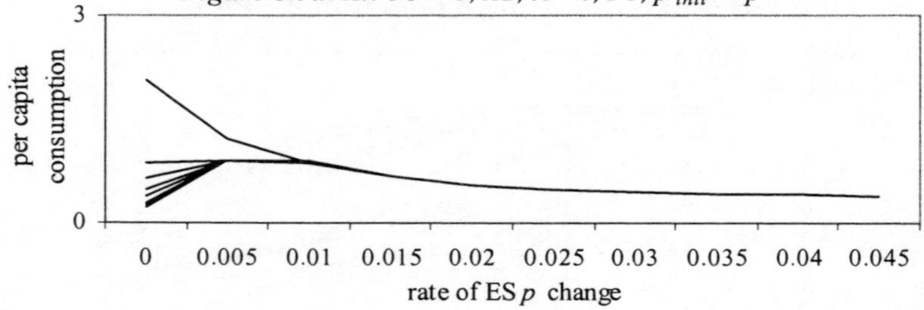

Figure 8.5b: All cs = 1, A2, t1=7, P1, $p_{init} = p'$

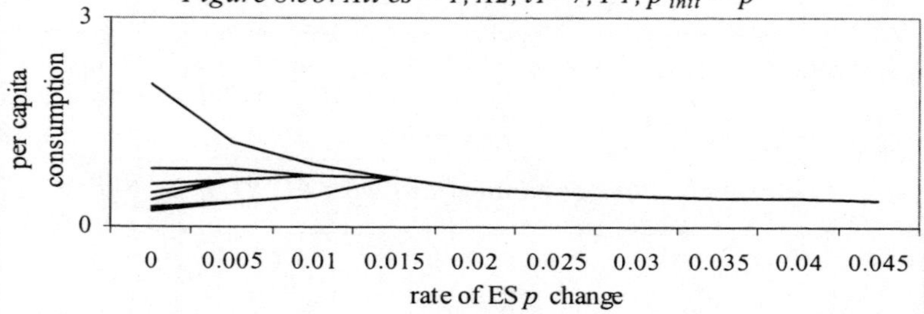

Figure 8.5c: All cs = 1, A2, t1=9, P1, $p_{init} = p'$

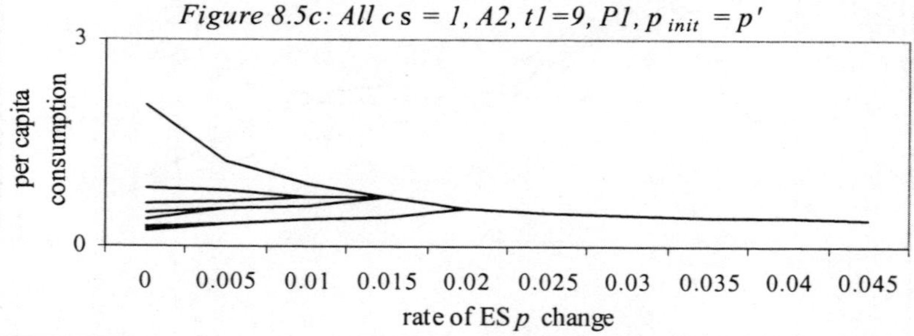

Figure 8.5d: All cs = 1, A2, t1=14, P1, $p_{init} = p'$

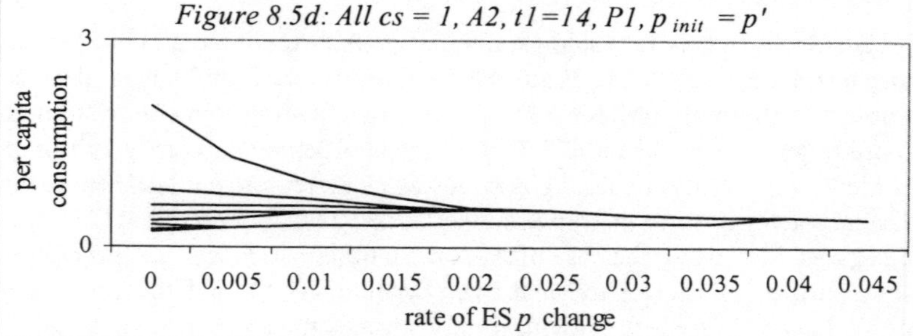

A1. The difference between A2 and A1 (as well as the analysis in Chapter 7) is more boldly illustrated by Figures 8.6a–d which consider values of *t1* in the range of [20, 1000]. Note now that even though a value of *t1* = 50 yields a wholly fractured subnetwork structure whereby only twins are paired, it is not until *t1* exceeds 100 that we see a pattern of per capita consumption similar to what we observe under A1 (Figure 8.3d). The explanation, of course, is that under A2 the less productive nodes are able to 'pull down' the consumption of the otherwise more productive nodes before subnetwork connections are broken.

It might seem, then, that we have at least two contradictory models and no empirical basis for choosing between them. But in fact there is an important implication here. The analysis in Chapter 7 is based on the assumption that nodes adjust strengths of connection on the basis of objectively correct criteria – the actual value of ES *p*. Assumption A2 assumes, in a sense, the opposite – that nodes adjust those strengths on the basis of the simplest observable: a nodes overall per capita consumption. Assumption A1 falls somewhere between these two extremes. Although it too adjusts connections on the basis of observable consumption, it nevertheless distinguishes between the consumption of high versus low *d*s. Unsurprisingly, then, our results under A1 fall between what we reported in Chapter 7 and the patterns we uncover under A2. The general implication, however, is this:

> *to the extent that the strengths of connection are based on simple observables as opposed to underlying objective parameter values, those strengths will be more readily maintained so as to allow less productive nodes to impact potentially more productive ones.*

A subsidiary implication, then, when all nodes are initially connected, is:

> *the evolution of networks will depend not simply on parameter values, but also on the extent to which there are underlying processes that render the evolutionary process itself dependent on objective values as opposed to simple observables.*

8.3 The Atomistic State

Rather than consider the intermediate cases of unpaired nodes and a partially complete initial connections matrix (C'), since we have a sense of how A1 and A2 change matters, it is more efficient to proceed directly from the case of a complete initial connections matrix to where $C = \mathbf{I}$ and where the initial vector of ES *ps* is the sequence of uniformly distanced values

P3 = (.05, .11, .17, .23, .29, .35, .41, .47, .53, .59, .65, .71, .77, .83, .89, .95)

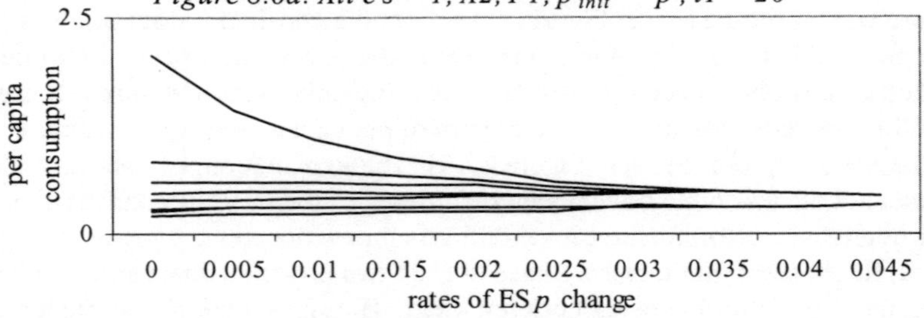

Figure 8.6a: All c s = *1, A2, P1,* $p_{init} = p', t1 = 20$

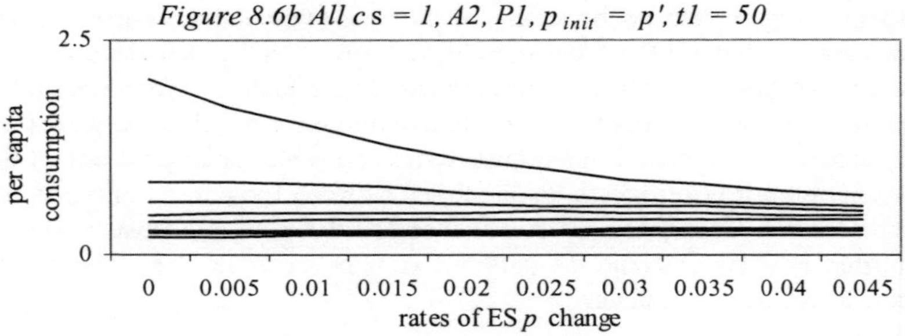

Figure 8.6b All c s = *1, A2, P1,* $p_{init} = p', t1 = 50$

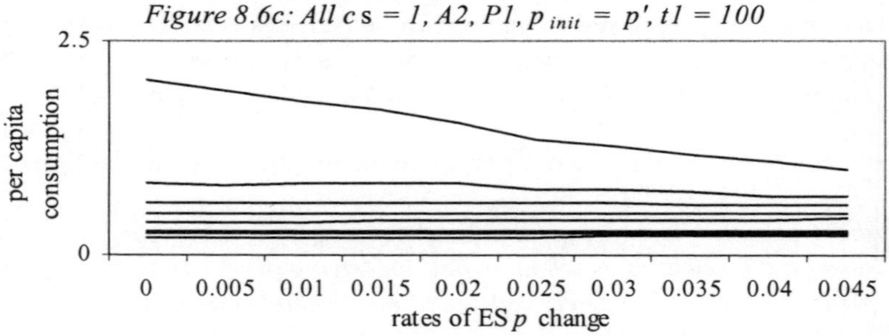

Figure 8.6c: All c s = *1, A2, P1,* $p_{init} = p', t1 = 100$

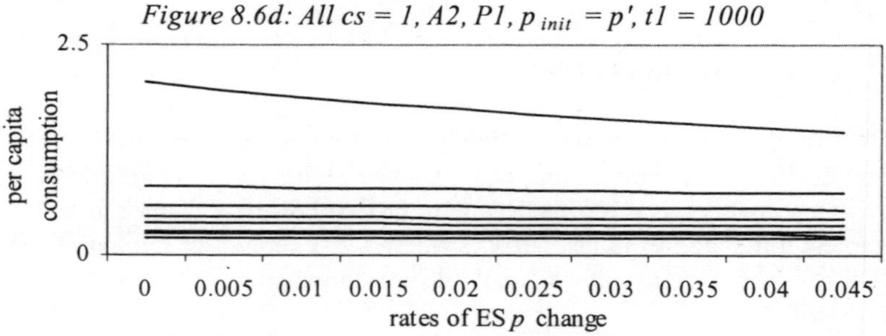

Figure 8.6d: All cs = 1, A2, P1, $p_{init} = p', t1 = 1000$

We can begin now by comparing what occurs under A1 and our analysis in Chapter 7 by setting *t1* = 16.7 and contrasting Figures 7.15a–c against Figures 8.7a–c. At first glance now it might seem as if there was little to distinguish between these two sets of figures. However, on closer inspection we can see a subtle yet important difference. Specifically, note in Figures 7.15 a–c that as t2 decreases, the less productive nodes are 'drawn up' to mimic the more productive ones. In the case of Figures 8.7a–c on the other hand, those lower nodes 'fall away' as *t2* decreases to form their own distinct subnetwork. If we look, for instance, at the subnetworks that form for two different values of ES p change rates we can see this difference perhaps a bit more clearly. Table 8.1, then, contrasts subgroups for the three values of *t2* considered in Figures 7.15a–c and 8.7a-c across three rates of ES *p* adjustment. In all cases we see the gradual consolidation of subnetworks. But the most telling column here is the last. In Chapter 7 a single subnetwork ultimately forms at the highest rates of ES *p* adjustment considered. In contrast, under alternative A1, as *t2* decreases, we see the development of a distinct subnetwork consisting of nodes 14 through 16.

Table 8.1: Comparison of subnetworks for Figures 7.15 and 8.7

	original model		
	ES *p* rate = 0.02	0.035	0.045
t2=.1	(1–12), (13), (14–16)	(1–14), (15, 16)	(1–16)
.05	(1–12), (13), (14–16)	(1–16)	(1–16)
.0125	(1–16)	(1–16)	(1–16)
	alternative A1		
.1	(1–10), (11), (12–16)	(1–14), (15-16)	(1–16)
.05	(1–9), (10), (11), (12), (13), (14–16)	(1–11), (12), (13), (14-16)	(1–15), (16)
.0125	(1–10), (11), (12), (13), (14–16)	(1–10), (11–12), (13), (14–16)	(1–12), (13), (14-16)

Before drawing any conclusions, though, we need to be cautious, for we have already seen a variety of discontinuities in subnetwork development as parameters change. Although it treats a hierarchy, Table 7.5, for instance, illustrates a situation in which as *t2* decreased, subnetworks suddenly shift from forming among the higher numbered nodes to the lowest numbered ones. To see if there are any such reversals here, suppose we consider greater rates of ES *p* change; namely, in the range of [0, 0.45]. In fact, what we see here is that at

Figure 8.7a: C = I, P3, t1 = 16.7, A1, t2 = 0.10

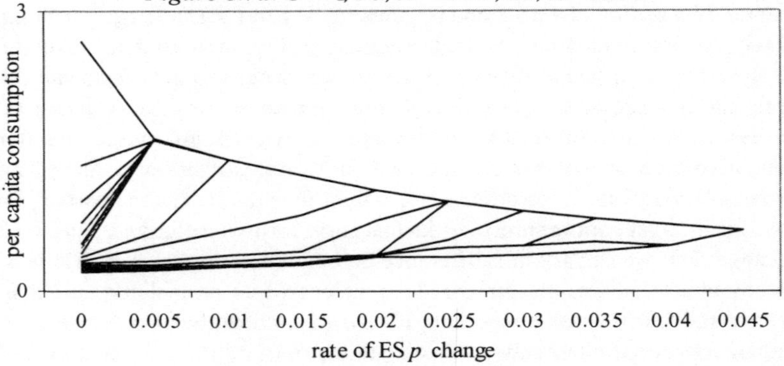

Figure 8.7b: C = I, P3, t1 = 16.7, A1, t2 = 0.05

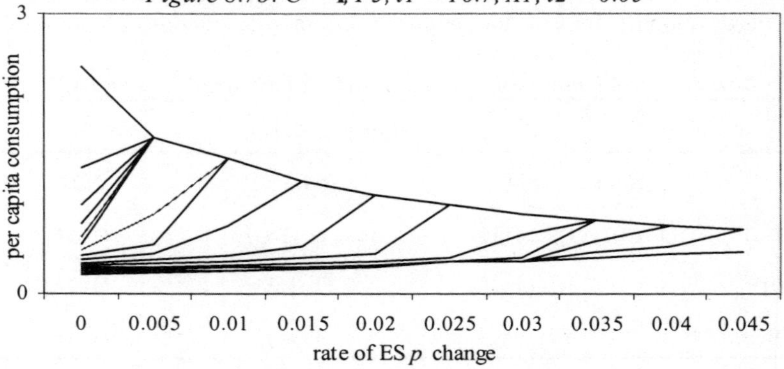

Figure 8.7c: C = I, P3, t1 = 16.7, A1, t2 = 0.0125

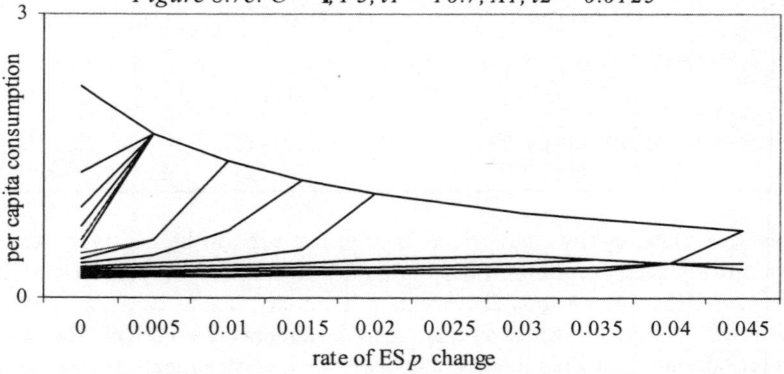

these higher rates, a single subnetwork forms under A1 in which all nodes are drawn down to the least productive nodes (#16) level of consumption. In this sense, then, our simulations at higher rates are a continuation of what we see in the second half of Table 8.1. For our original model, on the other hand, this 'draw down' occurs in the same way, but seemingly only for $t2$ = 0.1 and 0.05. On the other hand, when we make the formation of new connections more difficult by setting $t2$ = 0.0125, again a single complete subnetwork forms but – and as the continuation of Figure 7.15c – one in which all nodes are drawn UP to node 1's consumption level. In fact, though, there is no discontinuity here. Consider Figures 8.8a and b, where 8.8a returns once again to the model employed in Chapter 7 but which, like 8.8b, considers rates of ES p change in the intermediate interval of [0, 0.225].

Figure 8.8a: C = I, P3, t1 = 16.7, t2 = 0.05, Cp 7 model

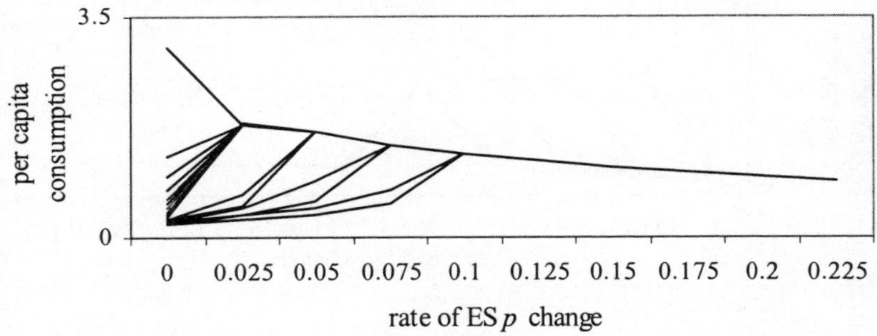

Figure 8.8b: C = I, P3, t1 = 16.7,t2 = 0.10, A1

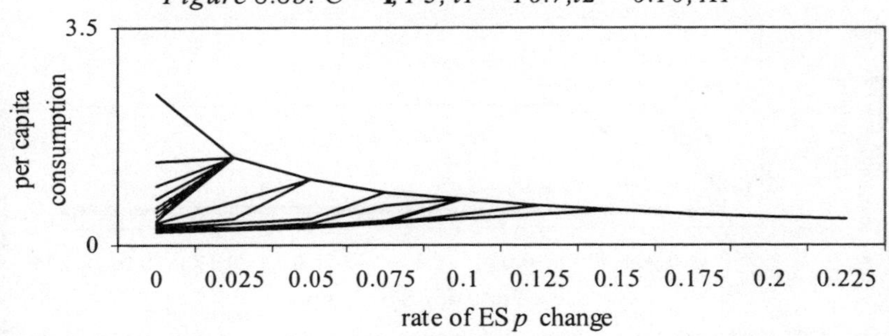

We see here that there is, in fact, little to choose from between the model from Chapter 7 and A1. It is the case that per capita consumption is greater under the assumption that strengths of connection depend on objective ES p values. But the appearance of a different structure to convergence to an equilibrium at lower rates of ES *p* adjustment is more an illusion than anything else – even the (14, 15, 16) subnetwork eventually joins the remaining nodes once the rate of ES *p* adjustment is great enough. Figures 8.8a and b, then, underscore the lesson from the previous section, namely that A1 seems less sensitive to parameter values than does the analysis in the preceding chapter, but that otherwise, our conclusions remain largely unchanged. There is, though, one parameter that requires consideration here – the vector of initial proportions of high and low *d* types. Figures 8.9a and b, then, employ the same parameter values as the preceding two figures, except that now we set $p_{init} = p'$ – the vector that matches initial proportions of types to initial ES *p* values.

Figure 8.9a: C = I, P3, t1 = 16.7, t2 = 0.05, Cp7 model, $p_{init} = p'$

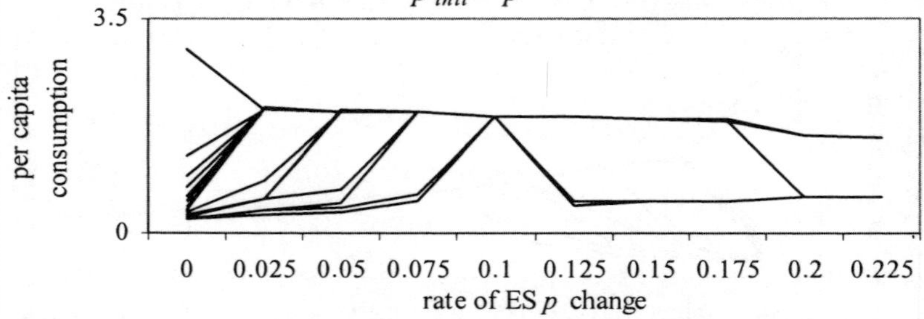

Figure 8.9b: C= I, P3, t1=16.7, t2=.05, A1, $p_{init} = p'$

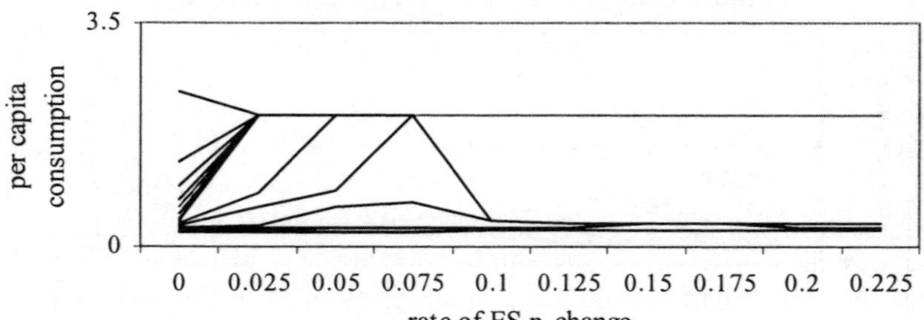

Figures 8.9a and b offer an interesting contrast. First, despite appearances, Figures 8.9a and 8.8a are quite similar, at least at the lower rates of ES *p* adjustment. If there is a change it is when ES *p* values are allowed to adjust relatively rapidly (.125 or higher), in which case those nodes that begin with a high proportion of low *d*s in accord with their initial (and high) ES *p* values

remain unproductive and fail to develop network connections with other nodes that might cause them to lower their ES *p* value and, correspondingly, invest for future consumption. However, it is when we substitute assumption A1 that we see the most dramatic change associated with the substitution of $p_{init} = p'$ for p. The principal change is that now some nodes (those with the highest initial ES *p* values) never join other nodes (except those that are similar to them initially), regardless of the rate of change of ES *p*.

This is not to say, of course, that an identical pattern would be observed for all values of *t1* or *t2*. Nor can we say that the way in which we make strengths of connection endogenous in Chapter 7 renders equilibria invariant with p_{init}. But the dual contrast between Figures 8.9a versus 8.8a and between 8.9b and 8.8b once again illustrates the specific sensitivity of outcomes to p_{init} when A1 dictates endogenous connection strengths.[1] At a minimum then, we see here that any attempt to understand the evolutionary development of social networks requires attention to a number of things, including those things that might impact the rate of evolutionary development and initial conditions such as p_{init} and initial ES *p* values.

8.4 The Hierarchy

We conclude in the previous chapter that '*if new connections are relatively easily made, it matters little who is at the top and who is at the bottom of a network hierarchy*' and that '*the social welfare implications of hierarchical networks in terms of who stands at the top of the hierarchy depends on the ease with which new connections can be made*' (pp. 201, 203). To see whether these broad brush stroke conclusions are maintained when we change the basis for adjusting connection strength, recall first the vector of initial ES *p*s considered in Chapter 7:

$$P4 = (.05, .11, .17, .23, .29, .35, .41, .47, .53, .59, .65, .71, .77, .83, .89)$$

and recall as well that we began our exploration of hierarchies with the initial connections matrix *C1*, which assumes that nodes at the same level in the hierarchy are not initially connected. We then presented a series of eight figures (Figures 7.20a–h) that detailed the dependence of subnetwork structure and per capita consumption for a range of values of *t2* when *t1* = 16.7. However, our analysis in Chapter 7 uncovers a discontinuity in the progression of subnetworks

[1] And once again, A2 offers the extreme. For the parameter values considered in Figures 8.9a–b, the plot of per capita consumption against rate of ES *p* change with $p_{init} = p'$ looks much like Figure 8.5a.

as *t2* decreased and established the critical dependence of those subnetworks on *t2*s value as well as the threshold *S*, above which a node takes into account the consumption of another node in the adjustment of its ES *p* value.

The source of that discontinuity is the particular way in which connection strengths change as a function of the parameter *t2*. Briefly, in Chapter 7 the strengths between numerically adjacent nodes slowly increase from 0 as *t2* increases until they cross the threshold *S*, at which point they become 1.0 in equilibrium At this point, of course, a node takes the consumption of the other into account in its own internal adjustments. The question, then, is whether a similar discontinuity can arise when we impose A1. To answer this question, Figures 8.10a–b graph per capita consumption against rate of ES *p* change for essentially the same parameter values as in Figures 7.20a–h, but consider a wider range of values for *t2*. And at first glance it might seem that these figures are equivalent in form to Figures 7.20a–h. However, there is a critical difference. Specifically, unlike what we see saw in Chapter 7, as *t2* decreases – as it becomes more difficult to form new connections – the system becomes increasingly atomistic. The difference between these figures and Figures 7.20a–h are well illustrated by Table 8.2 and the contrast it provides to Table 7.5. Table 7.5 reveals one part of the discontinuity as *t2* decreases – from subnetworks that from among the higher numbered (least initially productive) nodes, we suddenly move to a system whereby nodes 1 and 2 are separated from the rest. And we already know, it is only after *t2* is decreased further still that the system suddenly moves towards an atomistic structure. In Table 8.2, in contrast, the 'atomization' of the overall network structure follows a smoother progression as a function of *t2*. It is also evident, moreover, that as in our earlier simulations, subnetwork structure is less sensitive to *t2*s value under alternative A1 than it appears to be in Chapter 7.

Insofar as alternative A2 is concerned, although we do not present the results of any simulations here, it remains true that subnetwork structure is even less sensitive to *t2* under this contingency – but only because, for the range of parameters considered here, it encourages a single wholly connected sub-network. Thus, as before, A1 seems an intermediate model that lies between A2 and the one we explored earlier in Chapter 7. Thus, even in the case of an initial hierarchical system that places the most productive node at the top of the hierarchy,

> *our simulations under alternatives A1 and A2 – under the assumption that connection strengths depend on observable rates of consumption – remain less sensitive to the parameters t1 and t2 than do simulations that assume that strengths depend of on the objectively true values of ES p.*

Before proceeding further to see if this conclusion needs to be altered in any way when the least productive node (#15) stands at the top of the hierarchy, we

Figure 8.10a: C1, P4, t1 = 16.7, A1, t2 = 0.0000001

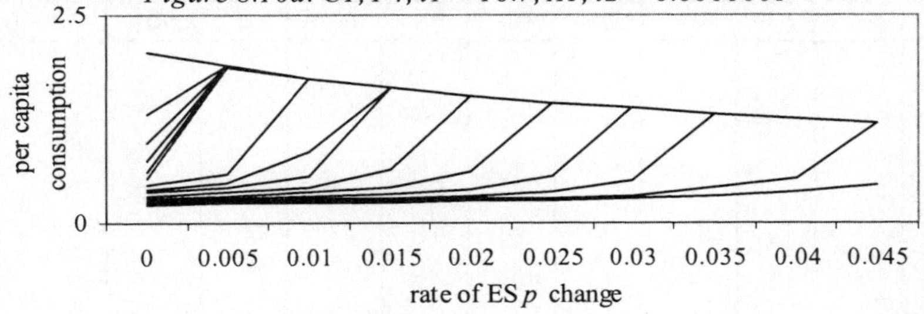

Figure 8.10b: C1, P4, t1 = 16.7, A1, t2 = 0.0001

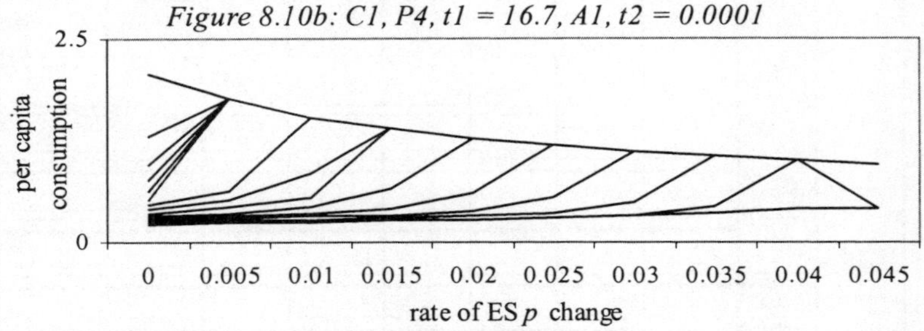

Figure 8.10c: C1, P4, t1 = 16.7, A1, t2 = 0.05

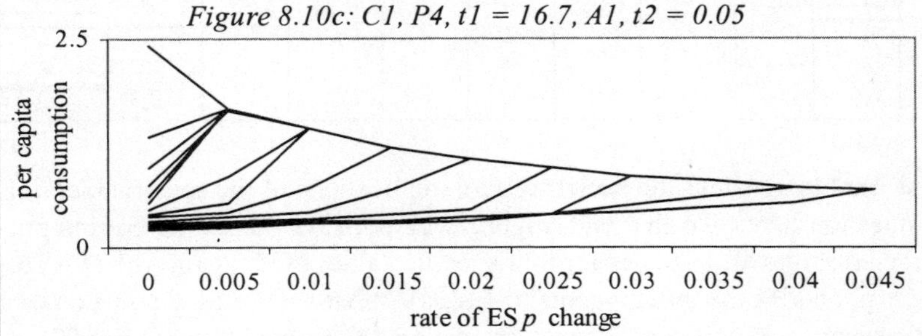

Figure 8.10d: C1, P4, t1 = 16.7, A1, t2 = 0.50

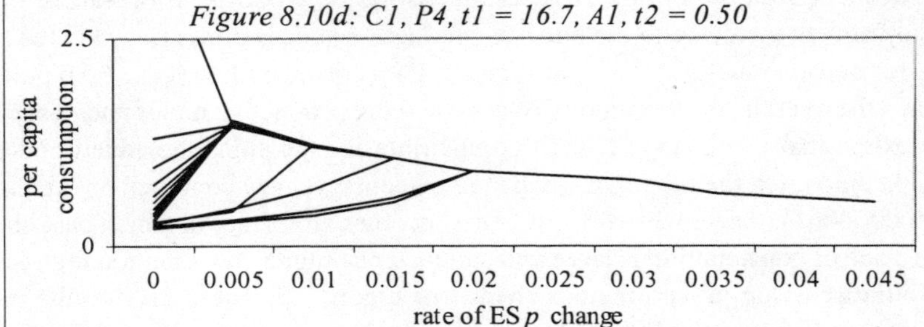

Table 8.2: Subnetwork boundaries vs. t2 with P4, t1=16.7, C1, A1

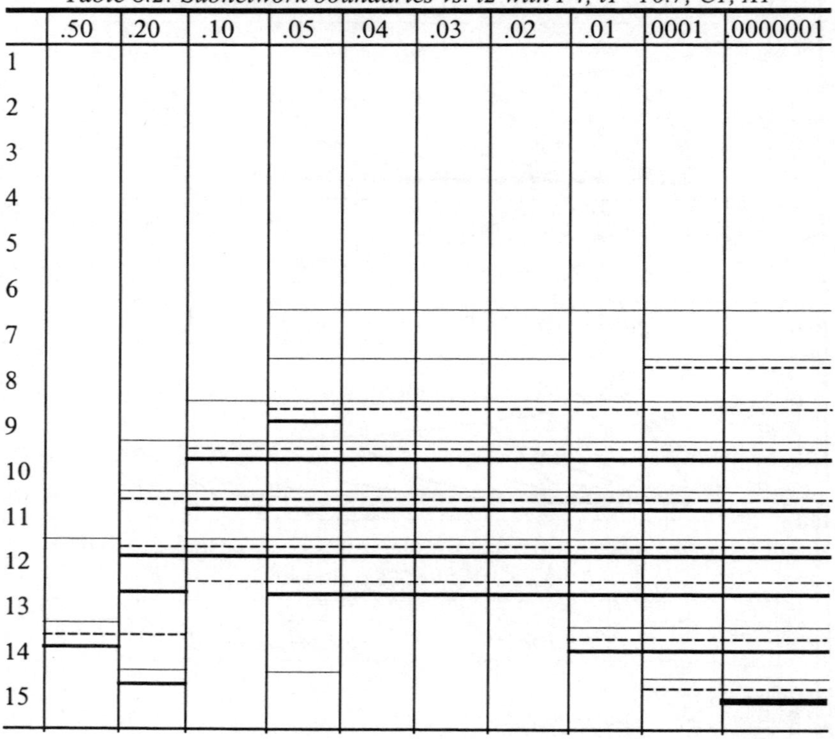

first want to consider the social welfare implications of the specific hierarchy in question here. To that end, Figure 8.11 portrays per capita consumption summed across all 15 nodes for four specific values of *t2* (again with *t1* = 16.7, p_{init} = p, and P4 the initial vector of ES *ps*). What we see here now comes as little surprise. When new connections are readily formed (e.g., *t2* = 0.50), the subsequent expansion of the largest subnetwork 'downwards' as the rate of ES *p* adjustment increases to include less productive nodes (from (1, ..., 11) to (1, ..., 13) and eventually at the highest rate of ES *p* considered, to (1, ..., 15)) pulls down the overall consumption of otherwise more productive nodes and to such an extent that any change fails to compensate for the gains the added nodes might enjoy. On the other hand, when *t2* discourages new connections (i.e., *t2* = 0.0000001), the subnetwork structure becomes atomistic, in which case the variance of consumption is large and summed per capita consumption high (as in out discussion in the previous chapter of Figure 7.3, Table 7.1 and the relationship between the World Bank's Human Development index and income diversity).

Figure 8.11: Summed per capita consumption, t1=16.7, P4, $p_{init} = p$, A1, C1

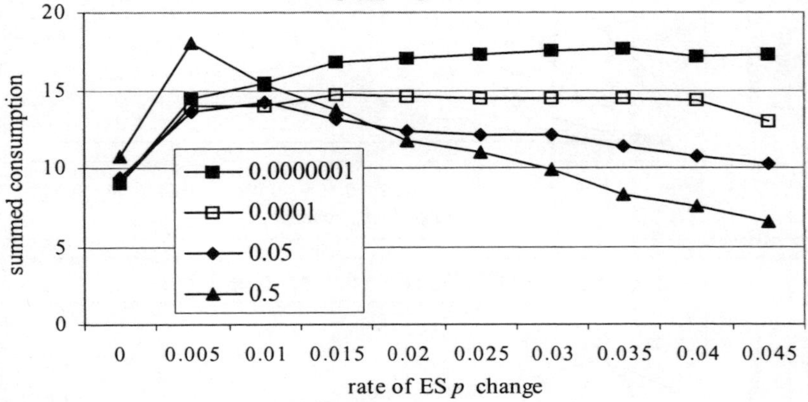

With these results as background let us now turn to the situation in which the least productive node (15) stands at the top of the hierarchy. Figures 8.12a–d, then, reproduce Figures 8.10a–d except that we substitute the initial connections matrix *C2* for *C1*. Perhaps the most evident comparison between these two sets of figures is that there does not appear to be much of a difference at all. Figures 8.10d and 8.12d look nearly identical, but there is one critical difference that is disguised by the format of these figures – the subgroups that form as a function of the rate of ES *p* change. To see what we mean, consider Table 8.3 which details those subgroups for *C1* and *C2* when *t2* = 0.05.

Table 8.3: Comparison of subnetworks for A1 under C1 and C2

ES *p* rate	*C1*	*C2*
0.005	(1– 6), {*i*}	(1 – 6), {*i*}, (13, 14, 15)
0.010	(1 – 7), {*i*}	(1 – 8), {*i*}, (13, 14), (15)
0.015	(1 – 9), {*i*}, (14, 15)	(1 – 9), {*i*}, (12 – 15)
0.020	(1 – 10), {*i*}, (14, 15)	(1 – 9), {*i*}, (12 – 15)
0.025	(1 – 11), {*i*}, (14, 15)	(1 – 10), (11), (12 – 15)
0.030	(1 – 12), (13), (14, 15)	(1 – 10), (11), (12 – 15)
0.035	(1 – 13), (14), (15)	(1 – 11), (12 – 15)
0.040	(1, ..., 14), (15)	(1, ..., 11), (12, ..., 15)
0.045	(1, ..., 13),(14, 15)	(1, ..., 12), (13, ..., 15)

Figure 8.12a: C2, P4, t1 = 16.7, A1, t2 = 0.0000001

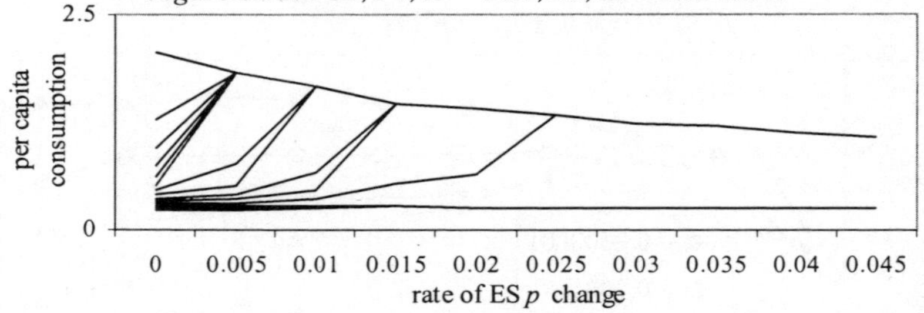

Figure 8.12b: C2, P4, t1 = 16.7, A1, t2 = 0.0001

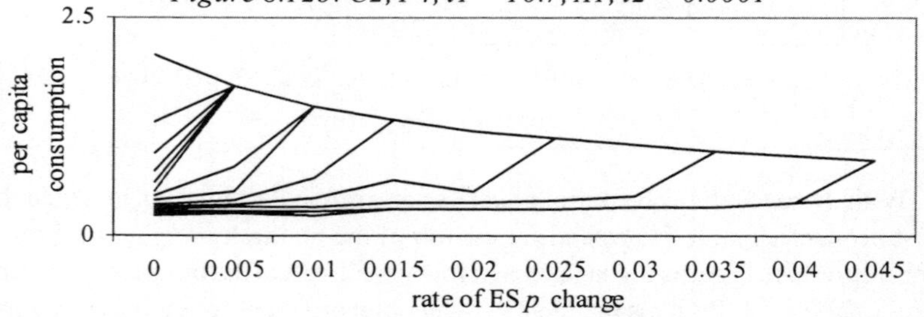

Figure 8.12c: C2, P4, t1 = 16.7, A1, t2 = 0.05

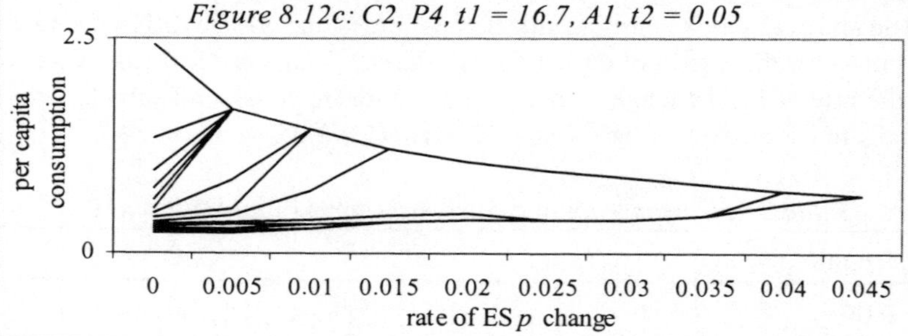

Figure 8.12d: C2, P4, t1 = 16.7, A1, t2 = 0.50

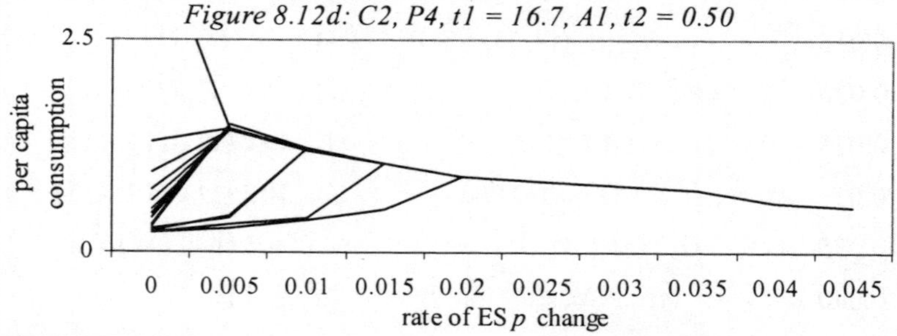

Table 8.3 tells a simple story: Under C1, the largest subnetwork, consisting of nodes at the top of the hierarchy and which are potentially the most productive (have the lowest initial ES *p* values), expands as the rates of adjustment of ES *p* increases at a faster rate than when C2 is the initial connections matrix. In contrast, when the potentially least productive nodes stand at the top of the hierarchy under *C2*, the four nodes with the highest ES *p* value (12, 13, 14 and 15) form a relatively stable subnetwork across a wide range of ES *p* adjustment rates. The differences here, of course, might not seem especially spectacular, but magnitudes here are of little consequence since there are no natural units to apply to any of our parameters. It is the relative order of things and not their magnitude that matters, and it also needs to be kept in mind that other initial parameter values (e.g., *t1*, *ar* and p_{init}) are likely to magnify this difference.

However, there is another way to see the difference between hierarchies for the parameter values under consideration. Figure 8.13 reproduces Figure 8.11 except that now we use *C2* as the input connection matrix. And once again, although differences can be detected between these two figures, one might wonder if we are seeing patterns when there are none. This is answered by Figure 8.14, which graphs the difference between summed consumption under *C2* less the summed consumption under *C1*.

Figure 8.13: Summed per capita consumption, t1=16.7, P4, p_{init} = p, A1, C2

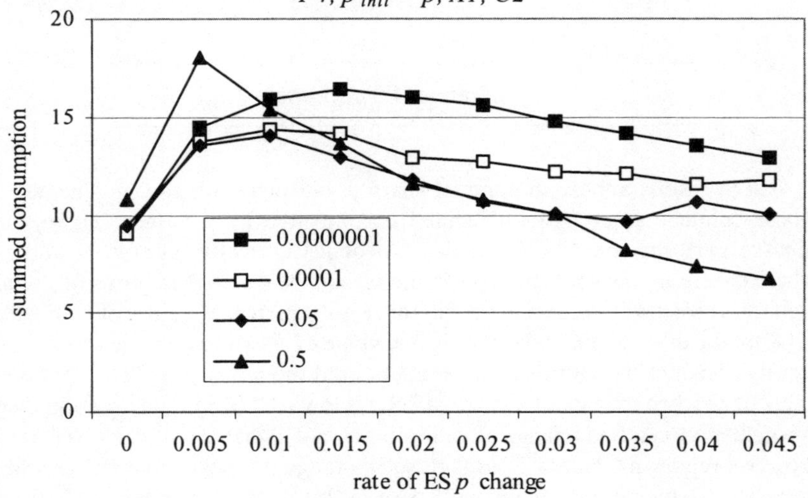

Figure 8.14 reveals the clear difference between *C1* and *C2* in terms of social welfare for the specific parameter values employed here; namely, except for the lowest rates of ES *p* adjustment considered, setting node 1 at the top of

the hierarchy yields a higher level of overall welfare (as measured by summed per capita consumption) than does setting node 15 at the top.

Figure 8.14: Differential per capita consumption, C2-C1, t1=16.7, P4, p init = p, A1

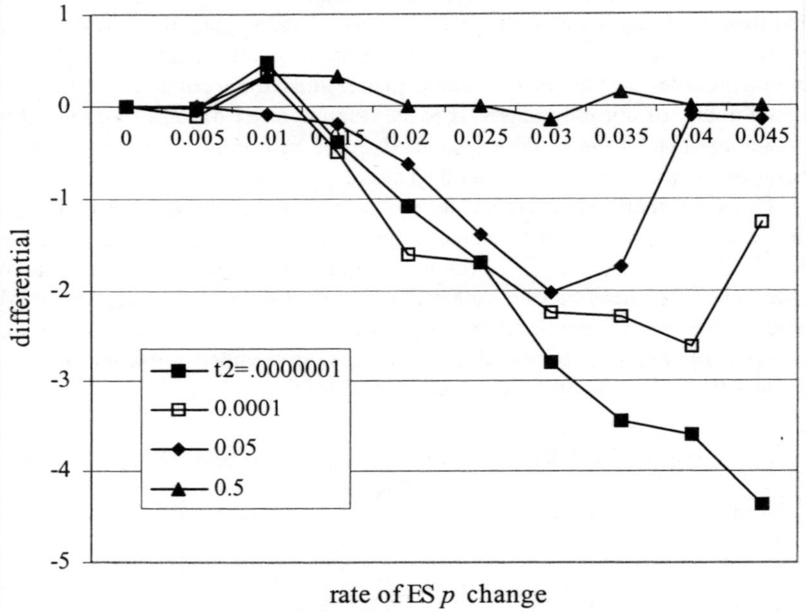

These results appear to contrast sharply with what we saw in Chapter 7 where connection strengths depended on objective ES *p* values. There, the relative performance of C2 versus C1 depends critically on both *t1* and *t2* whereas here it appears that C1 hold the advantage over C2 in terms of social welfare, at least with respect to the full range of 'reasonable' values of *t2*. Figure 8.14, on the other hand, looks at only one value of *t1*. Moreover, we should not wholly discount the fact that C2 seems to hold the advantage for all but one value of *t2* when the rate of change of ES *p* is low (≤ 0.015). The question then is whether we can find a range of values for *t1* and *t2* that give results similar to those we report in Chapter 7. In fact, such a range does exist, but unlike what we saw in Chapter 7, it is exceedingly narrow. Insofar as we can tell, aside from appearing when the rate of ES *p* is especially low as in Figure 8.14, it is largely limited to values of *t1* in the interval [40, 28] and then only for values of *t2* that are less than 0.000005 but greater than 0.

Figure 8.15 illustrates this range when we set *t1* = 30. What warrants emphasis here, however, is not the specific values that yield a higher overall average level of consumption when node 15 stands at the top of the hierarchy, but rather the fact that the circumstances under which this occurs are different from those we described in Chapter 7. First, as Table 7.7 illustrates, when strengths of connection depend on the objective values of ES p, we can describe the region where *C1* outperforms *C2* rather simply: As *t2* increases, *t1* must increase 'accordingly'. That is,

> *for the model in Chapter 7, C2 outperforms C1 if, when we decrease t2 so as to make the formation of new connections more difficult, we in a sense compensate by decreasing t1 so as to make their maintenance easier.*

Under alternative A1, in contrast, no simple rule of thumb makes itself evident to us. Instead,

> *under A1, C2 generally outperforms C1 only for a specific range of values for t1 and t2 – a range that seems to be considerably narrower than the one that operates in Chapter 7. This range, moreover, operates only when the rate of adjustment in ES p is sufficiently low.*

Figure 8.15: Differential per capita consumption
C2-C1, t1=30, P4, p init = p

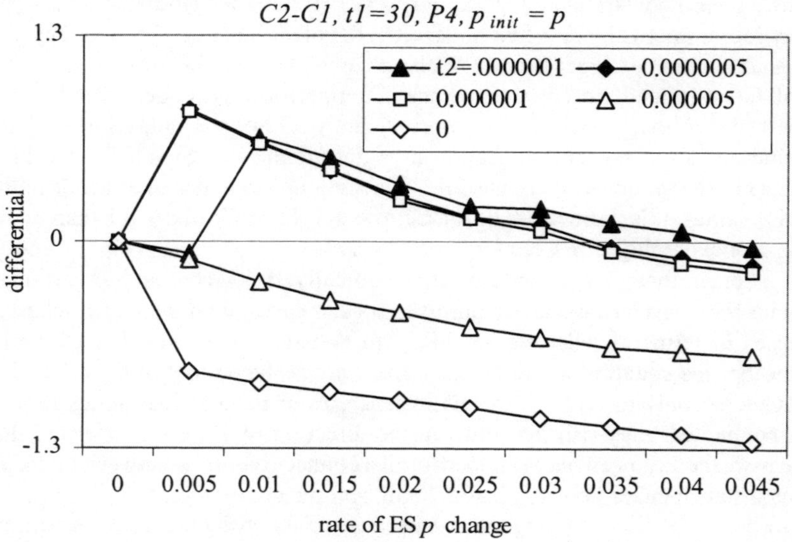

Referring again to Figure 8.15, we note that if *t1* equals 30, but *t2* either exceeds 0.0001 or is less than 0.0000001, then *C1* uniformly outperforms *C2*.

And the same is true *regardless* of *t2*s value, if *t1* is either less than 25 or greater than 40. As a qualitative conclusion, then, it appears that alternative A1 provides far fewer opportunities for a network headed by the least productive node to outperform a network headed by the most productive than does the model explored in Chapter 7. This observation, nevertheless, warrants a number of caveats. First, we should ask what substantive meaning we ought to give to the observation that it is only for relatively 'high' values of *t1* that *C2* performs better under A1. That is, what substantive meaning ought to be given to the fact that *C2* can outperform C1only for values of *t1* in the area of 30 or so when A1 dictates strength of connection whereas we have seen C2 outperform C1 even for values of *t1* as low as 6 in Chapter 7 (see Table 7.7). To answer this question, consider once again expression (7.1) and suppose $c_{old} = 1$. Let Δ denote whatever absolute difference is employed to dictate connection strength, in which case (7.1) becomes

$$c_{new} = (t2+1)/\Delta t1.$$

Hence, c_{new} exceeds the threshold $S = 0.6$ iff $\Delta t1 < 1.7t2 + 1.7$. Generally, we consider values of *t2* that are much less than 1. In Table 7.7, for example, we do not consider a value greater than .05, in which case the preceding expression becomes (approximately) $\Delta < 1.8/t1$. Thus, if *t1* is in the range of, say, 6, Δ cannot be bigger than 0.3 if a connection is to be sustained. This, though, is a fairly liberal constraint and, under P4, is largely satisfied (initially at least) by any four numerically adjacent nodes. On the other hand, if *t1* equals, say, 30, then Δ cannot be bigger than 0.06, else a connection will be broken and nodes will fail to influence each other. Since *C2* outperforms *C1* under A1 only when the order of magnitude of *t1* is approximately 30 but can outperform C1 for values as low as 6 when connection strengths are dictated by true ES p values, we can see that unless there is some 'bunching up' of nodes in terms of initial ES *p* values, it is more difficult for *C2* to outperform *C1* under A1 than under the model employed in Chapter 7.

There is, though, a second caveat. Specifically, the two hierarchical arrangements we consider here are themselves special cases, with their own inherent biases in terms of what we are likely to conclude generally. For example, consider the situation where nodes 1 and 2 are at the bottom of the hierarchy, connected only to node 7. In this instance, even if these two nodes remain disconnected, they will be drawn in the direction of 7's ES *p* in a way that narrows the ES *p* gap between them until a connection forms between them. In this instance either nodes 1, 2 and 7 will become connected, or nodes 1 and 2 will be while disconnecting from 7 because of the influence each has on the other. The tendency, then, will be either for 1 and 2 to be members of a subnetwork that excludes 7, or for all three to be members of a subnetwork that includes all intermediate nodes. This, of course, is precisely what we see in

Tables 7.5, 7.6, 8.2 and 8.3. In this sense, then, our results are an artifact of the specific structure of the hierarchies we have chosen to explore. And we can imagine a different set of results and different patterns with a different assignment of nodes to the vertices of Figure 7.19.

8.5 Appendix: A BASIC Program

The BASIC program that follows allows for iteration over one of five parameters:

- the initial proportion of low *d*s, p_{init},
- the rate of return, *r*, on investments,
- size of incremental adjustments, *ar*,
- initial ES *p*s, and
- rate of ES *p* adjustment.

The iteration parameter is taken to be equal for all nodes in the network. The program, when run, asks for three things: the iteration parameter, and the names of the input and output files. An example of the required input file and the variables it must contain is given at the end of this appendix. The output file contains either the record of all iterations (not recommended when, for instance, 5000 iterations are run as in all simulations reported in this and the last chapter) or the results of the last iteration. In this later case the output file reports the final per capita consumption for each node and at each value of the iteration parameter, the connection matrix at each value of the iteration parameter, and, for the record, the parameters values read off the input file.

```
 CLS
DIM apl!(10, 36), con!(10, 36), ch!(10, 36), cl!(10, 36), p!(10, 36), c!(10, 36, 36)
DIM r!(10), tot!(10), pch!(10, 36), pmax!(10, 36), idi%(10, 36), n%(36),cm1(10)
DIM pl!(10, 36), ih!(10, 36), trl!(10, 36), trh!(10, 36), incp!(10), po!(36)
DIM xxo!(36), tx!(36), ppx!(36), inx!(10), nlo!(36), nhi!(36), pt!(10,36),ctx!(36, 36),
 LOCATE 3, 10
 PRINT "Control parameters, including range and step size if not preset": PRINT
 PRINT " 1. po!(g) = initial proportion of low ds [0 to 1; step .1]"
 PRINT " 3. r  = rate of return on investments [0 to 1; step .1]"
 PRINT " 4. inx! = size of incremental adjustment [0 to .09; step .01]"
 PRINT " 5. p!(.,g)  = ES proportion of low ds [0 to 1 step .1]"
 PRINT " 6. incp!(.) = rate of adj. in ES ps [0 to .45 (.045) step .05 (.005)]":
PRINT: PRINT "  parameter for data set construction [enter 1, 3, 4, 5 or 6] ";
 INPUT parameter: PRINT
 INPUT "  Name of data file to be created:                    ", fle$: PRINT
 INPUT "  Name of data file for input:                        ", flein$
 OPEN fle$ FOR OUTPUT AS #1: OPEN flein$ FOR INPUT AS #7
```

```
REM  Create a data file with all iterations [y or n]?: INPUT #7, d$, iter$
REM  Input Maximum number of periods: INPUT #7, d$, t
REM  Input ESS ev adj based on percap [1] or abs consume [0]
 INPUT #7, d$, percap
REM  Input Number of groups: INPUT #7, d$, ngroups
REM  Input Supplemental income: INPUT #7, d$, kk
REM  Input Rate of return on investment: INPUT #7, d$, rr
REM  Input Group connections matrix: INPUT #7, d$
 FOR jc1 = 1 TO ngroups: FOR jc2 = 1 TO ngroups
 INPUT #7, c!(1, jc1, jc2): ctx!(jc1, jc2) = c!(1, jc1, jc2)
 FOR xx = 2 TO 10: c!(xx, jc1, jc2) = c!(1, jc1, jc2): NEXT xx: NEXT jc2
 IF ngroups = 36 THEN GOTO 101
 FOR jcc2 = ngroups + 1 TO 36: INPUT #7, zx: NEXT jcc2
101  NEXT jc1
 IF ngroups = 36 THEN GOTO 102
 FOR jcc1 = ngroups + 1 TO 36
 FOR jcc2 = 1 TO 36: INPUT #7, zx: NEXT jcc2
 NEXT jcc1
102  INPUT #7, d$
 FOR jc1 = 1 TO ngroups
REM  Input Size of groups: INPUT #7, n%(jc1): NEXT jc1
REM  Input Initital resource endowment vector: INPUT #7, d$, ko%: INPUT #7, d$
 FOR jc1 = 1 TO ngroups
REM  Input Initial values of ESS parameters: INPUT #7, ppx!(jc1): NEXT jc1
REM  Initialize ESS p for every iteration category
 FOR gg = 1 TO ngroups
 FOR xx = 1 TO 10:  p!(xx, gg) = ppx!(gg):  NEXT xx
 NEXT gg
  INPUT #7, d$
  FOR jc1 = 1 TO ngroups
REM  Input initial proportion of low ds: INPUT #7, po!(jc1)
  NEXT jc1
  INPUT #7, d$
REM  Input evolutionary rate: INPUT #7, inxx
REM  Input scale of change for parameter = 6 [10 or 1] and adj rate when par. not 6
 INPUT #7, d$, scle, incpp
REM  Input connection transformation parameters: INPUT #7, d$, t1, t2
REM  Input connection strength required for mimicking: INPUT #7, d$, similar
REM  Input number of connection matricies to be outputted
 INPUT #7, d$, nconp1
REM  Input iteration #s for printing of connection matricies
 FOR kook=1 TO nconp1: input #7, cm1(kook): NEXT kook
REM  ih = total investment of high ds
REM  ill = total investment of low ds
REM  cl = total consumption of low ds
REM  ch = total consumption of high ds
REM  pch=percap total consumption
REM  apl = prop. of consumption attributable to low ds
```

```
REM  trh = total resources of highds after investment
REM  trl = total resources of lowds after investment (if any)
REM   p = ES proportion of low ds
REM   pl = actual proportion of low ds
REM   con = conversion proportion in accord with ES p
REM   incp = adj rate for ES ps
 CLS
REM   Proceed to Calculations
 FOR g = 1 TO ngroups: FOR jj = 1 TO 10
 IF parameter = 1 THEN pl!(jj, g) = jj / 10 ELSE pl!(jj, g) = po!(g)
 NEXT jj: NEXT g
 FOR q = 1 TO 10: para = q / 20
 FOR g = 1 TO ngroups
  IF parameter = 3 THEN r!(q) = para ELSE r!(q) = rr
  IF parameter = 4 THEN inx!(q) = (para-.05)/5 ELSE inx!(q)=inxx
  IF parameter = 5 THEN p!(q, g) = para
  IF parameter = 6 THEN incp!(q) = (para - .05) / scle ELSE incp!(q) = incpp
  NEXT g: NEXT q
  FOR j = 0 TO t: FOR q = 1 TO 10
  FOR g1 = 1 TO ngroups: pmax!(q, g1) = 0: idi%(q, g1) = 0: NEXT g1
  FOR g = 1 TO ngroups
  nlo!(g) = n%(g) * pl!(q, g): nhi!(g) = n%(g) - nlo!(g)
  IF j > 0 THEN GOTO 103
REM   assume that only high ds invest, if at all
  ih!(q, g) = (1 - r!(q)) * ko% * nhi!(g)
  ch!(q, g) = r!(q) * ko% * nhi!(g): cl!(q, g) = ko% * nlo!(g)
103  pch!(q, g) = 0
   IF n%(g) * percap > 0 THEN pch!(q, g) = (ch!(q, g) + cl!(q, g)) / n%(g)
   IF percap = 0 THEN pch!(q, g) = ch!(q, g) + cl!(q, g)
   NEXT g
   FOR g = 1 to ngroups
   FOR g2 = 1 TO ngroups
   IF c!(q, g, g2) <= similar THEN GOTO 104
   IF pch!(q, g2) > pmax!(q, g) THEN pmax!(q, g) = pch!(q, g2): idi%(q, g) = g2
104  NEXT g2
   NEXT g
REM   compute ESS-relevant weight of low ds in each groups
   FOR xx11 = 1 TO ngroups: xxo!(xx11) = 0: tx!(xx11) = 0: NEXT xx11
   FOR xx11 = 1 TO ngroups: FOR xx22 = 1 TO ngroups
   xxo!(xx11) = xxo!(xx11) + c!(q, xx11, xx22) * cl!(q, xx22)
   tx!(xx11) = tx!(xx11) + c!(q, xx11, xx22) * (cl!(q, xx22) + ch!(q, xx22))
   NEXT xx22
   apl!(q, xx11) = xxo!(xx11) / tx!(xx11)
   NEXT xx11
REM   too many or too few ds in group g?
   FOR gg = 1 to ngroups: pt!(q,gg)=p!(q,gg)
   NEXT gg
   FOR g = 1 TO ngroups
```

```
  IF pch!(q,g)<pmax!(q,g) THEN p!(q,g)=p!(q,g)+incp!(q)*(pt!(q,idi%(q,g))-p!(q,g))
  IF apl!(q, g) <= p!(q, g) THEN 105: REM  too few lows
  IF apl!(q, g) > p!(q, g) THEN 106: REM  too many lows
105 REM    TOO FEW LOWS
REM   con(q,g)=prop. of highds that, if converted, sets consumption prop(g)=p(g)
  con!(q, g) = (p!(q, g) * (ch!(q, g) + cl!(q, g)) - cl!(q, g)) / ch!(q, g)
  IF con!(q, g) < 0 THEN con!(q, g) = 0
  IF con!(q, g) > 1 THEN con!(q, g) = 1
REM  adjust transfer by incremental evolutionary rate
  con!(q, g) = con!(q, g) * inx!(q)
REM new proportion of low ds
  pl!(q, g) = pl!(q, g) + con!(q, g) * (1 - pl!(q, g))
  trl!(q, g) = con!(q, g) * (1 + r!(q)) * ih!(q, g) + n%(g) * kk * pl!(q, g)
  trh!(q, g) = (1 - con!(q, g)) * (1 + r!(q)) * ih!(q, g) + n%(g) * kk * (1 - pl!(q, g))
REM   calculate investment and consumption for next period
  ih!(q, g) = (1 - r!(q)) * trh!(q, g): ch!(q, g) = r!(q) * trh!(q, g): cl!(q, g) = trl!(q, g)
  GOTO 107
106 REM    TOO MANY LOWS
  con!(q, g) = 1 - p!(q, g) * (cl!(q, g) + ch!(q, g)) / cl!(q, g)
  IF con!(q, g) < 0 THEN con!(q, g) = 0
  IF con!(q, g) > 1 THEN con!(q, g) = 1
  con!(q, g) = con!(q, g) * inx!(q): pl!(q, g) = (1 - con!(q, g)) * pl!(q, g)
  trl!(q, g) = n%(g) * kk * pl!(q, g)
  trh!(q, g) = (1 + r!(q)) * ih!(q, g) + n%(g) * kk * (1 - pl!(q, g))
REM  calculate investment and consumption for next period
  ih!(q, g) = (1 - r!(q)) * trh!(q, g)
  ch!(q, g) = r!(q) * trh!(q, g): cl!(q, g) = trl!(q, g)
107 REM
  NEXT g: NEXT q
  IF iter$ = "n" AND j < t THEN GOTO 108
  FOR gg = 1 TO ngroups: FOR w = 1 TO 10
  PRINT #1, USING "#########.##"; (cl!(w, gg) + ch!(w, gg)) / n%(gg);
  NEXT w
  PRINT #1, dum$: NEXT gg
  PRINT #1, dum$
108 REM
  FOR k1 = 1 TO ngroups: FOR k2 = 1 TO ngroups
  FOR k3 = 1 TO 10: IF c!(k3,k1,k2)+t2 = 0 THEN GOTO 110
  kv1 = p!(k3,k1)
  kv2 = p!(k3,k2)
  IF kv1 = kv2 THEN c!(k3, k1, k2) = 1: GOTO 110
  c!(k3, k1, k2) = (c!(k3, k1, k2) + t2) / (t1 * ABS(kv1-kv2))
109  IF c!(k3, k1, k2) < 0 THEN c!(k3, k1, k2) = 0: GOTO 110
  IF c!(k3, k1, k2) > 1 THEN c!(k3, k1, k2) = 1
110  NEXT k3: NEXT k2: NEXT k1: NEXT j
  FOR ncon1=1 TO nconp1
  cmat = cm1(ncon1)
  FOR jja = 1 TO ngroups: FOR jjb = 1 TO ngroups
```

```
PRINT #1, USING "#####.######"; c!(cmat, jja, jjb);
NEXT jjb
PRINT #1, dumm$
NEXT jja
print #1, dumm$
NEXT ncon1
PRINT #1, dumm$: PRINT #1, USING "##########.#"; t:
PRINT #1, USING "#########.##"; percap
PRINT #1, USING "##########.#"; ngroups
PRINT #1, USING "#########.##"; kk: PRINT #1, USING "#########.##"; rr
FOR jc1 = 1 TO ngroups: FOR jc2 = 1 TO ngroups
PRINT #1, USING "########.###"; ctx!(jc1, jc2);
NEXT jc2
PRINT #1, dum$: NEXT jc1
FOR jc1 = 1 TO ngroups: PRINT #1, USING "#########.##"; n%(jc1);
NEXT jc1
PRINT #1, dum$: PRINT #1, USING "########.###"; ko%
FOR jc1 = 1 TO ngroups: PRINT #1, USING "########.###"; ppx!(jc1);
NEXT jc1
PRINT #1, dum$
FOR jc1 = 1 TO ngroups: PRINT #1, USING "########.###"; po!(jc1);
NEXT jc1
PRINT #1, dum$
FOR jc1 = 1 TO 10: PRINT #1, USING "########.###"; inx!(jc1);
NEXT jc1: PRINT #1, dum$
PRINT #1, USING "#######.####"; scle; incpp
PRINT #1, USING "##.#########"; t1; t2
PRINT #1, USING "#########.##"; similar
PRINT #1, USING "#########.##"; cmat
CLOSE : END
```

A typical input data file looks as follow (Note that, in this example, the input connections matrix corresponds to the hierarchy with node 1 on top. Note also that the initial connections matrix, regardless of the actual number of nodes, must be 36 x 36. However, with N nodes, only the relevant NxN submatrix will be read as data):

```
"print all iterations? (y or n)"  n
"number of iterations" 5000
"per cap [1] or total consumption [0]" 1
"number of nodes" 15
"supplemental income" .2
"rate of return on investments" .05
"connections matrix"
1 0 0 0 0 0 0 0 0 0 0 0 0 0 0 0 0 0 0 0 0 0 0 0 0 0 0 0 0 0 0 0 0 0 0 0
1 1 0 0 0 0 0 0 0 0 0 0 0 0 0 0 0 0 0 0 0 0 0 0 0 0 0 0 0 0 0 0 0 0 0 0
1 0 1 0 0 0 0 0 0 0 0 0 0 0 0 0 0 0 0 0 0 0 0 0 0 0 0 0 0 0 0 0 0 0 0 0
0 1 0 1 0 0 0 0 0 0 0 0 0 0 0 0 0 0 0 0 0 0 0 0 0 0 0 0 0 0 0 0 0 0 0 0
```

```
0 1 0 0 1 0 0 0 0 0 0 0 0 0 0 0 0 0 0 0 0 0 0 0 0 0 0 0 0 0 0 0 0 0 0 0
0 0 1 0 0 1 0 0 0 0 0 0 0 0 0 0 0 0 0 0 0 0 0 0 0 0 0 0 0 0 0 0 0 0 0 0
0 0 1 0 0 0 1 0 0 0 0 0 0 0 0 0 0 0 0 0 0 0 0 0 0 0 0 0 0 0 0 0 0 0 0 0
0 0 0 1 0 0 0 1 0 0 0 0 0 0 0 0 0 0 0 0 0 0 0 0 0 0 0 0 0 0 0 0 0 0 0 0
0 0 0 1 0 0 0 0 1 0 0 0 0 0 0 0 0 0 0 0 0 0 0 0 0 0 0 0 0 0 0 0 0 0 0 0
0 0 0 0 1 0 0 0 0 1 0 0 0 0 0 0 0 0 0 0 0 0 0 0 0 0 0 0 0 0 0 0 0 0 0 0
0 0 0 0 1 0 0 0 0 0 1 0 0 0 0 0 0 0 0 0 0 0 0 0 0 0 0 0 0 0 0 0 0 0 0 0
0 0 0 0 0 1 0 0 0 0 0 1 0 0 0 0 0 0 0 0 0 0 0 0 0 0 0 0 0 0 0 0 0 0 0 0
0 0 0 0 0 1 0 0 0 0 0 0 1 0 0 0 0 0 0 0 0 0 0 0 0 0 0 0 0 0 0 0 0 0 0 0
0 0 0 0 0 0 1 0 0 0 0 0 0 1 0 0 0 0 0 0 0 0 0 0 0 0 0 0 0 0 0 0 0 0 0 0
0 0 0 0 0 0 1 0 0 0 0 0 0 0 1 0 0 0 0 0 0 0 0 0 0 0 0 0 0 0 0 0 0 0 0 0
1 1 1 1 1 1 1 1 1 1 1 1 1 1 1 1 1 1 1 1 1 1 1 1 1 1 1 1 1 1 1 1 1 1 1 1
1 1 1 1 1 1 1 1 1 1 1 1 1 1 1 1 1 1 1 1 1 1 1 1 1 1 1 1 1 1 1 1 1 1 1 1
1 1 1 1 1 1 1 1 1 1 1 1 1 1 1 1 1 1 1 1 1 1 1 1 1 1 1 1 1 1 1 1 1 1 1 1
1 1 1 1 1 1 1 1 1 1 1 1 1 1 1 1 1 1 1 1 1 1 1 1 1 1 1 1 1 1 1 1 1 1 1 1
1 1 1 1 1 1 1 1 1 1 1 1 1 1 1 1 1 1 1 1 1 1 1 1 1 1 1 1 1 1 1 1 1 1 1 1
1 1 1 1 1 1 1 1 1 1 1 1 1 1 1 1 1 1 1 1 1 1 1 1 1 1 1 1 1 1 1 1 1 1 1 1
1 1 1 1 1 1 1 1 1 1 1 1 1 1 1 1 1 1 1 1 1 1 1 1 1 1 1 1 1 1 1 1 1 1 1 1
1 1 1 1 1 1 1 1 1 1 1 1 1 1 1 1 1 1 1 1 1 1 1 1 1 1 1 1 1 1 1 1 1 1 1 1
1 1 1 1 1 1 1 1 1 1 1 1 1 1 1 1 1 1 1 1 1 1 1 1 1 1 1 1 1 1 1 1 1 1 1 1
1 1 1 1 1 1 1 1 1 1 1 1 1 1 1 1 1 1 1 1 1 1 1 1 1 1 1 1 1 1 1 1 1 1 1 1
1 1 1 1 1 1 1 1 1 1 1 1 1 1 1 1 1 1 1 1 1 1 1 1 1 1 1 1 1 1 1 1 1 1 1 1
1 1 1 1 1 1 1 1 1 1 1 1 1 1 1 1 1 1 1 1 1 1 1 1 1 1 1 1 1 1 1 1 1 1 1 1
1 1 1 1 1 1 1 1 1 1 1 1 1 1 1 1 1 1 1 1 1 1 1 1 1 1 1 1 1 1 1 1 1 1 1 1
1 1 1 1 1 1 1 1 1 1 1 1 1 1 1 1 1 1 1 1 1 1 1 1 1 1 1 1 1 1 1 1 1 1 1 1
1 1 1 1 1 1 1 1 1 1 1 1 1 1 1 1 1 1 1 1 1 1 1 1 1 1 1 1 1 1 1 1 1 1 1 1
1 1 1 1 1 1 1 1 1 1 1 1 1 1 1 1 1 1 1 1 1 1 1 1 1 1 1 1 1 1 1 1 1 1 1 1
1 1 1 1 1 1 1 1 1 1 1 1 1 1 1 1 1 1 1 1 1 1 1 1 1 1 1 1 1 1 1 1 1 1 1 1
1 1 1 1 1 1 1 1 1 1 1 1 1 1 1 1 1 1 1 1 1 1 1 1 1 1 1 1 1 1 1 1 1 1 1 1
1 1 1 1 1 1 1 1 1 1 1 1 1 1 1 1 1 1 1 1 1 1 1 1 1 1 1 1 1 1 1 1 1 1 1 1
1 1 1 1 1 1 1 1 1 1 1 1 1 1 1 1 1 1 1 1 1 1 1 1 1 1 1 1 1 1 1 1 1 1 1 1
1 1 1 1 1 1 1 1 1 1 1 1 1 1 1 1 1 1 1 1 1 1 1 1 1 1 1 1 1 1 1 1 1 1 1 1
"size of groups"
100 100 100 100 100 100 100 100 100 100 100 100 100 100 100
"initial endowment"  1
"initial ES ps" .05 .11 .17 .23  .29 .35   .41 .47  .53 .59 .65 .71 .77 .83   .89
"initial proportion of low ds"
.05 .05 .05 .05 .05 .05 .05 .05 .05 .05 .05 .05 .05 .05 .05
"initial ar, rate of change"  .025
"scale for ES p change and reversion ES p rate of change"  10  .025
"t1 and t2"  16.7  0.0000001
"S threshold"  .6
"# of connection matricies to be printed and which matricies"
9  2  3  4  5  6  7  8  9  10
```

9 Summary

These new replicators, which I call memes to distinguish them from genes, can propagate themselves from brain to brain ... (Dawkins 1987: 158)

9.1 Empirical Summary

This volume begins with two observations. First, within the rational choice paradigm, cooperation generally requires a 'shadow of the future' – individuals who are concerned sufficiently with the promise of future rewards or the threat of future losses or punishments in order to forestall defecting from cooperative arrangements today for immediate gain. Even if we assume, for example, that preferences and social norms conducive to cooperative action are somehow 'hard wired' by evolutionary development, that shadow must be a part of the preferences that evolve (Bendor and Swistak 2001, Gintis 2003). But second, this shadow, as usually modeled – as a private time discount rate – is taken as a fixed and exogenously determined parameter (and is even assumed at times to be uniform across populations and equal to objectively determined economic discount rates). It is this second observation that, we argue, corresponds to an untenable assumption that can distort our understanding of important social processes. If, for example, socialization and learning are modeled within the paradigm, then in lieu of assuming that people act out of habit (Pollak 1970, 1976) or simply develop a 'taste' for cooperation (see Hansson and Stewart 1990, Guttman 2000, 2003, Bisin and Verdier 2001, Bisin et al. 2004), it must be that what changes are people's parameters of decision making – preferences and beliefs. And, as the first observation tells us, one critical parameter is the weight people give to the future.

Although individual time discounts correlate with a number of socio-economic circumstances and variables (see Godoy et al. 2004 for a general survey), empirical instances of individual variation in time preferences suggest that this parameter of choice is itself a core concept and needs to be modeled as such. This book attempts to move us in that direction. We do so, moreover, with the assumption that, aside from variations in perceived uncertainty, time preference is a learned and changeable parameter and thereby a function of social interaction. That is, just as people 'learn' core preferences on the basis of experience and observation, we assume the same is true with their preferences for time. Thus, for example, if youth in a 'ghetto' interact primarily with their peers for whom immediate rewards take precedence over any potential long term goals, that immediacy will be learned and sustained as a characteristic of the network of which they are a part. And if crime and a propensity to drug addiction are, as we argue, likely consequences of high time discounts, then we can see one of the great difficulties with breaking the 'cycle' of crime and delinquency in central cities.

Additional support for the hypothesis of dynamically changing individual time preferences is offered by Bickel and Johnson (2003), who summarize and reanalyze much of the experimental research linking addiction with time discounting (smoking, alcohol, and opiates). For example, in Bickel et al.'s (1999) study, ex-smokers were found to discount money in a fashion identical to non-smokers while Petry (2001) reports that the discounting of recently abstinent alcohol-dependent individuals is intermediate between current alcoholics and a control population. Here we can also cite Vuchinich and Simpson's (1998: 292) conclusion that 'heavy social drinkers ... and problem drinkers ... both showed greater temporal discounting than light social drinkers'; Madden et al.'s (1997: 259) finding that 'the rate at which the subjective value of monetary rewards decreased with increasing delays (k) was significantly higher for opioid-dependent participants than for control participants'; Allen et al's (1998: 144) finding that 'subjects with a past history of drug dependence evidence more impulsivity in both behavioral and self-reporting measures when compared to subjects with no illicit drug use history'; Kirby et al's (1999: 78) summary conclusion that 'on average, heroin addicts' discount rates were twice those of controls'; and Odum et al.'s (2000: 259) conclusion that 'persistent needle sharing may be related to the relative inability of delayed outcomes to impact current behavior.' These findings, as well as Breteville-Jensen's (1999) study of ex-injectors of heroine and amphetamines, and Allen et al.'s (1998) study of ex-drug-dependent individuals, show that the time discounting can be modified if one applies a comprehensive strategy of making changes in their environment and behavioral choices. Thus, as Giordano et al. (2002: 174-5) summarize matters, 'a rapidly growing literature indicates that individuals with substance dependencies might be generally more susceptible to behavioral problems that may stem from diminished sensitivity to delayed outcomes ... drug-dependent individuals are controlled to a greater extent by smaller more immediate consequences at the expense of larger more delayed consequences' (see also Baket et al. 2003 ans Bickel and Johnson for surveys of the literature).

The empirical evidence we consider in Chapter 2 is consistent with these studies. Although we emphasize again that selection issues cloud definitive conclusions, the data we offer with respect to participants in Cocaine Anonymous as well as the differences observed in our samples of Ukrainian students are consistent with the hypothesis that time preference is an endogenous parameter. Both length of time in a recovery program and duration of one's stay in the West correlate positively with an apparent willingness to value the future. But such observations are consistent as well with the hypothesis that only those who value the future are successful in a CA program and primarily (but not exclusively) those who minimally discount the future emigrate from the former Soviet Union to study in the West. But in both cases, the mere suggestion that time preference is endogenous has potentially profound implications that warrant further investigation. In the case of CA and similar programs, an explicit appreciation of the impact such programs might have on time preference and the corresponding reduction in the likelihood of drug addiction suggests not only how such programs fundamentally alter people's preferences and decision criteria, but also point us in the direction of sharpening our focus on the treatment of addiction. We are not so bold as to suggest specific changes in the structure and operation of such programs, since, as we argued in Chapter 2, they already appear to be implicitly designed to alter time preference. We can only hope that by rendering that objective more explicit, those who operate such programs can sharpen their focus and techniques.

In the case of students emigrating to the West, to the extent that time preferences change as a part of their education, we can see here an important component of that education that extends beyond the things formally learned. It is often argued, in defending the funding for exchange programs of various types, that one benefit is introducing foreign students to democratic principles. The data we offer in Chapter 2 suggests that something more fundamental is also learned; namely, valuing the future, which is itself critical to sustaining the various forms of cooperation that render a democratic market-based society viable. And again, as with drug addiction, understanding fully what it is that is being learned might be of assistance to those who design and operate these exchange programs.

Of course, to reiterate, much follow-up research is required before definitive conclusions can be reached. At a minimum, it would be useful to conduct panel studies whereby the dynamics of individual time preference can be studied. For example, is the five-year mark that appears to be critical in our study of drug addiction a relatively uniform constant, or do individuals of different types or with different addictions require different lengths of participation before time preference changes definitively? Similar issues and questions arise with respect to our study of Ukrainian students. Do students who study in the West exhibit a lower time discount because they now perceive less uncertainty or because of a process of peer group socialization? Can we discriminate theoretically and then empirically between these two sources of change? In this instance it would be useful to develop an appropriately large sample of respondents who have studied in the West and then returned to their native countries and ask: Do the values they place on the future revert to pre-'immigration' values or do they remain higher, on average, than their peers? And if the impatience that eroded while studying in the West reemerges, does it do so because respondents perceive reduced opportunities for self-advancement or because they have reinserted themselves in a social group that values the future less?

9.2 Theoretical Summary and Extensions

Overview: Turning now to the theoretical component of this book, we cannot suppose that the analysis we offer is anything but an initial foray into the determinants of endogenous time preference. At the same time, it is also true that the empirical research Chapter 2 reviews does not test that analysis. It merely provides a basis for believing that individual time discount rates are subject to endogenous change. Our theoretical analysis, in contrast, seeks to explore one potential mechanism of such change with the understanding that, except in a few instances (e.g., modeling strength of ties), the literature on endogenous preferences and time discounting provides little theoretical guidance. Consequently, we have adapted what seems a tractable approach – an evolutionary perspective – wherein the parameters of individual discounting are 'learned' in the sense of being the surviving parameter values in an environment in which all potential values must 'compete'.

Our approach parallels the linkage between learning and replicator dynamics that has emerged in the field of evolutionary game theory. However, rather than suppose that evolutionary processes operate over biological entities, we follow Dawkins's (1976) suggestion that the objects of replicator dynamics can be memes, 'defined as cognitive or behavioral patterns that can be transmitted from one individual to another one by learning and imitation' (Heylighen 1992: 1–2). That is, the 'replicating entity is the meme, loosely defined as an idea that may be transmitted via mimicry' (Blok 2002: 1). In our analysis, time preference is a meme so that the evolutionary process we model is consistent with the prescription that 'in memetic evolution ... the less successful group simply imitates the more successful one' (Heylighen 1992: 6). Here we assume that the competing memes within an individual or the memes that characterize an individual adapt or evolve in the direction of an evolutionarily stable proportion (denoted ES p). The value of ES p, in turn, changes via a process of mimicry whereby the group (or individual) observes the performance of other groups or individuals, and adjusts its ES p in the direction of the seemingly best-performing group (individual). This notion of mimicry, then, parallels the concept of learning as originally introduced by Bush and Mosteller (1951: see also Cross 1973) and subsequently incorporated into evolutionary game theory (see, for instance, Harley 1981, Fudenberg and Tirole 1991, Weibull 1995, Borgers and Sarin 1997, Fudenberg and Levine 1998).

It is useful to distinguish here between two implicit modes of learning, the first of which occurs strictly within a group or network node and corresponds to the usual treatment of replicator dynamics. The second occurs across nodes. With respect to the first, whereby the proportion of high discount and low discount memes are assumed to converge incrementally to the ES p that corresponds to that node, our analysis allows for two subsidiary interpretations. In one we can view a node (the single 'group' analyzed in Chapters 3 and 4) as a collection of discrete decision making entities, each with a specific meme or time discount factor. Alternatively, we can interpret a node as an individual in which there are competing memes or 'ideas' (competing time preferences) whereby the proportion of types adjusts to an exogenously set ES p. With either interpretation, we can interpret change (either of the individual or of the proportion of meme types within an individual) as learning (Borgers and Sarin 1997, Borgers et al. 2002; see also Tromsdorf and Lamm 1975).

Note, however, that both interpretations follow the traditional analysis of replicator dynamics and, thereby, are subject to the usual critique of such analyses whether applied to biological or psychological replicators: namely, that the ES p to which a system converges is a function of a set of relative fitness parameters, the determinants of which are exogenous to the analysis. It is here that we introduce the innovation of embedding our evolutionary model in a model of social networks that allows for a second level of learning whereby a node's ES p can mimic other relatively well-performing groups. Specifically, in any one period decisions are made by, implicitly, drawing time preferences at random from the population that describes a node and an average (per capita) consumption calculated. Depending then on the performance of that entity relative to others to which it is connected, we incrementally adjust the node's ES p in the direction of the better-performing node. Thus, with adjustments occurring both within and across nodes, we can say that the replicator dynamics of our analysis correspond to a two-stage process of learning.

The final innovation, which constitutes the analysis in Chapter 5, is to make connections within the network – the groups or nodes each specific group or node mimics – endogenous as well.

The Simple Investment/Consumption Model: Insofar as the limitations and extensions of the model offered are concerned, we can begin with Chapter 3 and the alternative substantive scenarios considered there that we link to an evolutionary perspective of endogenous discounting. Each of those scenarios, which corresponds to a traditional 2 × 2 evolutionary game model, presents a potential (but as yet not wholly developed) substantive interpretation of the analysis that follows. That is, if our analysis is to have full empirical content that is subject to empirical application, those scenarios need to be developed further, and the conclusions we offer later need to be redirected 'back' to them. Specifically, the core input variables in these scenarios are the payoffs (normally interpreted as fitness parameters in evolutionary replicator dynamics) we employ to derive evolutionary stable proportions of low and high discount rates. And it is these variables that need to be modeled and measured so as to give our theoretical analysis full empirical content. Moreover, when, in later chapters, we allow evolutionary stable proportions to adjust within the nodes of a network, it is again the parameters of a more basic and substantively defined evolutionary game that implicitly change. It is the fuller elaboration of this change, set in specific substantive contexts, that warrant further investigation. Nevertheless, the scenarios in Chapter 3 do highlight the general relevance of our analysis. They suggest, among other things, that the study of such matters as juvenile crime and drug abuse, along with political/economic corruption in transitional societies, need to be augmented with a fuller treatment of time preference and how those preferences are determined in specific social contexts.

However, even if we limit our attention to the theoretical concerns set forth here and adhere to the approach taken, there are any number of alternative assumptions that might be employed in developing a model that can be tied more tightly to specific empirical applications. For example, the model of investment and consumption offered in Section 4.1 is especially simple and a non-linear capital stock perspective may, for some applications, be more appropriate (see, for example, Boylan et al. 1991). Admittedly, the choice of model here is driven more by analytical tractability. We suspect that similar conclusions will hold for other models, but we cannot be certain until a complete comparative analysis is done, however, note that one consequence of this move to analytic tractability is that the *quantitative* value of an individual's discount rate loses significance – what matters is simply whether it is 'high' or 'low'. That is, what matters is whether one's discount is sufficiently high to induce full consumption or sufficiently low to induce investment, where the rate of investment depends only on the rate of return of those investments. In a capital stock model, in contrast, the rate of investment will depend as well on the quantitative value of a person's discount rate (see, for example, Klochko 2004).

Chapters 3 and 4 are but a preliminary step in the direction of modeling endogenous time preferences in networks. Nevertheless, we should illuminate several assumptions that warrant treatment in subsequent research. The first is the assumption that rates of return on investments are uniform across the population, as is supplemental income. Hence, our analysis assumes, in effect, that all persons share equivalent opportunities. This assumption is perhaps not restrictive if we choose to interpret the 'individuals' here as memes, since in this instance they can be construed as all 'operating' in the head of a single decision-making entity. On the other hand, if the 'individuals' in the analysis are taken to be discrete entities, our assumption is clearly restrictive. It also needs to be kept in mind that although we make time discounts endogenous, we do so in Chapter 4 only by making a variety of other variables exogenous – notably, the rate of return on investments, the rate of evolutionary change, and the initial value of ES p. Admittedly, the assumption that rate of return is exogenous is not especially restrictive since we are not attempting to model the full dynamics of a society or economy. The assumption of a fixed rate of return is restrictive only if we are attempting to model the consequences of all interactions within a society and is no less restrictive than the assumption, for example, that individual consumers cannot affect price. Similarly, an exogenously fixed rate of adjustment is not especially problematical since we can do comparative statics to assess the consequences of different rates.

In contrast, the assumption of an exogenously set initial ES p, although characteristic of the literature on replicator dynamics, is restrictive in that it merely pushes the problem back a step or two. Also exogenous and in a trivial way is network structure: if we interpret the group in Chapter 4 as a collection of individuals, each with a high or low discounting meme, the implicit assumption is that all nodes are connected to all others with a strength of 1. Indeed, the absence there of any explicit consideration of networks necessarily renders that analysis merely preliminary. If, as we assume, time preference changes in an evolutionary adaptive way, then the study of endogenous time preference requires formalization of 'social context'. If, for instance, one's propensity to drug addiction, crime, or street gang membership is a consequence of one's private time preference, then we cannot understand those propensities and how they might be determined and change in different social environments without examining how time preferences change in those environments.

The Exogenously Structured Network: Chapters 5 and 6 begin to generalize this analysis in a direction more consistent with what we know about learning – namely, embedding a model of investment and consumption in a social network. Here we allow for a 3-node network, and although network structures are taken as exogenous inputs (with strengths of connection equal either to 1 or 0), we begin the process of generalizing our theoretical development by allowing evolutionary stable proportions of high and low discounters to itself be endogenous. Of course, here again we see a variety of alternative ways to implement endogeneity with respect to these proportions. The model we choose here is one in which a node in the network 'looks' at all other nodes to which it is connected and incrementally (at an exogenously fixed rate) mimics the node that appears to be doing best (in terms of per capita consumption). This is the simplest formulation of learning or adaptation (again, see Bush and Mosteller 1951, Cross 1973) and seems a reasonable starting point for any initial inquiry.

A number of abstract conclusions are developed in this somewhat more elaborate model of endogenous time discounting, such as the finding that if the rate of return on investments is 'sufficiently high' so as to encourage consumption rather than investment, network structure becomes irrelevant to group or individual performance. From a sociological perspective, however, the more substantively interesting conclusion is that if the rate of mimicry is 'sufficiently high', a small cadre of visible individuals (high consumers with low rates of saving) can change the behavior of the mass of the population and that it is only when that rate of adjustment in ES *p* is low that that mass can exert a decisive positive influence on this smaller cadre. We can, then, speculate as to the sorts of things that accelerate or decelerate rates of adjustment. For example, if the manifestations of corruption – wholly visible 'extravagant consumption' and a class of corrupt officials that seemingly flaunt their status as occurred in Russia and Ukraine – we might suppose that the rate of mimicry is increased over what might occur in a society whereby the more corrupt elements of the population are more isolated from the rest of society (e.g., Mexico). Similarly, we can speculate that closed neighborhoods with only incidental external contacts, such as might characterize the environment of inner-city youth, would accelerate the process of replicator dynamics and thereby increase the impact of the values and beliefs of small gang cadres on those who might not otherwise participate in gang membership. The suggestion here, then, is that policies and programs that slow this rate (e.g., summer sports programs and the establishment of youth centers) can have a profound influence on the overall social environment of otherwise 'vulnerable' populations.

Our analysis is also consistent with those elements of game theory that model learning via mechanisms of fictitious play (see Fudenberg and Tirole 1991: 23-9) and which conclude that rapid adjustments often preclude convergence to any equilibrium. Here we find that social networks will operate in 'more understandable' ways to the extent that the changes within them are not 'too abrupt' whereas if change is rapid – if the rate of convergence within a node to its ES *p* is high or if nodes 'too readily' mimic others that appear to be performing better – an impression of incoherence may prevail.

Finally, we see that there need not be a monotonic relationship between overall social welfare and the connectedness of networks. It is not the case that connectedness necessarily yields a higher or lower overall level of consumption in equilibrium. Rather, that relationship is mediated by rates of adjustment and the relative sizes of groups. More generally, these observations tell us that in any reasonably complex model of endogenous time preference, we should not anticipate monotonic relationships among parameters. That is, if non-monotonicity appears in the relatively simple linear system that comprises the model here, we can suppose that even more complex relationships might prevail in models that employ non-linear structures (e.g., a capital stock model).

There is, however, one feature of our analysis that warrants further discussion. Much of what we see in Chapters 5 and 6 is the immediate consequence of our formulation of mimicking whereby one group contrasts its level of consumption with other groups to which it is connected is that groups or nodes that discount the future greatly and thereby consume rather than invest, have a greater influence on groups or nodes that invest rather than vice versa.

That is, when assessing the relative performance of a group or node, a group that invests will consume little and thereby not be mimicked as readily as a group that wholly consumes. It is only until a group with an initially high valuation of the future saves and earns a return on its investment that it can consume at a rate that will lead others to mimic its ES p. Note, though, that if the rate of adjustment is 'sufficiently' high, that group can be 'dragged down' quickly and never be afforded the opportunity to influence a group that initially consumes all its resources. It is for this reason, then, that the internal rate of mimicry is an important parameter in our analysis.

The point we wish to make here is that any conclusions we might offer about the influence of one group on another is an immediate consequence of our linear investment-consumption model. A node that places little weight on the future will consume all of its resources and will be replenished only by its supplemental income rather than by any return on investments. A model in which the proportion of resources consumed is a smoother function of individual time discounts is likely to produce different conclusions about the influence of groups of different types and sizes and perhaps different conclusions about rates of change in parameters. This is not to say that the analysis we offer is 'wrong', but only that it need not apply universally and that other models of investment and consumption need to be considered as well.

Chapters 5 and 6 also offer two alternative assumptions about how the nodes of a network adjust their ES p values. One alternative follows from the framework of Chapter 4 whereby nodes act as if they take as their estimate of the ES p's of other nodes the proportion of consumption attributable to the consumption of high versus low d's. That is, the proportion of types at each node is weighted by the consumption of that type. In Chapter 6, in contrast, we in effect assume that each node knows the objective value of every other node's ES p value. Once again, we cannot say which assumption is the more viable empirically and to that extent Chapters 5 and 6 can be taken as an assessment of the sensitivity of our analysis to this assumption. And although some differences are observed, they are 'minor', at least compared to the sensitivity of outcomes to other parameters such as speeds of adjustment and network structure.

The Endogenous Network: Despite the numerous parameters that can be explored in the context of the model offered in Chapters 5 and 6, there remains one critical exogenously fixed 'parameter' – network structure – that, as an assumption, seems especially untenable. It is, then, this parameter that is the focus of our final two chapters. Moreover, unlike earlier attempts to render network structure endogenous, we allow the strengths of connection to be a continuous parameter that varies between 0 and 1. Our assumption here is that the strength of connection between two network nodes is a monotonic function of the similarity of those nodes in terms of ES p. Thus, if two groups are similar in this respect, the strength of connection will move towards 1, whereas if they are dissimilar, it will move towards 0. Here, however, we encounter two additional alternative formulations: one in which strengths of connection depend on the objective values of ES p and another in which connection strength depends on the similarity of rates of consumption (i.e., where the presumed ES p is taken to be the proportion of consumption attributable to low ds). Chapter 7, then, explores the first alternative while Chapter 8 assumes that strengths of connection depend on the values of ES p inferred from patterns of consumption.

Beyond this, the extension to endogenous networks is critical to our understanding of social systems generally. Specifically, one phenomenon that cannot be addressed with the model in Chapters 5 and 6 is *clustering* – the propensity of people to form distinct subnetworks and associate primarily with those of their own type. Only by rendering network structures endogenous can we ascertain whether clustering hinders or facilitates social welfare, whether it hinders or retards the influence of extreme types within a society, and whether rapid of slow rates of adjustment hinder or encourage clustering.

Given the complexity of the resulting model and the number of relevant input parameters (i.e., the rate of return on investment, the rate of evolutionary adaptation, the rate of mimicry, the supplemental income per period, an n-element vector of initial evolutionary stable proportions, an $n \times n$ matrix that describes the initial network structure, and the two parameters linking strength of connection to similarity of nodes), only a small fraction of parameter configurations can be considered here. Indeed, even if we fix all other parameters, the potential variation in initial network structures and initial values for ES p across nodes is daunting. Thus, there is little doubt that a great many more patterns remain to be discovered over and above the ones we reveal here. Nevertheless, several of the patterns we do uncover are substantively suggestive. For example, although it is commonplace to assert that social welfare declines as heterogeneity increases – as income disparities increase – we find the opposite in our model. And indeed, if we divide the World Bank's data on income inequality and economic development into subcategories (developed 'Western' states, the former Soviet bloc, and 'other') we find evidence that supports this implication (see Section 7.4).

This is not to say, of course, that the opposite relationship might not exist for a different constellation of initial parameter values. But minimally, our analysis brings into question the universality of one presumed relationship. Insofar as other patterns are concerned, Chapter 7 offers the following. First, when all nodes are initially connected, the presence of a twin (in terms of the vector of initial ES ps) renders sub-network structure somewhat less sensitive to initial parameter values, at least in the case where those values allow for low rates of adjustment. At low adjustment rates a node can resist mimicking those nodes that consume and do not save until its twin builds up its investments so as to achieve high rates of consumption in the long run, at which point, even those initially unproductive nodes will themselves begin to mimic the more productive ones. Thus, although it seems reasonable to suppose with respect to, say, street gang membership or drug addiction, that a person who initially prefers to invest in the future through education would be better able to resist the short-term pressures of conformity if there are other similar persons with whom it can connect, our analysis adds the important proviso that this influence is likely to be productive only if there are role models of success who demonstrate the long term value of such an investment strategy.

Second, the absence of gaps in the initial vector of ES *p*s yields a more homogeneous population overall in terms of per capita consumption. As is only reasonable, initial gaps in the distribution of ES *p*s, in contrast, encourage clustering: the dissimilarities in investment and consumption that result initially from those gaps erode whatever network connections might exist at first, as 'birds of a feather flock together'. One can conclude, then, that societies that begin with a continuum of types are more likely to become homogeneous than are ones that begin with discrete clusters. Thus, again with respect to drug addiction among youth, we can speculate that if the initial population of, say, a high school is divided among subpopulations of hard drug users, abstainers and students who vary in their willingness to experiment, the ability of abstainers to resist experimenting and eventually becoming addicts will be less than if the only two sub-populations are hard users and abstainers.

Perhaps an especially important and general conclusion is that to understand the connection that arises between any pair of nodes in a network, it is not sufficient to simply examine those two nodes in isolation. Rather, we need to study the structure of interactions throughout the entire system taken as a whole. Thus, almost as a justification for Sociology as a discipline itself, in a very real sense we cannot understand fully the complexity of social systems except as a system. A number of additional patterns are revealed if we next consider an initial input vector of connections that defines an incomplete network. Quite reasonably, if network connections are difficult to form, it need not be the case that members of a sub-network all achieve the same level of per capita consumption, whereas if those connections are easily formed, per capita consumption tends to equalize within sub-networks. For a potential empirical manifestation of this phenomenon we can contrast a small close-knit community in which people readily interact face to face against a larger community where it is simply physically impossible for such intense interaction to occur. One can suppose, of course, that incomes will be more varied in the larger community than in the smaller one – that there is a 'natural' homogeneity in small groups. The question, however, is whether the absence of homogeneity in the larger group exceeds what we might assume will prevail by chance – by a process of random selection. Is it the case that any correlation between, say, a Gini index of inequality and community size exceeds what we might expect to find if we assume that income types are a random draw from a distribution that describes the overall population?

Finally, if we allow for initially atomistic systems or for asymmetric connections (e.g., hierarchies so that node *j* can be connected to node *k*, but not vice versa, which in the current context, means that node *j* 'looks at' and can mimic node *k* in adjusting its ES *p* but node *k* does not take any account directly of node *j*), then regardless of whether the most or least advantaged node (in terms of the ES *p*) is set atop an initially hierarchical network, overall social welfare is greater than if the initial network is complete. Thus, at least on average, it is socially advantageous to allow social connections to evolve from the 'neutral' starting point of a minimally connected or hierarchical system than to have it begin with a system in which those who give little value to the future are fully connected to all others. Again, if we attempt to draw substantive conclusions from this pattern, and if we consider again a closed neighborhood as might pertain to youth in the inner city, we can see why a thorough mixing of the population that includes youths initially predisposed to gang membership or drugs with youth predisposed otherwise can be especially disadvantageous in the sense that it will be the first category of youths who will exert the primary influence that an initially complete network will provide.

Of course, any implication we might attempt to draw from our analysis inference needs to be tempered by the observation that we consider only one type of hierarchy (see Figure 7.19), whereby node 1 is asymmetrically connected to nodes 2 and 3, node 2 is asymmetrically connected to nodes 4 and 5, node 3 asymmetrically connected to nodes 6 and 7, and so on. There are also those forms in which, for instance, node j is asymmetrically connected to all nodes $k > j$. It remains to be seen whether a complete network is necessarily inferior in terms of overall social welfare to these other possibilities. And only until we can answer all questions that pertain to such things can we offer, with any confidence, predictions about the spread of corruption, inner city crime and drug addiction as a function of initial social conditions.

If there is, however, a single general and qualitative theoretical conclusion that summarizes our analysis of endogenous networks in Chapter 7 it is that *those models of endogenous network formation that allow only for a dichotomous structure of connections will necessarily fail to reveal the potential complexity of network evolution.* Substantively, then, we should not be concerned simply with whether or not individuals coexist in the same general network, but also with the possibility that strengths of connection might vary. It need not be the case, for instance, that we should attempt to somehow disrupt connections with the goal of wholly eliminating the influence of gangs in a neighborhood or corrupt individuals in a society, but rather that we engage in policies that merely mute these connections and slow this influence.

Once again, though, we need to recognize that Chapter 7 considers only one of a multitude of possibilities. Specifically, our measure of the similarity of nodes is calculated on the basis of differences in each node's proportion of high and low discount types to which that node is converging. Chapter 8 considers one alternative, whereby it is the proportion of consumption attributable to low *d* types that dictates similarity. However, rather than review the implications of that analysis here since Chapter 8 attempts to summarize the various differences observed, we note that still other formulations are equally plausible. For example, we could just as easily assume that similarity and strength of connection depend on per capita rates of consumption. Indeed, the number of alternatives here, especially if we also consider alternative functional forms in that relationship, are nearly endless.

If there is, then, any general conclusion to be reached about the analysis this book offers it is that a great many parameters and functional relationships must be specified, usually on an *ad hoc* basis, before any formal analysis can be assumed to provide a complete picture of the social processes that concern us. This fact, of course, can be taken to mean that we have somehow poorly conceptualized endogenous time preferences and networks and that some alternative conceptualization might yield a more coherent picture of those processes in the context of closed-form results. Alternatively, the complexity our analysis reveals – including the often non-linear behavior of systems – can also be taken to mean that the phenomena under consideration are sufficiently complex that analytically tractable models with closed-form implications cannot capture the essential character of that phenomena. Self-contained models that yield straightforward conclusions may be adequate for revealing or illustrating isolated theoretical possibilities or alerting us to relationships only dimly perceived, but it may also be true that an understanding of social processes sufficient for policy formation and prediction requires the application of more complex methods. We are reminded, for example, of theoretical results such as the Median Voter Theorem or the characterization of equilibria in abstract markets that serve as the theoretical core of the respective disciplines of political science and economics. Surely these results point the way to a deeper theoretical understanding of democracy and market economies. However, modeling a polity or an economy so that we can predict the consequences of a specific policy generally requires a more complex (and often more *ad hoc*) approach. Similarly, in the natural sciences, we have a variety of theoretical results about fluid flow dynamics and thermodynamics, each of which constrains what we can expect from real systems. Nevertheless, modeling a specific system, even relatively simple ones, generally requires experimentation and computer simulation. The same is almost certainly true when it comes to social learning and the dynamics of networks.

Bibliography

Agarval A. and K.K. Tripathi 1980. 'Temporal orientation and deprivation,' *Journal of Psychological Researches* 24 (3): 144–52.

Ainslie, George. 1992. *Picoeconomics: the Strategic Interaction of Successive Motivational States within the Person.* Cambridge: Cambridge University Press.

Ainslie, George. 1991. 'Derivation of "Rational" Economic Behavior from Hyperbolic Discount Curves,' *American Economic Review* 81 (2): 334–40.

Ainslie, George and Varda Haendel. 1982. 'The Motives of the Will,' in *Etiologic Aspects of Alcohol and Drug Abuse,* ed. by E. Gottheil, et al. Springfield (Ill.): C. C. Thomas, pp 119–40.

Akers, Ronald L. 1973. *Deviant Behavior; a Social Learning Approach.* Belmont, Calif., Wadsworth

Akers, Ronald L. 1985. *Deviant Behavior*. Wadsworth

Akers,Ronald L., Marvin D. Krohn, Lonn Lanza-Kaduce, and Marcia Radocevich. 1979. 'Social Learning and Deviant Behavior: A Specific Test of a General Theory,' *American Sociological Review* 44: 636–55.

Allen, Terry J., F. Gerard Moeller, Howard M. Rhoades, and Don R. Cherek. 1998. 'Impulsivity and History of Drug Dependence,' *Drug and Alcohol Dependence* 50:137–45.

Allender, Eric. 2001. 'Basic Complexity,' in *Aspects of Complexity* ed. by Rod Downey and Denis Hirschfeldt. Berlin: Walter de Gruyter Pub.

Altman, Irwin and Taylor, Dalmas. 1973. *Social Penetration: The Development of Interpersonal Relationships*, New York: Holt, Rinehart & Winston

Andvig, J.C. and K.O. Moene. 1990. 'How Corruption May Corrupt,' *Journal of Economic Behavior and Organization* 13: 63–76.

Axelrod, Robert M. 1997. *The Complexity of Cooperation: Agent-Based Models of Competition and Collaboration*. Princeton: Princeton University Press.

Axelrod, Robert M. 1984. *The Evolution of Cooperation*. New York: Basic Books,

Baumol, W.J. 1970. 'On the Discount Rate for Public Projects,' in *Public Expenditures and Policy Analysis* ed. by R.H. Haveman and J. Margolis. Chicago: Markham, pp. 272–90.

Becker, Gary S. 1996. *Accounting for Tastes.* Cambridge, Mass.: Harvard University Press.

Becker, Gary S. 1976. 'Altruism, Egoism, and Genetic Fitness: Economics and

Sociobiology,' *Journal of Economic Literature* 14 (3): 817–26.

Becker, Gary S. and Robert J. Barro. 1988. 'A Reformulation of the Economic Theory of Fertility,' *Quarterly Journal of Economics* 103 (1): 1–25.

Becker, Gary S. and Casey B. Mulligan 1997. 'The Endogenous determination of Time Preference,' *Quarterly Journal of Economics (*August): 729–58.

Becker, Gary S. and Kevin M. Murphy. 1988, 'A Theory of Rational Addiction,' *Journal of Political Economy* 96 (4): 675–700.

Bendor, Jonathan and Piotr Swistak. 2001. 'The Evolution of Norms,' *American Journal of Sociology* 106 (6): 1493–1545.

Benzion, Uri, Amnon Rapoport and Joseph Yagil. 1989. 'Discount Rates Inferred from Decisions: An Experimental Study,' *Management Science* 35 (3): 270–84.

Bergadaa M. M. 1990. 'The role of time in the action of the consumer,' *Journal of Consumer Research* 17: 289–302.

Bergstrom, Theodore C and Oded Stark, 1993. 'How Altruism Can Prevail in an Evolutionary Environment,' *American Economic Review* 83 (2): 149–55

Bester, Helmut and Werner Guth. 1998. 'Is Altruism Evolutionary Stable?' *Journal of Economic Behavior & Organization* 34: 193–209

Bicchieri, Cristina. 1993. *Rationality and Coordination*. Cambridge: Cambridge University Press.

Bicchieri, Cristina and C. Rovelli, 1995. 'Evolution and Revolution: The Dynamics of Corruption,' *Rationality and Society* 7 (2): 201–24.

Bickel, Warren K and Matthew W. Johnson. 2003. 'Delay Discounting: A Fundamental Behavioral Process of Drug Discounting,' in *Time and Decision* ed. by G. Loewenstein, D. Read and R. Baumeister. New York: Russell Sage, pp. 419-40.

Bickel, Warren K., Amy L. Odum, and Gregory L. Madden. 1999. 'Impulsivity and Cigarette Smoking: Delay Discounting in Current, Never, and Ex-Smokers' *Psychopharmacology* 146: 447–54.

Bikhchandani, Suchil, David Hirshleifer and Ivo Welch. 1998. 'Learning From the Behavior of Others: Conformity, Fads, and Informational Cascades,' *Journal of Economic Perspectives* 12 (3): 151–70.

Bisin, Alberto and Thierry Verdier. 2001. 'The Economics of Cultural Transmission and the Dynamics of Preferences,' *Journal of Economic Theory* 97: 298–319.

Bisin, Alberto, Georgio Topa and Thierry Verdier. 2004. 'Cooperation as a Transmitted Cultural Trait,' *Rationality and Society* 16 (4): 447–507.

Blau, Peter 1964 *Exchange and Power in Social Life.* New York: Wiley

Blok, Hendrik J. 2002. 'Conditions for Memetic Driving,' unpublished ms., Center for Applied Ethics, University of British Columbia, Canada. Http://rikblok.cib.net/lib/blok(01).html

Blumberg, Boris F. 2001. 'Cooperation Contracts between Embedded Firms,' *Organization Studies* 22 (5): 825–52.

Borgers, Tilman and Rajiv Sarin, 1997. 'Learning Through Reinforcement and Replicator Dynamics,' *Journal of Economic Theory* 77, 1–14

Borgers, Tilman, Antonio J. Morales and Rajiv Sarin. 2002. 'Expedient and Monotone Learning Rules,' unpublished ms., Department of Economics, University College London, London, UK

Bourdieu, Pierre. 1985. 'The Social Space and the Genesis of Groups' (in French), *Sur les Sciences Sociales* 24 (2): 195–220.

Bowles, Samuel. 1998. 'Endogenous preferences: The Cultural Consequences of Markets and other Economic Institutions,' *Journal of Economic Literature* XXXVI: 75–111.

Bowles, Samuel and Herbert Gintis. 1998. 'The Moral Economy of Communities,' *Evolution and Human Behavior* 19: 3–25.

Boyd, Robert and Peter J. Richerson. 1985. *Culture and the Evolutionary Process*. Chicago; University of Chicago Press.

Boyd, Robert and Peter J. Richerson. 1992. 'Punishment Allows the Evolution of Cooperation (or Anything Else) in Sizeable Groups,' *Ethology and Sociology* 13: 171–95.

Boylan, Richard, John Ledyard, Arthur Lupia, Richard D. McKelvey and Peter C. Ordeshook, 1991. 'Political Competition in a Model of Economic Growth: An Experimental Study,' in *Laboratory Research in Political Economy* ed. by T. Palfrey, Ann Arbor: University of Michigan Press, pp 33–68.

Breteville-Jensen, Anne L. 1999. 'Addiction and Discounting.' *Journal of Health and Economics* 18: 393–407.

Brinton, Mary and Zun Tang. 2002. 'Network Embeddedness and Stratification in the Japanese Labor Market,' mimeo. Department Of Sociology. Cornell University.

Burkett, S.R. and E.L. Jensen. 1975. 'Conventional Ties, Peer Influence and Fear of Apprehension,' *Sociological Quarterly* 16: 522–33.

Burt, Ronald S. [1992], 1995. *Structural Holes: The Social Structure of Competition*. Cambridge, Mass.: Harvard University Press.

Bush, R. and F. Mosteller, 1951. 'A Mathematical Model for Simple Learning,' *Psychological Review* 58: 313–23.

Bush, R. and F. Mosteller. 1955. *Stochastic Models For Learning*. New York: Wiley.

Buskens, Vincent and Jeroen Weesie. 2000. ' An Experiment on the Effects of Embeddedness in Trust Situations,' *Rationality and Society* 12 (2): 227–53.

Camerer, Colin and George Loewenstein. 2004. 'Behavioral Economics: Past, Present, Future,' in *Advances in Behavioral Economics* ed. by Colin F. Camerer, George Loewenstein and Matthew Rabin. Princeton: Princeton University Press, pp. 3–52.

Chapman, Gretchen B. and Jennifer R. Winquist. 1998. 'The Magnitude Effect: Temporal Discount Rates and Restaurant Tips,' *Psychonomic*

Bulletin and Review 5 (1): 119–123.

Cohen, M., R. Riolo, and Robert Axelrod. 2001. 'The Role of Social Structure in the Maintenance of Cooperative Regimes,' *Rationality and Society* 13 (1): 5–32.

Coleman, James S. 1987. 'Norms as Social Capital,' in *Economic Imperialism: The Economic Approach applied Outside the Field of Economics* ed. by Gerard Radnitzky and Peter Bernholz, New York: Paragon House. pp. 133–53.

Coleman, James S. 1990. *Foundations of Social Theory*. Cambridge, Mass.: Harvard University Press.

Cook, K. and R. Hardin. 2001. 'Norms of Cooperativeness and Networks of Trust,' in *Social Norms* ed. by M. Hechter and K.D. Opp. New York: Russell Sage, pp. 327–47.

Cotterell, John. 1996. *Social Networks and Social Influences in Adolescence.* New York: Routledge.

Cropper, Maureen L., Sema K. Aydede, and Paul R. Portney. 1994. 'Preferences for Life Saving Programs: How the Public Discounts Time and Age,' *Journal of Risk and Uncertainty* 8: 243–65.

Cross, John G. 1973. 'A Stochastic Learning Model of Economic Behavior,' *Quarterly Journal of Economics* 87 (2): 239–66

Davids, Anthony and Bradley B. Falkof. 1975. 'Juvenile Delinquents Then and Now,' *Journal of Abnormal Psychology* 84 (2): 161–64.

Dawkins, R. 1976. *The Selfish Gene.* Oxford: Oxford University Press

Dawkins, R. 1987. *The Blind Watchman*. New York: Norton

Deutsch, Karl W. 1962. *Nationalism and Social Communication: An Inquiry into the Foundations of Nationality*. Cambridge, Mass.: MIT University Press.

Diamond, Jared. 2002. 'Lessons from Lost Worlds,' *Time*, August 26: A54–A55.

DiMaggio, Paul and Hugh Louch. 1998. 'Socially Embedded Consumer Transactions,' *American Sociological Review* 63: 619–37.

Dixit, Avinash and Susan Skeath. 1999. *Games of Strategy*. New York: Norton.

Duncan, Otis Dudley, David L. Featherman, and Beverly Duncan. 1972. *Socioeconomic Background and Achievement.* New York: Seminar Press

Elster, Jon. 2000. *Ulysses Unbound*. Cambridge: Cambridge University Press.

Ericksen Eugene P and William L. Yancey. 1979. 'The Antecedents of Community: The Economic and Institutional Subculture of Urban Neighborhoods,' *American Sociological Review* 44 (April): 253—2.

Ericksen, Eugene, and William L Yancey 1980 'The Locus of Strong Ties,' unpublished, Department of Sociology, Temple University, Philadelphia

Fernandez, Roberto, Emilio J. Castilla, and Paul Moore. 2000. 'Social Capital at Work,' *American Journal of Sociology* 105 (5): 1288–1356.

Festinger, Leon. 1957. *A Theory of Cognitive Dissonance*. Stanford, Calif.:

Stanford University Press.

Fisher, Irving. 1930. *The Theory of Interest*. New York: Macmillan.

Fraser, J.T. and N. Lawrence (eds). 1975. *The Study of Time II*. New York: Springer-Verlag

Frederick, Shane. 1999. 'Discounting, Time Preference, and Identity,' PhD Dissertation, Department of Social and Decision Sciences, Carnegie Mellon University, Pittsburgh.

Frederick, Shane. 2002. 'Measuring Intergenerational Time Preference: Are Future Lives valued Less,' working paper, MIT, Sloan School of Management, MIT, Cambridge, Mass.

Frederick, Shane. 2003. 'Time Preference and Personal Identity,' in *Time and Decision* ed. by G. Loewenstein, D. Read and R. Baumeister. New York: Russell Sage.

Frederick, Shane, George Lowenstein and Ted O'Donoghue. 2002. 'Time Discounting and Time Preference: A Critical Review,' *Journal of Economic Literature* 40 (June): 351–401.

Friedman, James N. 1977. *Oligopoly and the Theory of Games*. Amsterdam: North Holland.

Fudenberg, Drew and Jean Tirole. 1991. *Game Theory*. Cambridge (Mass.): MIT Press.

Fudenberg, Drew and D. Levine. 1998. *The Theory of Learning in Games*. Cambridge, Mass.: MIT Press

Gallopin, Gilberto, Al Hammond, Paul Raskin and Rob Swart. 1997. 'Branch Points: Global Scenarios and Human Choice,' Stockholm Environment Institute, report 7, Stockholm, Sweden.

Gibbons, Robert. 2001. 'Trust in Social Structures: Hobbes and Coase Meet Repeated Games,' in *Trust in Society* ed. by Karen Cook. New York: Russell Sage, pp. 332–53.

Gibson, James. 2001. 'Social Networks, Civil Society, and the Prospects for Consolidating Russia's Democratic Transition,' *American Journal of Political Science* 45 (1): 51–69.

Gintis, Herbert. 2003. 'Solving the Puzzle of Prosociality,' *Rationality and Society*. 15 (2): 155-87.

Giordano, Louis A., Warren K. Bickel, George Loewenstein, Eric A. Jacobs, Lisa Marsch and Gary J. Badger. 2002. 'Mild Opioid Deprivation Increases the Degree that Opioid-dependent Outpatients Discount Delayed Heroin and Money," *Psychopharmacology* 163: 174–82.

Glaeser, Edward L., Bruce Sacerdote, and Jose A. Scheinkman. 1996. 'Crime and Social Interactions,' *Quarterly Journal of Economics* 111(2): 507–48.

Godoy, Ricardo, Elizabeth Byron, Victoria Reyes-Garcia, William R. Leonard, Karishma Patel, Lilian Apaza, Eddy Perez, Vincent Vadez and David Wilkie. 2004. 'Patience in a Foraging-Horticultural Society: A Test of Competing Hypotheses,' *Journal of Anthropological Research* 60: 179–202.

Gordon, Robert M. 2000. 'Criminal Business Organizations, Street Gangs and 'wanna-be' groups: A Vancouver Perspective,' *Canadian Journal of Criminology* (January): 39–60.

Gottfredson, Michael R and Travis Hirschi. 1990. *A General Theory of Crime*. New York: Macmillian.

Gould, Roger V. 1995. *Insurgent Identities: Class, Community, and Protest in Paris from 1848 to the Commune*. London: University of Chicago Press.

Granovetter, M. 1995 [1974]. *Getting a Job: A Study of Contacts and Careers*. Chicago: University of Chicago Press.

Granovetter, M. 1973. 'The Strength of Weak Ties,' *American Journal of Sociology* 78: 1360–80.

Green, Leonard, Astrid F. Fry, and Joel Myerson. 1994. 'Discounting of Delayed Rewards,' *Psychological Science* 5 (1): 33–36.

Greenfeld, Liah. 1992. 'Kitchen Debate,' *The New Republic* 207 (13): 22–25.

Grief, Avner and Eugene Kandel. 1994. 'Contract Enforcement and Institutions: Historical Patterns and Current Status in Russia,' in *Economic Transition in Eastern Europe and Russia* ed. by E.P. Lazear. Stanford: Hoover Institution Press, 291–321.

Greif, Avner, Paul Milgrom and Barry R. Weingast. 1998. 'Coordination, Commitment, and Enforcement,' in *Explaining Social Institutions* ed. by Jack Knight and Itai Sened, Ann Arbor, Mich.: University of Michigan Press, 27–56.

Guth, Werner and Hartmut Kliemt. 1994. 'Competition or Cooperation: On the Evolutionary Economics of Trust, Exploitation and Moral Attitudes,' *Metroeconomica* 45 (2): 155—7 .

Guthrie, Douglas. 1998. 'The Declining Significance of Guanxi in China's Economic Transition,' *The China Quarterly* 154: 31– 62.

Guttman, Joel M. 2000. 'On the Evolutionary Stability of Preferences for Reciprocity,' *European Journal of Political Economy*. 16: 31–50.

Guttman, Joel M. 2003. 'Repeated Interaction and the Evolution of Preferences for Reciprocity,' *Economic Journal* 113 (July): 631–56

Hansson, Ingemar and Charles Stuart. 1990. 'Malthusian Selection of Preferences,' *American Economic Review* 80 (3): 529–44.

Hardin, Russel. 1982. *Collective Action*. Baltimore, Md.: Johns Hopkins University Press.

Hardin, Russel. 1989. 'Why a Constitution,' in *The Federalist Papers and the New Institutionalism* ed. by B. Grofman and D. Wittman., New York: Agathon, pp. 100–20.

Hardin, Russell. 1995. *One for All: The Logic of Group Conflict*. Princeton, N.J.: Princeton University Press.

Harley, C.B. 1981. 'Learning the Evolutionary Stable Strategy,' *Journal of Theoretical Biology* 81 (4): 611–33.

Heavner, D. Lee and Lance Lochner. 2002. 'Social Networks and the

Aggregation of Individual Decisions,' NBER working paper 8979: http://www.nber.org/papers/w8979.

Heylighen, Francis. 1992. 'Selfish Memes and the Evolution of Cooperation,' *Journal of Ideas,* 2 (4): 70–84.

Holcomb, James H. and Paul S. Nelson. 1992. 'Another Experimental Look at Individual Time Preference,' *Rationality and Society* 4(2): 199–220.

Homans, George. 1961. *Social Behavior: Its Elementary Forms*. New York: Harcourt, Brace & World.

Horowitz, John K. 1991. 'Discounting Money Payoffs: An Experimental Analysis,' in *Handbook of Behavioral Economics* (Volume 2B). Greenwich, Ct,: JAI Press, pp: 309-24.

Howell, James C. 1994. 'Recent Gang Research: Program and Policy Implications,' *Crime & Delinquency* 40 (4): 495–515.

Hummon, Norman P. 2000. 'Utility and Dynamic Social Networks,' *Social Networks* 22: 221–49.

Insead, Subramanian Rangan. 2000. 'The Problem of Search and Deliberation in Economic Action: When Social Networks Really Matter,' *Academy of Management Review* 25 (4): 813–28.

Jain, Arvind, K. 2001. 'Corruption: A Review,' *Journal of Economic Surveys* 15 (2): 71–121.

Johnson, Richard E., Anastasios C. Marcos and Stephen J. Bahr. 1987. 'The Role of Peers in the Complex Etiology of Adolescent Drug Use,' *Criminology* 25 (2): 323–40.

Johnstone, J.W.C. 1983. 'Recruitment to a Youth Gang,' *Youth and Society* 14: 281–300.

Kacelnik, Alex. 2003. 'The Evolution of Patience,' in *Time and Decision* ed. by G. Loewenstein, D. Read and R. Baumeister, New York: Russell Sage, pp. 115–38.

Kagel, John H., Leonard Green and Thomas Caraco 1986. 'When Foragers Discount the Future: Constraint or Adaptation,' *Animal Behavior* 34: 271–83.

Kandel, Denise (ed). 1978. *Longitudinal Research on Drug Use: Empirical Findings and Methodological Issues*. Washington: Hemisphere.

Kirby, Kris N., Nancy M. Petry and Warren K. Bickel. 1999. 'Heroin Addicts Have Higher Discount Rates for Delayed Rewards Than Non-Drug-Using Controls,' *Journal of Experimental Psychology* 128 (1): 78–87.

Klein, Malcolm W. 2002. 'Street Gangs: A Cross-National Perspective,' in *Gangs in America III* ed. by C. Ronald Huff, Thousand Oaks, Calif.: Sage Publications, pp. 237–56.

Klineberg Stephen L.1968. 'Future Time Perspective and the Preference for Delayed Reward,.' *Journal of Personality and Social Psychology* 8(3): 253–57.

Klochko, Marianna A. and Peter C. Ordeshook. 2003. 'Corruption, Cooperation and Endogenous Time Discount Rates,' *Public Choice* 115: 259–83.

Klochko, Marianna A. 2004. *Endogenous Time Preferences: An Evolutionary Approach.* PhD Dissertation, Department Of Sociology, Cornell University, Ithaca, NY.

Klochko, Marianna A 2005. 'A Note on Time Discounting and Sobriety: Evidence of the Endogenous Determination of Discount Rates,' mimeo, Ohio State University, Marion, OH.

Klochko, Marianna A. 2006. 'Time Preferences and Selection vs Learning: A Case Study of Ukrainian Students,' *Rationality and Society* forthcoming 18 (3 August)

Klyamkin, I. and L. Timofeev. 2000. 'Shadow Russia: Economic-Sociological Research,' *Tenevaya Rossiya*, Russian Humanitarian State University.

Kneen, Peter. 2000. 'Political Corruption in Russia and the Soviet Legacy,' *Crime, Law and Social Change* 34: 349–67.

Knight, Jack. 2001. 'Social Norms and the Rule of Law,' in *Trust in Society* ed. by Karen Cook. New York: Russell Sage, pp. 354–73.

Koopmans, Tallings C. 1960. 'Stationary Ordinal Utility and Impatience,' *Econometrica* 28: 287–309.

Krackhardt, A. 1992. 'The Strength of Strong Ties: The Importance of Philos in Organizations,' in *Networks and Organizations* ed. by Nitin Nohria and Robert G. Eccles. Boston: Harvard Business School Press, pp. 216–39.

Krahn, M. and Gafni, A. 1993. 'Discounting in the Economic Evaluation of Health Care Interventions,' *Medical Care* 31 (5): 403–18.

Laibson, David I and Howard Rachlin (eds). 1997. *The Matching Law: Papers in Psychology and Economics.* New York: Russell Sage.

Lawrance, Emily C. 1991. 'Poverty and the Rate of Time Preference: Evidence from Panel Data,' *Journal of Political Economy* 99 (1): 54–77.

Ledeneva, Alena V. 1998. *Russia's Economy of Favours: Blat, Networking and Informal Exchange.* New York: Cambridge University Press.

Lewis, David K. 1969. *Convention: A Philosophical Study.* Cambridge (Mass): Harvard University Press.

Lin, Nan, Paul W. Dayton, and Peter Greenwald. 1978. 'Analyzing the Instrumental Use of Relations in the Context of Social Structure,' *Sociological Methods and Research* 7:149–66.

Lin, Nan, Walter M. Ensel, and John C. Vaughn. 1981. 'Social Resources and Strength of Ties: Structural Factors in Occupational Status Attainment,' *American Sociological Review* 46 (4): 393–405.

Lomborg, Bjorn. 1996. 'Nucleus and Shield: The Evolution of Social Structure in the Iterated Prisoners' Dilemma,' *American Sociological Review* 61 (April): 278–307

Loewenstein, George. 1987. 'Anticipation and the Valuation of Delayed Consumption,' *Economic Journal* 97: 666–84.

Loewenstein, George. 1992. 'The Fall and Rise of Psychological Explanations in the Economics of Intertemporal Choice,' in *Choice over Time* ed. by G. Loewenstein and Jon Elster. New York: Russell Sage.

Loewenstein, George F. and Drazen Prelec. 1993. 'Preferences for Sequences of Outcomes,' *Psychological Review* 100 (1): 91–108.

Lott, Albert J. and Bernice E. Lott 1964. *Influence of Classroom Group Cohesiveness on Learning and Adherence to Standards*. Lexington, Ky.: Kentucky Research Foundation, University of Kentucky

Macy, Michael. 1996. 'Natural Selection and Social Learning in Prisoners' Dilemma: Co-adaptation with Genetic Algorithms and Artificial Neural Networks,' *Sociological Methods and Research* 25 (July): 103–37.

Macy, Michael W. and Yoshimichi Sato. 2002. 'Trust and Market Formation in the U.S. and Japan,' *Proceedings of the National Academy of Sciences* 99 (May 14): 7214–20.

Macy, Michael and John Skvoretz. 1998. 'The Evolution of Trust and Cooperation between Strangers: A Computational Model,' *American Sociological Review* 63 (5): 638–60.

Madden, Gregory J., Nancy M. Petry, Gary J. Badger and Warren K. Bickel. 1997. 'Impulsive and Self-Control Choices in Opioid-Dependent Patients and Non-Drug-Using Control Participants: Drug and Monetary Rewards,' *Experimental and Clinical Psychopharmacology* 5 (3): 256–62.

Mahoney, Paul G. and Chris Sanchirico. 2001. 'Competing Norms and Social Evolution,' *Social Science Research Network Electronic Paper Collection*, Paper No.00-15 papers.ssrn.com/ paper.taf'absract_id=229694

Mahoney, Paul G. and Chris Sanchirico. 2002. 'Norms, repeated Games, and the Role of Law,' mimeo.

Mauro, Paolo. 1997. 'The Effects of Corruption, Investment and Government Expenditure,' in *Corruption and the Global Economy* ed. by Kimberly A. Elliot, Washington D.C.: Inst. for International Economics, pp. 83–107.

Maxson, Cheryl and Monica L. Whitlock. 2002 'Joining the Gang: Gender Differences in Risk Factors for Gang Membership,' in *Gangs in America III* ed. by C. Ronald Huff, Thousand Oaks(Calif.): Sage Publications, pp. 19–36.

Maynard-Smith, John. 1974. The Theory of Games and the Evolution of Animal Conflicts,' *Journal of Theoretical Biology* 132: 247–60.

Maynard-Smith, John. 1982. *Evolution and the Theory of Games*. Cambridge: Cambridge University Press.

McKelvey, Richard D. and Peter C. Ordeshook. 1990. 'Information and Elections: Retrospective Voting and Rational Expectations,' in *Information and Democratic Processes* ed. by J. Ferejohn and J. Kuklinski, Champaign-Urbana: University of Illinois Press.

Miller, W.B. 1975. Violence by Youth Gangs and Youth Groups as a Crime Problem in Major American Cities (report to the National Institute for

Juvenile Justice and Delinquency Prevention). Washington, DC.

Nagel, Thomas. 1970. *The Possibility of Altruism*. Princeton: Princeton University Press.

Nagin, Daniel S., Greg Pogarsky. 2001. 'Integrating Celerity, Impulsivity, and Extralegal Sanction Threats into a Model of General Deterrence: Theory and Evidence,' *Criminology* 39 (4): 865–92.

Nee, Victor and Peng Lian. 1994. 'Sleeping with the Enemy: A Dynamic Model of Declining Political Commitment in State Socialism,' *Theory and Society* 23: 253–96.

O'Donoghue, Ted and Matthew Rabin. 1999. 'Doing it Now or Later,' *American Economic Review* 89 (1): 103–24.

Odum, Amy L., Gregory J. Madden, Gary J. Badger and Warren K. Bickel. 2000. 'Needle Sharing in Opioid-dependent Outpatients: Psychological Processes Underlying Risk,' *Drug and Alcohol Dependence* 60: 259–66.

Oetting, E.R. and F. Beauvais. 1987. 'Common Elements in Youth Drug Abuse: Peer Clusters and Other Psychosocial Factors,' *Journal of Drug Issues* 17: 133–51.

Oevermann, Ulrich. 1969. 'Role Structure of the Family and its Implications for the Cognitive Development of Children,' in *Education in Europe: Sociological Research* ed. by M. A. Matthijssen and C. E. Vervoort, The Hague: Mouton. Pp. 103–22.

Ok, Effie and Y. Masatlioglu. 2003. 'A General Theory of Time Preference,' unpublished, Department Of Economics, New York University.

Olson, Mancur. Jr. 1965. *The Logic of Collective Action*. Cambridge (Mass.): Harvard University Press.

Ordeshook, Peter C. 1993. 'Some Rules of Constitutional Design,' *Social Philosophy and Policy* 10 (2): 198–232.

Orphanides, Athanasios and David Zervos. 1995. 'Rational Addiction with Learning and Regret,' *Journal of Political Economy* 103 (4): 739-58.

Orphanides, Athanasios and David Zervos. 1998. 'Myopia and Addictive Behavior,' *Economic Journal* 108 (January): 75–91.

Ostrom, Elinor. 1990. *Governing the Commons: The Evolution of Institutions for Collective Action.* Cambridge: Cambridge University Press.

Petersen, Trond, Ishak Saporta, and Marc-David L. Seidel. 2000. 'Offering a Job: Meritocracy and Social Networks,' *American Journal of Sociology* 106 (3): 763–816.

Petry, Nancy M. 2001. 'Delay Discounting of Money and Alcohol in Actively Using Alcoholics, Currently Abstinent Alcoholics, and Controls,' *Psychopharmacology* 154: 243–50.

Podolny, Joel M. 2001. 'Networks as the Pipes and Prisms of the Market,' *American Journal of Sociology* 107 (1) : 33–60.

Podolny, Joel M., Toby E. Stuart, and Michael T. Hannan. 1996. 'Networks, Knowledge, and Niches: Competition in the Worldwide Semiconductor

Industry, 1984-1991,' *American Journal of Sociology* 102: 659–89.
Pollak, Robert A. 1970. 'Habit Formation and Dynamic Demand Functions,' *Journal of Political Economy* 78 (4): Part 1: 645–763.
Pollak, Robert A. 1976. 'Habit Formation and Long-Run Utility Functions,' *Journal of Economic Theory* 13: 272–297.
Posner, Eric A. 2000. *Law and Social Norms*. Cambridge, Mass.: Harvard University Press.
Putnam, Robert D. 2001. *Bowling Alone: The Collapse and Revival of American Community*. New York: Touchstone.
Raiser, Martin. 1999. 'Trust in transition,' Working paper # 39 of European bank for Reconstruction and Development.
Raub, Werner and Jerome Weesie. 1990. 'Reputation and Efficiency in Social Interactions: An Example of Network Effects,' *American Journal of Sociology* 96: 626–54.
Read, Daniel. 2001. 'Is Time Discounting Hyperbolic or Subadditive,' *Journal of Risk and Uncertainty* 23 (1): 5–32
Remington, Thomas. 1981. 'The Mass Media and Public Communication in the USSR,' *Journal of Politics* 43: 803–17.
Roberts, Russell D. 1991. 'Myopic Discounting: Empirical Evidence,' in *Handbook of Behavioral Economics*. Greenwich, Ct.: JAI Press. Pp. 342–45.
Rogers, Alan R. 1994. 'Evolution of Time Preference by Natural Selection,' *American Economic Review* 84 (3): 460–81.
Rooks, Gerrit, Werner Raub, Robert Selten and Frits Tazelaar. 2000. 'How Inter-firm Co-operation De[ends on Social Embeddedness: A Vignette Study,' *Acta Sociologica* 43: 123–37.
Rose-Ackerman, Susan. 1999. *Corruption and Government*. Cambridge: Cambridge University Press.
Rubin, Paul H and Chris W. Paul II. 1979. 'An Evolutionary Model of Taste for Risk,' *Economic Inquiry*. XVII (October): 585–96
Ruble, N. And W.L. Turner. 2000. 'A Systemic Analysis of the Dynamics and Organization of Urban Street Gangs,' *American Journal of Family Therapy* 28 (2): 117–32.
Sally, David. 2000. 'A General Theory of Sympathy, Mind-reading, and Social Interaction, with an Application to the Prisoners' Dilemma,' *Social Science Information* 39 (4, December): 567–634.
Samuelson, Paul. 1937. 'A Note on Measurement of Utility,' *Review of Economic Studies* 4: 155–61.
Samuelson, Larry. 1997. *Evolutionary Games and Equilibrium Selection*. Cambridge, Mass.: MIT Press.
Schachter, Stanley. 1959. *The Psychology of Affiliation*. Stanford, Calif.: Stanford University Press.
Schneider, L. and S. Lysgaard,1953. 'The Deferred Gratification Pattern,'

American Sociological Review 18: 142–59.

Shapiro, Jonathan. 2002. Notes on Presentation for Complex Systems Summer School, Santa Fe, New Mexico.

Shleifer, Andrei and Robert W. Vishny. 1993. 'Corruption,' *Quarterly Journal of Economics* 108 (3): 599–618.

Sober, Elliott. 1984. *The Nature of Selection: Evolutionary Theory in Philosophical Focus.* Chicago: University of Chicago Press.

Sorokin, Pitirim. 1914. 'Crime and Punishment, Heroism and Reward,' originally published in *Individual, Civilization, Society* ed. by A. Sogomonov, Moscow: Publishers of Political Literature, pp. 25–155.

Spergel, Irving A. 1990. 'Youth Gangs: Continuity and Change,' in *Crime and Justice: A Review of Research* ed. by Michael Tonry and Norval Morris, Chicago: University of Chicago Press, 12: 171–275.

Stigler, George J. 1984. *The Theory of Price.* New York: Macmillan.

Stolle, Dietlind 2001 'Clubs and Congregations: The Benefits of Joining an Association,' in *Trust in Society.* ed. By K.S. Cook, New York: Russel Sage, pp. 202–44.

Strang, David and Michael W. Macy. 2000. 'In Search of Excellence: Fads, Success Stories and Adaptive Emulation,' unpublished working paper, Department of Sociology, Cornell University, Ithaca, NY.

Strotz, R.H. 1955. 'Myopia and Inconsistency in Dynamic Utility Maximization,' *Review of Economic Studies* 23 (3): 165–80.

Sutherland, Edwin. 1939. *Principles of Criminology*, Chicago: J. B. Lippincott.

Sutton, Richard S. and Andrew G. Barto. 1998. *Reinforcement Learning: An Introduction*. Cambridge, Mass.: MIT Press.

Taylor, Michael. 1976. *Anarchy and Cooperation*. New York: Wiley.

Taylor, Peter D. 1979. 'Evolutionary Stable Strategies With Two Types of Player,' *Journal of Applied Probability* 16: 76–83.

Teahan, John E. 1963 'Parental Attitudes and College Success,' *Journal of Educational Psychology* 54 (April):104–09.

Thomas, L.C. 1984. *Games, Theory and Applications.* New York: Halstead Press.

Thornberry, Terence P., Marvin D. Krohn, Alan J. Lizotte and Deborah Chard-Wierschem. 1993. 'The Role of Juvenile Gangs in Facilitating Delinquent Behavior,' *Journal of Research in Crime and Delinquency* 30 (1): 55–87.

Tomasello. M. 2004. 'Learning Through Others,' *Daedalus*. 133 (January 1): 51–8.

Trice, Harrison M. 1966. *Alcoholism in America*. New York: McGraw Hill

Trommsdorf G. and Lamm H. 1975. 'An Analysis of Future Orientation and Some of its Social Determinants,' in *The Study of Time* (vol II) ed. by J. T. Fraser and N. Lawrence, New York: Springer-Verlag,

Tybout, Alice M and Nancy Artz, 1994. 'Consumer Psychology,' *Annual*

Review of Psychology 45: 131–70.

Unger, Donald G. and Abraham Wandersman. 1985. 'The Importance of Neighbors: The Social, Cognitive, and Affective Components of Neighboring,' *American Journal of Community Psychology* 13 (2): 139–69.

Uzzi, B. 1996. 'The Sources and Consequences of Embeddedness for Economic Performance of Organizations: The Network Effect,' *American Sociological Review* 61: 674–98.

Volker, Beate and Henk Flap. 2001. 'Weak Ties as a Liability: The.Case of East Germany,' *Rationality and Society* 13 (4): 397–428.

Von Weizsacker, Carl Christian. 1971. 'Notes on Endogenous Change of Tastes,' *Journal of Economic Theory* 3: 345–72.

Vuchinich, Rudy E. and Cathy A. Simpson. 1998. 'Hyperbolic Temporal Discounting in Social Drinkers and Problem Drinkers,' *Experimental and Clinical Psychopharmacology* 6 (3): 292–305.

Wank, David L. 1996. 'The Institutional Process of Market Clientelism,' *China Quarterly* (September): 820–39.

Wasserman, Stanley and Katherine Faust. 1994. *Social Network Analysis*. Cambridge: Cambridge University Press.

Watts, Duncan J., P.S. Dodds and M.E.J. Newman. 2002. 'Identity and Search in Social Networks,' *Science* 296 (5571), 1302–05.

Weibull, Jorgen W. 1995. *Evolutionary Game Theory*. Cambridge, Mass.: MIT Press.

Weibull, Jorgen W. 1998. 'What Have We Learned from Evolutionary Game Theory So Far,' unpublished, Stockholm School of Economics.

Wilson, Drew and Lloyd Donaldson. 1996. *Russian Etiquette & Ethics in Business*. Lincolnwood, Ill.: NTC Business Books.

Wilson, David Sloan. 1997. 'Incorporating Group Selection into the Adaptionist Program,' in *Evolutionary Social Psychology* ed. by J. Simpson and D.T. Kenrick, Hillsdale, NJ.: Lawrence Erlbaum, pp. 345–86.

Wilson, David Sloan. 2002. *Darwin's Cathedral: Evolution, Religion, and the Nature of Society*. Chicago: University of Chicago Press.

Wilson, James Q. and R. J. Hernstein. 1985. *Crime and Human Nature*. New York: Simon & Schuster.

Winfree, Thomas L. Jr., Christine S. Sellers, and Dennis L. Clason. 1993. 'Social Learning and Adolescent Deviance Abstention: Toward Understanding the Reasons for Initiating, Quitting, and Avoiding Drugs,' *Journal of Quantitative Criminology* 9 (1): 101–25.

Winfree, Thomas L. Jr., Teresa Vigil Backstrom and G. Larry Mays. 1994. 'Social Learning Theory, self-reported Delinquency, and Youth Gangs: A New Twist on a General Theory of Crime and delinquency,' *Youth & Society* 26 (2): 147–77.

Winston, Gordon C and Richard G. Woodbury. 1991. 'Myopic Discounting: Empirical Evidence,' *Handbook of Behavioral Economics*. Vol. 2B:

325–41.

Yamagishi, Toshio. 2001. 'Trust as a Form of Social Intelligence,' in *Trust in Society* ed. by Karen S. Cook. New York: Russell Sage, pp. 121–47.

Yamagishi, Toshio and Midori Yamagishi. 1994. 'Trust and Commitment in the United States and Japan,' *Motivation and Emotion* 18 (2): 129–166.

Yamagishi Toshio and Karen S. Cook. 1993. 'Generalized Exchange and Social Dilemmas,' *Social Psychology Quarterly* 56: 235–48.

Yamagishi, Toshio, Karen S. Cook and Motoki Watabe. 1998. 'Uncertainty, Trust, and Commitment Formation in the United States and Japan,' *American Journal of Sociology* 104 (1): 165–94.

Zeggelink, Evelien. 1995. 'Evolving Friendship Networks: An Individual-oriented Approach Implementing Similarity,' *Social Networks* 17: 83–110.

Zwimgle, Erla. 1993. 'Ogallala Aquifer: Wellspring of the High Plains,' *National Geographic* 183 (March): 80–109.

Name Index

Subject Index